THE STUDENT GUIDE TO

MINITAB®

RELEASE 14

John McKenzie would like to dedicate this book to his four brothers and sisters, Bruce, Ellen, Laurie, and David, and to their families.

Robert Goldman would like to dedicate this book to his wife, Maryglenn, and their daughter, Katy, with love.

THE STUDENT GUIDE TO

MINITAB®

RELEASE 14

John D. McKenzie, Jr.
Babson College

Robert Goldman
Simmons College

PEARSON
Addison
Wesley

Boston   San Francisco   New York
London   Toronto   Sydney   Tokyo   Singapore   Madrid
Mexico City   Munich   Paris   Cape Town   Hong Kong   Montreal

Publisher: Greg Tobin
Executive Editor: Deirdre Lynch
Editorial Assistant: Sara Oliver
Marketing Manager: Phyllis Hubbard
Project Manager: Cindy Johnson, *Publishing Services*
Managing Editor: Karen Wernholm
Senior Manufacturing Buyer: Evelyn Beaton
Senior Media Buyer: Ginny Michaud
Composition: Cindy Johnson, *Publishing Services*
Cover Design: Andrea Menza
Cover Photo: © Stockbyte

Minitab at Work and data set credits appear on p. 541, which constitutes an extension of the copyright page.

Portions of MINITAB Statistical Software input and output contained in this book are printed with permission of Minitab, Inc.

Library of Congress Cataloging-in-Publication Data
McKenzie, John, 1944–
    The student guide to Minitab Release 14 / John D. McKenzie, Robert Goldman.
        p. cm.
    Includes index.
    ISBN 0-321-11312-8 — ISBN 0-201-77469-0
    1. Mathematical statistics—Data processing. 2. Minitab for Windows. I. Goldman, Robert N.,
    1944– II. Title.

QA276.4.M299 2004
519.5'0285'536—dc22

                                                2004052227

Trademarks: Many of the designations used by manufacturers and sellers to distinguish their products are claimed as trademarks. Where those designations appear in this book, and Addison-Wesley was aware of a trademark claim, the designations have been printed in initial caps or all caps.

The programs and applications presented in this book have been included for their instructional value. They have been tested with care but are not guaranteed for any particular purpose. The publisher does not offer any warranties or representations. Nor does it accept any liabilities with respect to the programs or applications.

ISBN 0-201-77469-0   (Manual/Software Package)
ISBN 0-321-11312-8   (Manual)
ISBN 0-321-11313-6   (Software)

4 5 6 7 8 9 10 —DOC—070605

Welcome to *The Student Guide to MINITAB Release 14.*

This book is designed to guide the user step-by-step from the basics to the more sophisticated features of MINITAB Release 14. Written for users of both the Professional and Student versions of MINITAB, the guide is suitable for students of introductory statistics, those using statistics in later courses, and for users interested in learning MINITAB on their own. This guide may be purchased with a CD-ROM containing MINITAB Student Release 14 or as a standalone guide to be used with MINITAB Release 14 purchased from Minitab, Inc.

The Student Guide to MINITAB Release 14 is organized into 16 interactive tutorials that expose readers to the full power of the software through a careful, step-by-step explanation of the necessary procedures. All the tutorials include case studies from business, psychology, education, engineering, health care, and the life and physical sciences. These case studies give students the opportunity to immediately apply their knowledge of the software in an interesting and real-life context, using real data provided for this purpose. As students perform steps in the tutorial, screen views are shown in the book to teach and illustrate MINITAB concepts and features with as little additional outside instruction as possible. Throughout the book students are shown how to take advantage of MINITAB's excellent Help capability. Each tutorial ends with a summary and review section including a MINITAB command summary and a section that tests mastery of the material through exercises and practice problems. Each tutorial requires approximately 60 minutes to complete (not including end-of-chapter material).

Features of the Student Guide

Organized largely by statistical topic, the tutorials in this book closely match current teaching practice. Students complete a thorough introduction to data analysis, including contingency tables, correlation, and regression, before moving on to inferential methods. This organization makes the guide a perfect supplement for many introductory statistics texts.

- Minitab techniques are illustrated with more than 90 real data sets from a wide variety of application areas. One-third of the data sets are new to this edition.
- Each tutorial includes a set of Matching and True/False problems designed to help students become familiar with the features of MINITAB.
- A series of up to 12 problems in each tutorial reinforce the techniques covered in the tutorial by asking the user to apply them to realistic problems and real data. Many of these problems require the student to interpret MINITAB output or draw a conclusion from the analysis.

- Each tutorial concludes with an "On Your Own" project in which the student is invited to tackle a more open-ended problem.
- In selected tutorials, a "Minitab at Work" feature illustrates how MINITAB is used in the workplace.

New in This Edition

- Graphs are used throughout the book as a basic tool for exploring data before inference and as an essential tool for examining assumptions.
- Each tutorial emphasizes the importance of checking the validity of the assumptions in statistical inference. The guide takes full advantage of MINITAB's built-in graphs to demonstrate the importance of checking assumptions with each inferential technique.
- Techniques for importing and exporting data and output, including graphs, are integrated carefully throughout the book.
- Many new problems and an "On Your Own" project have been added to each tutorial. Answers to the Matching and True-False problems are included at the end of each tutorial.
- The Web site for the book (*www.aw-bc.com*, keyword: MINITAB) contains solutions to selected end-of-tutorial problems, a list of alternative pathways through the book that correspond to popular statistics textbooks, and errata.

About MINITAB Release 14

MINITAB is a powerful tool originally developed at The Pennsylvania State University as a teaching aid for introductory statistics courses. It is now one of the most widely used statistical analysis packages for college instruction, currently used at over 4,000 colleges and universities worldwide. MINITAB is also the most popular software employed in Advanced Placement Statistics courses. Learning MINITAB is a valuable asset for future careers. Across the globe, distinguished and successful corporations choose MINITAB for quality and process improvement and for general data analysis. In fact, many Fortune 500 companies rely on Minitab's award-winning software as part of their formula for success.

MINITAB Student Release 14

MINITAB Student Release 14 is a streamlined version of MINITAB Release 14 designed for use in introductory statistics and business statistics courses. It contains the core features of the Professional version. MINITAB Student Release 14 now allows users to work with worksheets of up to 10,000 cells of data (twice as many as in the last release).

Features of MINITAB Release 14

- Comprehensive statistical capabilities, including descriptive statistics, simulations and distributions, elementary inferential statistics, analysis of variance, regression, analysis of categorical data, nonparametric methods, time series analysis, and control charts and other quality management tools

- Presentation-quality graphics, which enable users to produce a comprehensive array of professional graphs
- A menu-driven interface providing easy access to MINITAB's statistical, graphical, and data management capabilities
- Alternative means of issuing commands such as typing Session window commands, clicking toolbar buttons, and using the right-hand mouse button
- Data windows that permit data entry, editing, and browsing in a spreadsheet-like display
- An on-line Help facility using the Microsoft Windows Help system to provide context-specific help at the click of a mouse button
- The ability to import and export data, including data from Microsoft Excel and Microsoft Word

New in MINITAB Release 14

State-of-the-Art, Presentation-Quality Graphs

- Users can now update graphs automatically, customize graphs and save preferences, place several graphs in the same window, obtain data tips, and graph summary statistics.
- A new pictorial gallery and simplified and unified graph dialog boxes make selecting the most appropriate graph quick and easy.
- Users can graph subsets of their data without first manipulating the worksheet.
- Double-clicking on a graph component now leads directly to the dialog box for editing it.

More Basic Statistics

- Release 14 offers the user a wider collection of descriptive statistics.
- Users can enter summarized data for most basic statistics methods.
- One-sided confidence intervals can be produced.

Spreadsheet-Like Features

- New features such as Autofill, column hiding, and find-and-replace add power to worksheet management.
- Excel workbooks can now be imported as separate worksheets.

Ease of Use and Support

- Comprehensive HTML-based Help system makes how-to information easier to find and print.
- Enhanced Project Manager, StatGuide, and ReportPad features help the user logically organize data and graphs, explain output, and generate reports.

Overview of the Guide

The fundamentals of MINITAB are covered in the first three tutorials. Tutorial 0 includes a concise explanation of the conventions used throughout the manual, the system requirements needed to run the software, and information regarding technical support. It also contains a sample MINITAB session designed to give students an introduction to the software and the tutorial approach. Tutorials 1 and 2 cover the entering, saving, printing, and manipulation of data.

Tutorials 3 through 6 illustrate how MINITAB performs data analysis and handles probability distributions and random data. Tutorial 3 covers data analysis for a single variable; Tutorials 4 and 5 show how to use MINITAB to examine the relationship between two variables. All of these tutorials make extensive use of graphs to provide insight into the analysis. Tutorial 6 illustrates MINITAB's excellent capabilities in working with probability distributions and random data.

Tutorials 7 through 10 cover the basic inference techniques seen in introductory, applied statistics texts. Tutorials 7 and 8 illustrate one-sample and two-sample techniques and include examples of sample size determination and power calculations. Tutorial 9 introduces techniques for one-way and two-way ANOVA. The fundamentals of inference in simple, quadratic, and multiple regression are covered in Tutorial 10.

Tutorials 11 through 15 cover more specialized topics. Tutorial 11 introduces dummy variables in multiple regression, collinearity, the best-subsets technique for model construction, and logistic regression. Chi-square tests of fit and for independence are covered in Tutorial 12. Tutorial 13 illustrates the use of nonparametric methods. The last two tutorials, 14 and 15, cover time series and total quality management tools.

Pathways Through the Guide

Students unfamiliar with Minitab should complete Tutorials 0 through 2 before moving on to the other tutorials. Most students will complete, in order, Tutorials 3 through 8. The remaining tutorials (9 through 15) can be completed in any sequence.

Appendices

Appendix A contains descriptions of all of the real data sets used in the tutorials and practice problems. It also contains descriptions of other data sets that students may want to use for additional practice.

Appendix B contains images of each of the distinct menus and toolbars in MINITAB Student Release 14. These menus form the core of, and in many cases are identical to, the corresponding menus in the Professional version of the software. (Professional version users may activate the Student version menus by using the Tools > Manage Profiles > Manage command.)

Supplements

Visit *www.aw-bc.com*, keyword: MINITAB, for access to:

- Solutions to selected end-of-tutorial problems
- A list of alternative pathways through the book that correspond to popular statistics textbooks
- Errata

Reviewers

We would like to thank the following reviewers who assisted in the development of this edition.

Major William K. Farmer, *United States Military Academy, West Point*
Stergios B. Fotopoulos, *Washington State University*
Lance Hemlow, *Raritan Valley Community College*
Timothy H. Husband, *Siena Heights University*
Farid Islam, *Utah Valley State College*
Colleen Kelly, *San Diego State University*
Ampalavanar Nanthakumar, *State University of New York at Oswego*
Stephen Russell, *Weber State University*
Doug Sizemore, *Covenant College*
James Smart, *Santa Fe Community College*
Kai-Sheng Song, *Florida State University*

J. M.
R. G.

ABOUT THE AUTHORS

John McKenzie has been teaching applied statistics courses to undergraduate and MBA students at Babson College since 1978. He also provides statistical advice to the college's faculty and staff. He has used statistical software in his teaching and consulting for over 30 years, and served as co-editor of the Minitab User's Group Newsletter. He has held a number of positions with the American Statistical Association, among them the representative of the Council of Chapters on its Board of Directors. He is a Fellow of the Association, from whom he received a Founders Award. In 2004 he was the chair of the Mathematical Association of America's Special Interest Group on Statistics Education. He has also been a member of the Planning Committee of the Making Statistics More Effective in Schools of Business conferences. He has an A.B. in mathematics from Amherst College and, from the University of Michigan, master's degrees in mathematics and statistics and a Ph.D. in statistics.

Robert Goldman has been teaching a wide variety of statistics courses at Simmons College since 1972. He is the co-author of the statistics textbook *Statistics: An Introduction* and has consulted widely for a number of states and the Federal government. He is a past-president of the Boston Chapter of the American Statistical Association. His current interests include creating and improving an online statistics course. He has a B.Sc. degree in statistics from the London School of Economics and a master's degree and a Ph.D. in statistics from Harvard University.

BRIEF TABLE OF CONTENTS

Minitab Fundamentals

Tutorial 0   Getting Started With Minitab   1

Tutorial 1   Working with Data   22

Tutorial 2   Summarizing, Transforming, and Manipulating Data   59

Data Analysis, Distributions, and Random Data

Tutorial 3   Data Analysis for One Variable   92

Tutorial 4   Data Analysis: Comparing Groups   132

Tutorial 5   Examining Relationships Between Two Quantitative Variables   158

Tutorial 6   Distributions and Random Data   199

Inference

Tutorial 7   Inferences from One Sample   228

Tutorial 8   Inferences from Two Samples   255

Tutorial 9   Comparing Population Means: Analysis of Variance   283

Tutorial 10   Fundamentals of Linear Regression   312

Special Topics

Tutorial 11   Building Regression Models   337

Tutorial 12   Analyzing Qualitative Data   378

Tutorial 13   Analyzing Data with Nonparametric Methods   403

Tutorial 14   Time Series Analysis   426

Tutorial 15   Total Quality Management Tools   456

Appendix A   Data Sets   483

Appendix B   Minitab Menus and Toolbars   521

CONTENTS

Tutorial 0:
Getting Started with Minitab   1

0.1   Introduction to Minitab   2
      What's in This Book?   *3*
      Typographical Conventions   *3*
      Technical Support   *4*
      Installing Minitab   *4*
      The Minitab Menu   *5*
      Types of Data Files   *6*
0.2   Touring Minitab and Using Help   6
      Opening Minitab   *6*
      Worksheets and the Data Window   *9*
      Issuing Minitab Commands   *10*
      Context-Sensitive Help   *10*
      Using the Help Menu   *11*
      Getting Help with Data Sets   *12*
      Exiting Minitab   *14*
      Web Site for the Book   *14*
0.3   Sample Minitab Session   14
      Getting Started   *14*
      Exploring an Existing Worksheet   *15*
      **CASE STUDY: GEOLOGY—
      DURATION OF "OLD FAITHFUL" ERUPTIONS**   15
      Getting Information About a Data Set   *16*
      Creating Graphs   *19*
      Printing a Graph   *20*
      Exiting Minitab   *21*
Review and Practice   21

Tutorial 1:
Working with Data   22

1.1   Opening Minitab   23
      **CASE STUDY: HEALTH MANAGEMENT—
      NUTRITIONAL ANALYSIS**   23
1.2   Entering Data   27
1.3   Correcting Mistakes   28
1.4   Text and Numeric Data   28
1.5   Saving Data   29

1.6   Naming Columns   30
1.7   Printing from the Data Window   31
1.8   Moving in the Data Window   32
1.9   Leaving Minitab   33
1.10 Retrieving Data   33
1.11 Manipulating Data Using the Calculator
      Command   34
1.12 Copying Columns   36
1.13 Opening a Minitab Stored Data File   40
      **CASE STUDY: MANAGEMENT—
      SALARY STRUCTURE**   40
      Getting Information About a Data Set   *41*
1.14 Deleting and Inserting Rows   42
1.15 Deleting Columns   45
1.16 Viewing Information About the Data Set   45
1.17 Exporting Data to Excel   46
1.18 Importing Data from Excel   48
Minitab Command Summary   50
Review and Practice   52

Tutorial 2:
**Summarizing, Transforming, and
Manipulating Data**   59

2.1   Summarizing Cases and Rounding   60
      CASE STUDY: EDUCATION—CLASS EVALUATION   60
2.2   Summarizing Columns   65
2.3   Using Session Commands   66
2.4   Coding Data   69
2.5   Ranking Data   71
2.6   Sorting Data   72
2.7   Standardizing Data   74
2.8   Creating Subsets   75
2.9   Combining Data Using the Stack Option   76
2.10 Separating Data Using the Unstack Option   78
2.11 Printing the Results of Your Analysis   80
2.12 Creating a Text File   83
Minitab Command Summary   85
Review and Practice   86

Tutorial 3:
Data Analysis for One Variable 92

3.1 Creating a Project 93
   CASE STUDY: PUBLIC HEALTH—
   INFANT NUTRITION 93
3.2 Summarizing Qualitative Variables 95
3.3 Creating Bar Charts 97
   Naming Graphs 98
   The Project Manager 99
   Bar Graphs Using Summarized Data 100
   Saving Graphs 101
   Percentages Using Summarized Data 102
3.4 Creating Pie Charts 103
3.5 Quantitative Variables: Creating
   Histograms 105
3.6 Creating Stem-and-Leaf Displays 107
3.7 Creating Dotplots 111
3.8 Creating Individual Value Plots 112
3.9 Creating Boxplots 113
3.10 Quantitative Variables: Summarizing Data
   Numerically 115
3.11 Using StatGuide and the Glossary 119
3.12 Constructing Other Descriptive Statistics 120
3.13 Copying Text Output into ReportPad and
   Microsoft Word 121
3.14 Saving and Reopening a Project 123
   MINITAB AT WORK: EDUCATION 124
Minitab Command Summary 125
Review and Practice 126

Tutorial 4:
Data Analysis: Comparing Groups 132

4.1 Contingency Tables 133
   CASE STUDY: PUBLIC HEALTH—INFANT NUTRITION
   (CONTINUED) 133
4.2 Cluster and Stack Bar Charts 136
4.3 Comparing Dotplots 140
   CASE STUDY: SPORTS—BASEBALL STADIUMS 140
4.4 Comparing Individual Value Plots 141
4.5 Comparing Boxplots 142
4.6 Describing Subgroups 143
4.7 Using Charts to Display Descriptive
   Statistics 146
4.8 Exporting Graphs 149
Minitab Command Summary 151
Review and Practice 152

Tutorial 5:
Examining Relationships Between Two
Quantitative Variables 158

5.1 Creating Scatterplots 158
   CASE STUDY: SPORTS—BASEBALL STADIUMS
   (CONTINUED) 159
   Using Crosshairs 164
5.2 Adding a Grouping Variable to a
   Scatterplot 165
   Paneling a Scatterplot 167
5.3 Viewing the History Folder 168
5.4 Creating Marginal Plots 170
5.5 Computing Covariance 171
5.6 Computing Correlation 172
5.7 Computing the Least Squares/Regression
   Line 174
5.8 Displaying the Least Squares/Regression
   Line 175
5.9 Creating Plots on Which X Represents
   Time 178
   CASE STUDY: BUSINESS—COMPETITION 178
5.10 Overlaying Plots 180
5.11 Exporting Data 182
   CASE STUDY: SPORTS—BASEBALL STADIUMS
   (CONTINUED) 182
   Exporting Formatted Data 184
   Exporting Data to an Excel Spreadsheet 186
5.12 Importing Data 187
   Renaming a Worksheet in Minitab 188
   Importing Formatted Text Data Files 188
   Importing an Excel File 190
Minitab Command Summary 191
Review and Practice 193

Tutorial 6:
Distributions and Random Data 199

6.1 Calculating Binomial Probabilities 200
   CASE STUDY: BIOLOGY—BLOOD TYPES 200
   Individual Binomial Probabilities 200
   Cumulative Binomial Probabilities 204
6.2 Generating Random Data from a
   Discrete Distribution 207
   CASE STUDY: MANAGEMENT—
   ENTREPRENEURIAL STUDIES 207
6.3 Generating Random Data from a Normal
   Distribution 210
   CASE STUDY: PHYSIOLOGY—HEIGHTS 211

6.4 Checking Data for Normality 212
  The Normal Probability Plot 213
6.5 Determining Cumulative Probabilities and Inverse Cumulative Probabilities for the Normal Distribution 215
6.6 Sampling from a Column 218
  MINITAB AT WORK: PUBLIC SAFETY 220
Minitab Command Summary 220
Review and Practice 223

Tutorial 7:
  Inferences from One Sample 228

7.1 Testing a Hypothesis About μ When σ is Known 229
  CASE STUDY: SOCIOLOGY—AGE AT DEATH 229
7.2 Computing a Confidence Interval for μ When σ Is Known 234
7.3 Sample Size for Estimating μ When σ is Known 235
7.4 Inferences about μ When σ Is Unknown 237
  The t-Test and Confidence Interval with Summarized Data 239
7.5 Inferences About a Population Proportion 241
  CASE STUDY: HEALTH CARE— WORK DAYS LOST TO PAIN 241
7.6 Computing the Power of a Test 245
  MINITAB AT WORK: RETAILING 249
Minitab Command Summary 249
Review and Practice 250

Tutorial 8:
  Inferences from Two Samples 255

8.1 Comparing Population Means from Two Independent Samples 256
  CASE STUDY: SOCIOLOGY— AGE AT DEATH (CONTINUED) 256
  Two-Sample t-tests Using Stacked Data 256
  Using Help to Find a Formula 260
  Two-Sample t-Tests Using Unstacked Data 262
  Two-Sample t-Tests Using Summarized Data 264
  Obtaining a 95% Two-Sided Confidence Interval for $\mu_F - \mu_M$ 264

8.2 Inference on the Mean of Paired Data 265
  CASE STUDY: HEALTH CARE— CEREAL AND CHOLESTEROL 265
8.3 Sample Size and Power for Comparing the Means of Two Independent Samples 269
  CASE STUDY: WELFARE REFORM— ERRORS IN GRANT DETERMINATION 269
8.4 Comparing Population Proportions from Two Independent Samples 272
  CASE STUDY: HEALTH CARE— WORK DAYS LOST TO PAIN (CONTINUED) 272
  Comparing Two Proportions Using Summarized Data 274
8.5 Sample Size and Power for Comparing Two Independent Proportions 275
  MINITAB AT WORK: SCIENTIFIC RESEARCH 277
Minitab Command Summary 277
Review and Practice 278

Tutorial 9:
  Comparing Population Means: Analysis of Variance 283

9.1 Comparing the Means of Several Populations 284
  CASE STUDY: CHILD DEVELOPMENT— INFANT ATTENTION SPANS 284
9.2 Checking the Assumptions for a One-Way ANOVA 289
9.3 Performing Tukey's Multiple Comparisons Test 291
9.4 Comparing the Means of Several Populations with Responses in Separate Columns 294
9.5 Performing a Two-Factor Analysis of Variance 296
  CASE STUDY: PSYCHOLOGY— MEASURING DEPTH PERCEPTION 296
  The Two-Way ANOVA 297
  Checking the Assumptions for the F-tests in a Two-Way ANOVA 301
  The Interactions Plot 303
  Creating Factor Levels 304
Minitab Command Summary 306
Review and Practice 307

Tutorial 10:
Fundamentals of Linear Regression 312

10.1 Fitting a Straight Line to Data: Simple Linear Regression 313
    CASE STUDY: INSTITUTIONAL RESEARCH—TUITION MODELING 313
    A Confidence Interval for the Population Slope 317
    Obtaining Residuals 318
    The Fitted Line Plot 320
10.2 Computing Response Variable Estimates 321
10.3 Performing a Quadratic Regression 324
    Using Transformations 326
10.4 Performing Multiple Linear Regression 326
10.5 Obtaining Multiple Linear Regression Response Variable Estimates 328
    MINITAB AT WORK: HUMAN RESOURCES 331
Minitab Command Summary 331
Review and Practice 332

Tutorial 11:
Building Regression Models 337

11.1 The Importance of Graphs in Regression 338
    CASE STUDY: DATA ANALYSIS—IMPORTANCE OF GRAPHS 338
11.2 Identifying Collinearity 345
    CASE STUDY: INSTITUTIONAL RESEARCH—TUITION MODELING (CONTINUED) 345
11.3 Verifying Linear Regression Assumptions 349
    Storing the Residuals 349
    Checking the Normality Assumption 351
    Using Scatterplots to Verify the Homoscedasity Assumption 351
    Checking the Independence Assumption with a Time Series Plot 353
11.4 Examining Unusual Observations 355
    Cook's Distance 357
11.5 Incorporating an Indicator (Dummy) Variable into a Model 357
    Interpreting the Regression Coefficient for an Indicator Variable 361
11.6 Performing Best Subsets Regression 362
11.7 Performing a Binary Logistic Regression 368
Minitab Command Summary 372
Review and Practice 373

Tutorial 12:
Analyzing Qualitative Data 378

12.1 Comparing an Observed Distribution of Counts to a Hypothesized Distribution 379
    CASE STUDY: MANAGEMENT—ENTREPRENEURIAL STUDIES (CONTINUED) 379
12.2 A Minitab Exec Macro for a Chi-Square Goodness-of-Fit Test 384
12.3 The Chi-Square Test for Independence for Two Qualitative Variables in a Contingency Table 388
    CASE STUDY: HUMAN RESOURCES—EMPLOYMENT STATISTICS 388
12.4 The Chi-Square Test for Independence for Two Qualitative Variables Using Raw Data 394
Minitab Command Summary 398
Review and Practice 399

Tutorial 13:
Analyzing Data with Nonparametric Methods 403

13.1 The Runs Test for Randomness 404
    CASE STUDY: METEOROLOGY—SNOWFALL 404
13.2 Testing a Hypothesis About the Population Median Using the Sign Test 408
    CASE STUDY: HEALTH CARE—CEREAL AND CHOLESTEROL (CONTINUED) 408
13.3 Estimating the Population Median with the 1-Sample Sign Confidence Interval Estimate 410
13.4 Testing Hypotheses About the Population Median Using the Wilcoxon Test 411
13.5 Estimating the Population Median with the Wilcoxon Confidence Interval Estimate 412
13.6 Comparing the Medians of Two Independent Populations 413
    CASE STUDY: SOCIOLOGY—AGE AT DEATH (CONTINUED) 413
13.7 Computing the Medians of K Independent Populations Using the Kruskal-Wallis Test 417
    CASE STUDY: CHILD DEVELOPMENT—INFANT ATTENTION SPANS (CONTINUED) 417
    MINITAB AT WORK: MEDICAL DIAGNOSTICS 420
Minitab Command Summary 421
Review and Practice 422

Tutorial 14:
   Time Series Analysis   426

14.1 Performing a Trend Analysis of a
     Time Series   427
     CASE STUDY: ENVIRONMENT—
     TEMPERATURE VARIATIONS   427
14.2 Performing a Classical Decomposition of a
     Time Series   430
14.3 Autocorrelation and Partial Autocorrelation
     Plots   435
     Autocorrelation   435
     Partial Autocorrelations   437
14.4 Transforming a Time Series   439
     Lagging Data   439
     Computing Differences   440
14.5 Performing a Box-Jenkins ARIMA Analysis of a
     Time Series   441
     Constructing a Seasonal Model   444
14.6 Forecasting with ARIMA   447
     Plotting the ARIMA Forecasts   448
14.7 Comparing the Two Forecasting Models   449
     MINITAB AT WORK: STOCK MARKET   451
Minitab Command Summary   451
Review and Practice   452

Tutorial 15:
   Total Quality Management Tools   456

15.1 Creating a Cause-and-Effect Diagram   457
     CASE STUDY: EDUCATION—FACULTY SURVEY   457
15.2 Creating a Pareto Chart   459
15.3 Constructing an Xbar Chart   462
     CASE STUDY: PRODUCTION—
     QUALITY CONTROL CHARTS   462
15.4 Constructing a Range Chart   466
15.5 Constructing an Individuals Chart   469
15.6 Constructing a Moving Range Chart   470
15.7 Constructing a Proportion Chart   473
     MINITAB AT WORK: QUALITY MANAGEMENT   476
Minitab Command Summary   477
Review and Practice   478

Appendix A:
   Data Sets   483

Academe.mtw   484
AgeDeath.mtw   484
Assess.mtw   485
Baby.mtw   485
Backpain.mtw   486
BallparkData.mtw   486
BodyTemp.mtw   486
Candya.mtw – Candyc.mtw   487
     Candya.mtw   487
     Candyb.mtw   487
     Candyc.mtw   487
Carphone.mtw   488
Chol.mtw   488
CollMass.mtw and *CollMass2.mtw*   488
Compliance.mtw   489
CongressSalary.mtw   490
Cotinine.mtw   490
CPI2.mtw   490
Depth.mtw   490
DJC20012002.mtw, DJC20012002a.mtw
     and DJC20012002b.mtw   491
Donner.mtw   492
Drive.mtw   492
DrivingCosts.mtw   492
DrugMarkup.mtw   493
Election2.mtw   493
EMail.mtw   493
EmployeeInfo.mtw   493
Endowment.mtw   494
ExamScores.mtw   494
Fja.mtw   494
Force.mtw   495
GasData.mtw   495
Height.mtw   495
Homes.mtw   496
Infants.mtw   496
Jeans.mtw   497
Lakes.mtw   498
Lotto.mtw   499
Marathon2.mtw   499
Marks.mtw   499
MBASurvey.mtw   500
Mercedes.mtw   500
MLBGameCost.mtw   500
MnWage2.mtw   501
MonthlySnow   501
Movies.mtw   501

Murders.mtw and *Murderu.mtw*   502
    Murders.mtw   502
    Murderu.mtw   502
MusicData.mtw   503
NHL2003.mtw   503
Note02.mtw   504
OldFaithful.mtw   504
OpenHouse.mtw   504
PayData.mtw   505
PhoneRates.mtw   505
Pizza2.mtw   505
Process.mtw   506
Prof.mtw   506
Pubs.mtw   507
PulseA.mtw   507
Radlev.mtw   508
RandomIntegers.mtw   508
Rivera.mtw – Rivere.mtw and *Rivers.mtw*   509
Riverc2.mtw   510
Salary02.mtw   510
SBP.mtw   511
SchoolsData.mtw   511
Sleep.mtw   512
SP5002.mtw   512
SPCarData.mtw   512
SpeedCom.mtw   513
Stores2.mtw   513
Survey.mtw   514
TBill2.mtw   514
Temco.mtw   514
Textbooks.mtw   515
Top25Stars.mtw   516
Tvhrs.mtw   516
TwinsYankees.mtw   517
TwoTowns.mtw   518
UGradSurvey.mtw   518
USAArrivals.mtw   519
USDemData.mtw   519
WastesData.mtw   519
YearlySnow.mtw   520
YogurtData.mtw   520

Appendix B:
   Minitab Menus and Toolbars   521
Menus   522
Toolbars   525

Index   527

Credits   541

Tutorial

Getting Started with Minitab

Welcome to *The Student Guide to Minitab!* As the name suggests, this book is designed to guide Minitab users through the software. This first tutorial will introduce you to Minitab and take you through a brief analysis. The more specific objectives of this tutorial are set out below.

OBJECTIVES

In this tutorial, you learn how to:

- Become familiar with the organization and conventions used in this book
- Navigate around Minitab and use online Help to get information about the software and the statistical procedures it performs
- Open and explore an existing worksheet
- Describe the characteristics of a variable by using basic statistics
- Create a professional-quality graph based upon data in the worksheet

0.1 Introduction to Minitab

This section will introduce you to the features and organization of this book and to some of the important features of Minitab.

Minitab is an easy-to-use, general-purpose statistical computing package for analyzing data. It is a flexible and powerful tool that was designed from the beginning to be used by students and researchers new to statistics. It is now one of the most widely used statistics packages in the world. Minitab produces accurate and professional quality graphs and performs tedious computations almost instantly. This power frees the user to focus on the exploration of the structure of the data and the interpretation of the output.

This book describes the features of MINITAB Release 14. There are two versions of MINITAB Release 14. You can have up to 200 graphs open at once (though the default number is only 100) and up to 4,000 columns of data in a worksheet in both versions.

1. MINITAB Release 14 contains all of Minitab's commands. The size of the worksheets permitted in this version are limited only by the memory available on the computer.

   MINITAB Release 14 may be purchased or leased from Minitab. Students (and their instructors) may rent this Professional version for limited use. Go to Minitab's Web site *www.minitab.com* for information about these options.

2. MINITAB Student Release 14 is an educational form of MINITAB Release 14. It includes all but a limited number of the features and functions of MINITAB Release 14. In MINITAB Student Release 14, for example, worksheets may contain up to 10,000 cells and no more than five worksheets may be open at a time.

   You can purchase this book in one of two ways: with MINITAB Student Release 14 or as a standalone guide. (MINITAB Student Release 14 is also available bundled with a variety of textbooks.) The Minitab Web site (*www.minitab.com*) contains more information about acquiring MINITAB Student Release 14 as well as a complete comparison of the features of the two versions of the software.

   Whether you have access to MINITAB Release 14 or MINITAB Student Release 14, this book will take you step by step through the software. Both versions of the software contain the data files that you will need to complete the tutorials in this book.

   To run MINITAB Release 14, regardless of version, your computer system must meet the following requirements:

- Windows® 98, ME, NT 4, 2000, or XP (Windows 2000, XP recommended)
- 300 MHz processor (500 MHz recommended)
- 64 MB RAM (160 MB recommended)
- CD-ROM required for installation
- 85 MB hard disk for full installation

   It is important to note that MINITAB Release 14 will not run on Windows 95 or on earlier Windows operating systems. In addition, Minitab does not support a version for the Macintosh or the Linux operating systems.

◆ **Note** This book is intended for users of both MINITAB Release 14 and MINITAB Student Release 14. Users of MINITAB Release 14 (the Professional version) should ignore references to the Student version. ■

What's in This Book?

This book uses a series of hands-on tutorials to lead the new user through the most widely used Minitab features. The tutorials apply all of the techniques found in the standard, applied, introductory statistics course. This first tutorial introduces you to the book and the software.

Regardless of your previous experience with the Windows operating system and with online Help, you should read through the remainder of this introductory tutorial (Tutorial 0).

Tutorials 1 through 15 form the heart of the book. Beginning with the simplest features of Minitab, the tutorials introduce the major features of the software in a step-by-step, hands-on format. The extensive exercises and practice problems at the end of each tutorial reinforce the procedures you have learned. Almost all of the tutorials and practice problems use the data sets that come as part of Minitab.

Here is a rough breakdown of the contents of these tutorials.

- In Tutorials 1 and 2 you will learn the fundamentals of entering, saving, organizing, manipulating, and retrieving data in Minitab.
- Tutorials 3 through 7 cover the use of Minitab for graphical and numerical data analysis, for working with probability distributions, and for basic inference.
- The remaining tutorials, 8–15, cover the use of Minitab to perform inferential procedures found in most introductory, applied statistics books. These include two-sample inferences for means and proportions, analysis of variance, simple and multiple regression, chi-squares tests, nonparametric methods, time series analysis, and quality methods.

Some of the later tutorials continue case studies introduced in earlier ones, but you can still do these problems independent of their earlier use.

There are two appendices at the back of the book. Appendix A contains a detailed description of each of the data sets used in this book as well as other data sets on which you can practice. Appendix B is a complete pictorial list of all the menus and toolbars associated with the Student version of Minitab.

Typographical Conventions

Typographical conventions and symbols are used throughout this book to make the material easier to use.

Any action that you should perform appears on a line by itself, indented, and preceded by a square bullet (■). Boldface type indicates letters, numbers, and symbols that you are to enter using the keyboard. Also in boldface are elements on the screen, including commands that you are to choose or click. For example:

■ Choose **Stat > Basic Statistics > Display Descriptive Statistics**
■ Click the Minitab Help window's **Close** button ⊠ and then click on the **Cancel** button

Here, *Click* indicates that you depress the left-hand mouse button.

References to clicking the right-hand mouse button are indicated explicitly, as below.

- Hold your cursor over the title, click the **right-hand mouse button**, and choose **Edit Title** from the shortcut menu

Here, the term *mouse* is used to refer to both the device with that name and the touch pad on a laptop computer.

Quotation marks set off labels that identify sections of the dialog box in which you must enter data on the screen.

Italicized type introduces a new statistics or Minitab term or concept. Such type is also used with file names and when issuing commands to obtain graphs.

Choose tells you to make a menu selection by using either the keyboard or the mouse.

Type instructs you to enter text using the keyboard. It is important to type this information exactly as shown, including any punctuation or spacing.

Press indicates that you should press the specified key on the keyboard, such as Tab or ◄┘ (the Enter key).

Instructions for menu selections will ordinarily be displayed in an abbreviated form using the greater than symbol (>). For example, instructions to open the Stat menu, choose Basic Statistics, and then choose Display Descriptive Statistics, will be written as follows:

- Choose **Stat > Basic Statistics > Display Descriptive Statistics**

You can invoke commands in Minitab in a variety of ways. You should follow the tutorial instructions exactly so that you will gain experience with the different ways of issuing commands. As you become familiar with them, you can use the approach that is most comfortable for you. Be aware also that the contents of some menus change depending on which window is active.

Technical Support

If you encounter difficulty operating Minitab, please check the Troubleshooting section of Minitab's online help (see Section 0.2 for instructions on using Help). Additional tips and technical support are available on the Minitab Web site at *www.minitab.com* and the Addison-Wesley Web site at *www.aw-bc.com*, keyword: MINITAB.

Those who purchase or lease MINITAB Release 14 are entitled to telephone support from Minitab. Registered instructors who have adopted this product for their students are also entitled to such support. However, neither Minitab, Inc., nor Addison-Wesley Publishing Company, Inc., provides telephone support to users of MINITAB Student Release 14. Students should report any problems they encounter with the Minitab version to their instructors. Be sure to note exactly what action you were performing when the problem occurred, as well as the exact error message received, if any.

Installing Minitab

Whether you are downloading Minitab onto your hard drive or loading it from a CD-ROM, the authors recommend that you install your Minitab folder in your Programs folder (this is the default) and that you create a Minitab 14 icon on your desktop. The Minitab installation program will do this for you automatically, but you may consult the Minitab Web site at *www.minitab.com* for help with any questions that arise during installation.

▶ **Note** MINITAB Student Release 14 cannot be installed on a network server. ■

▶ **Note** Because the Minitab CD-ROM contains compressed files, you must use the installation program to load the files onto your hard disk. You will not be able to use these files if you copy them directly from the CD-ROM to your hard drive. ■

The Minitab Menu

If you installed Minitab to the default location, you will find a Minitab folder in the Programs folder on your Start menu (if you did not, your folder may be located elsewhere on your Start menu). To see the Minitab menu options:

■ Choose **Start** > **Programs** > **MINITAB 14** (or **MINITAB 14 Student**) (Recall that this means click Start, then click on Programs, and then click on Minitab.)

For the Professional version of Minitab, you should see the following eight items on the Minitab menu, as shown in Figure 0-1. (The items may be in a slightly different order)

FIGURE 0-1

The Minitab menu for the Professional version on the Start menu

1. "Meet MINITAB" is a self-contained, basic guide to using Minitab.
2. "MINITAB 14" launches your copy of Minitab.
3. "MINITAB Help" is available here and also during a Minitab session from its menu bar. (You will learn about Minitab Help in the next section.)
4. "MINITAB Session Command Help" is available here and also during a Minitab session from its menu bar. (You will learn about this component of Minitab Help in Tutorial 2.)
5. The "MINITAB Tutorials" are self-paced introductions to various aspects of Minitab (not to be confused with the tutorials in this book). Several of these Minitab tutorials are recommended during the course of the book.
6. The "ReadMe" file includes technical information that may be useful if you encounter installation difficulties.
7. "StatGuide" is Minitab's guide to interpreting statistical output. It is introduced in Tutorial 3 in this book.
8. "What's New" provides a listing of the features that are new to Release 14 of Minitab.

If you performed a complete installation using the default options and do not see these items on your menu, repeat the installation process.

Types of Data Files

Moving through the tutorials in this book, you will work with two different sets of data files:

1. The data files that come with Minitab:

These data files do not appear on the Minitab menu but are stored in *data folders*. Almost all the data sets used in this book are in a folder named Studnt14. This should be the default folder if you are using the Student version of Minitab. A large number of other data sets may be found in a folder appropriately called Data. This should be the default folder if you are using the Professional version of Minitab. As you work through the tutorials, you will be asked to open files from one of these two folders and work with the data they contain. If you are working on a lab computer, your instructor may have installed the data files in a different location. If so, substitute that location each time you are asked to open a file from one of these folders.

2. The files that you will create as you work through the tutorials:

You should save these files to a folder *other* than Studnt14 or Data so that you don't accidentally overwrite an existing file that you will need later. If you are working in a school lab, you will want to save your work to a floppy disk, to a CD-ROM, or to a removable disk drive. (You shouldn't count on saving your work on the school's computer.) If you are working on your own computer or have an account on the school's network, create a different folder in which you will save all of your work. In the tutorials you will be directed to save files to the location where you are saving your work.

0.2 Touring Minitab and Using Help

Opening Minitab

You can open Minitab in one of two ways:

- Double-click the **MINITAB 14** (or **MINITAB 14 Student**) icon 🖥 on the desktop

  or

- Choose **Start > Programs > MINITAB 14** (or **MINITAB 14 Student**) > **MINITAB 14** (or **MINITAB 14 Student**)

  When you open Minitab for the first time, a First Time Alert notice appears.

- Click the **Close** button on the alert window

The main Minitab window opens as shown in Figure 0-2. The Minitab menu bar appears near the top of the window. Each menu lists groups of related commands that you use to operate Minitab. Below the menu bar is the toolbar. The buttons on the toolbar invoke the most commonly used commands and make them available with a single click of the mouse.

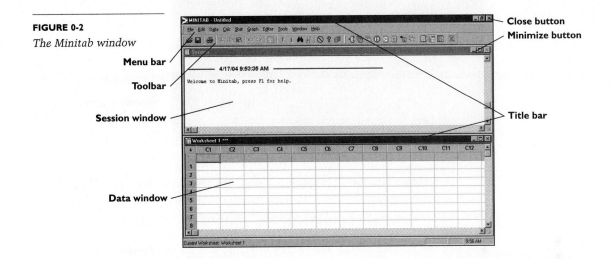

You see two of Minitab's open windows: the Session window and the Data window (labeled Worksheet 1 ***). Each window has its own title bar, and can be restored, moved, sized, minimized, and maximized. The Student version's session window informs you that worksheets may only contain up to 10,000 cells.

One other window, the Project Manager window, is open but minimized. To see this window:

■ Minimize the Session window by clicking its **Minimize** button ▬ and the Data window by clicking its **Minimize** button ▬

Your screen should look as shown in Figure 0-3, with all three windows lying along the bottom of the screen.

FIGURE 0-3

The Minitab windows minimized

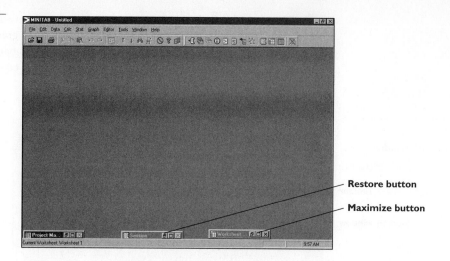

Restore button

Maximize button

To return a window to its previous size, you can click on its title bar and use the menu that appears to restore it. This menu also allows you to move the window, maximize it, or close it. You also can click its maximize, restore, or close buttons directly.

- Click the **Restore** button 🗗 , shown in Figure 0-3, for the **Session** window

Notice that each window, except the Data window, is identified by its name in the title bar. The Data window's title bar displays the name of the current data set, or worksheet, in this case the generic "Worksheet 1 ***".

- Click the Data window's **Maximize** button ⬜

When a window is maximized, its title is added to the Minitab window title bar. For the Professional version, for example, the window title is "Minitab - Untitled - [Worksheet 1 ***]" , as shown in Figure 0-4.

FIGURE 0-4

The Data window maximized

Active cell

Status bar

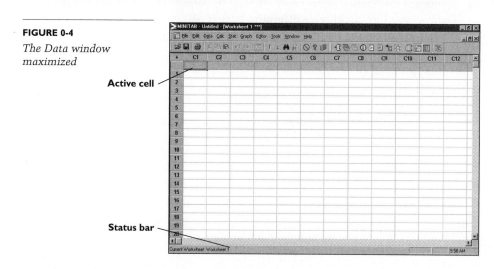

When one window is maximized, its title bar is not visible, but its mini-mize, restore, and close buttons appear directly below those for the main Minitab window. You can use these to change the size of the window or to close it.

To make a different window active when one is maximized, choose it by name from the Window menu on the menu bar:

■ Choose **Window > Session**

The Session window is now active and maximized, and its name appears in the Minitab title bar. To make the Data window active again:

■ Choose **Window > Worksheet 1 *****

The Data window is active again. Only one Minitab window can be active at a time. It is important to keep in mind that several of the Minitab menus change depending on which window in active.

Worksheets and the Data Window

The Data window presents a view of the data set stored in a *worksheet*. Each worksheet has its own Data window. With the Professional version of Minitab there is no limit on the number of worksheets—Data windows—you can have open at once. With the Minitab Student version you can have no more than five worksheets—Data windows—open at the same time.

Similar to a spreadsheet, a worksheet has *columns* and *rows*. However, unlike electronic spreadsheets, such as those in Microsoft Excel and Lotus 1-2-3, Minitab worksheets contain only numbers, text, and dates—not formulas. Minitab worksheets can also store single-number constants and matrices, although you can't see them in the Data window.

A *cell* is the intersection of a column and a row. Notice the outline around the cell in column C1 in Figure 0-4. The outline means that the cell is active. You can activate any cell in the worksheet by clicking in it. Minitab updates the Data window automatically whenever you make any changes to the worksheet.

Generally, each column lists data for one variable and each row contains a set of observations for an individual case. The worksheet may contain up to 4,000 columns. In the Professional version you may have as many cells as your computer's available memory allows. In the Student version the total number of cells must be less than or equal to 10,000. In Minitab columns are referred to by column number (C1, C2, and so on) or by name. Assigning names to the columns makes it easier to remember what they contain. (The columns in the worksheet shown in Figure 0-4 are unnamed.) Rows are referred to by number: 1, 2, and so on.

Minitab commands work on the current worksheet, which is identified in the *status bar* at the bottom of the screen. Worksheets (no more than five in the Student version) and the information associated with them (in the Session, Graph, and Project Manager windows) can be stored in a *project*. You will learn more about how Minitab stores data, about projects, and about Graph windows, as you work through Tutorials 1–3.

▶ **Note** Be careful to avoid opening the Student version two or more times simultaneously. If you try to perform any command in such a situation, Minitab will terminate the program. This problem should not occur with the Professional version of Minitab. ■

Issuing Minitab Commands

Virtually all Minitab commands can be issued in a variety of ways. Depending on the command, you may choose it from a menu on the menu bar, click a corresponding toolbar button, use a right-hand mouse click, or type its corresponding Session command in the Session window. Most menu commands open a dialog box that allows you to make further choices with regard to how the command will be carried out.

You'll learn more about the individual commands and the different ways to issue them in the first few tutorials in the book. To learn more about a particular command or to get help when you are in a dialog box, you can use Minitab's online Help feature.

Context-Sensitive Help

There are a variety of ways to access help in Minitab. One particularly useful method is available at the point of issuing a menu command. Here is an example.

- Choose **Stat > Basic Statistics**

    Notice that Minitab uses pictorial aids on many of the menu commands to illustrate the general intent of the command.

- Choose **Display Descriptive Statistics**

    The Display Descriptive Statistics dialog box appears. To find out more about the Display Descriptive Statistics command:

- Click the **Help** button in the lower left-hand corner of the dialog box

    Minitab Help displays—in the right-hand side of the window—the main Help entry for Display Descriptive Statistics, as shown in Figure 0-5. (Your window may be a different size when it opens. As with other windows, Help windows may be moved and resized.)

FIGURE 0-5

The Display Descriptive Statistics Help window

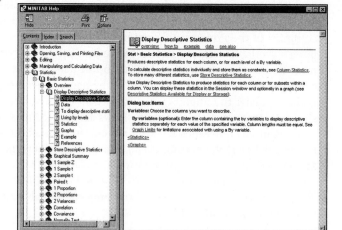

After a brief explanation of the function of this command, each of the options in the Display Descriptive Statistics dialog box is explained. Underlined words in the Help window are *links* to additional information. For most commands, Minitab Help offers the following links:

- "overview" introduces the command and the statistical concepts related to it
- "how to" presents step-by-step instructions for using the command
- "example" shows how the steps would be carried out with a particular data set and offers help with interpreting the results
- "data" describes conditions that must be met by the data set to be used with the command
- "see also" links to topics that relate to the command

To view an application of this command:

■ Click on **example**

You can see an example of the descriptive statistics and graphs available with this command.

■ Click the Minitab Help window's **Close** button ☒ and then click the **Cancel** button

Using the Help Menu

The Help menu, shown for the Professional version in Figure 0-6, offers other ways to get information about Minitab. (The corresponding menu for the Student version is only slightly different.)

FIGURE 0-6

The Help menu

The menu presents a number of areas in which to receive help. Here the focus is on Help.

■ Choose **Help > Help**

The main Minitab Help window opens as shown in Figure 0-7.

FIGURE 0-7

The Minitab Help window

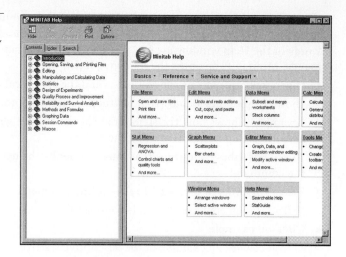

In the right-hand side of the window you can search through links related to each of the Minitab menus as well as links to Basics, Reference, and Service and Support. If, for instance, you need a definition for a term you encountered in Minitab, the Reference link will lead you to a Glossary.

For a second example:

■ Choose **Stat Menu** in the right-hand panel

Links to all of the commands on the Stat menu appear.

■ Choose **Basic Statistics**

Each item under Basic Statistics is also a link.

■ Choose **Display Descriptive Statistics**

The main Help window for this command appears as you saw it in Figure 0-5.

Getting Help with Data Sets

All of the data sets you will use in this book—and several others that are included with Minitab—are described in Minitab's Help. You can see a list of these data sets in alphabetical order by searching on the topic Data Sets. To search for a topic in Help, use the Search tab.

■ If necessary, click on the **Search** tab in the left-hand side of the Help window

■ Type **data sets** in the "Type in the word(s) to search for" text box and click the **List Topics** button

■ Highlight **sample data sets** in the list box and click the **Display** button

The Sample Data Sets topic appears, as shown in Figure 0-8. The alphabet buttons at the top of the window link you to a list of data sets with names that start with a given letter.

FIGURE 0-8

*The Sample Data Sets
Help window*

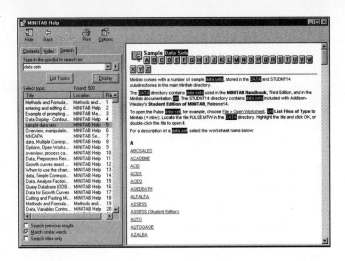

You can also type the name of the data set itself to go directly to its Help window.

■ Type **OldFaithful** in the "Type in the word(s) to search for" text box, click the **List Topics** button, and double-click ***OLDFAITHFUL.MTW*** in the list box

The description of the *OldFaithful.mtw* data set and the variables it contains appears as shown in Figure 0-9.

FIGURE 0-9

*The OldFaithful.mtw
Help window*

File location

Variable descriptions

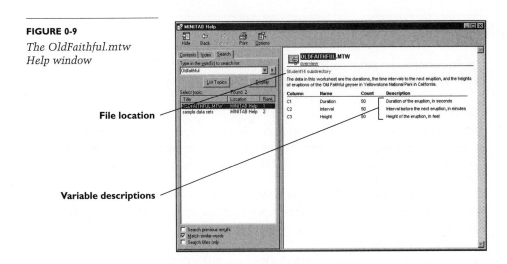

➧ **Note** All Minitab data files end with the file type *mtw*. ■

For each data set, Minitab Help tells you where the file is located; in this case, it is in the Studnt14 folder. After a brief description of the file, each variable is listed with its column number, variable name, number of observations, and description.

◆ **Note** In this book the term *folder* is used rather than the more traditional terms *directory* and *subdirectory*. Be aware, however, that the latter two terms are still used in some Minitab documentation. ■

When you have located a topic, you can print a copy of the information by clicking the Print button in the Help window.

    To leave Help:

■ Click the **Close** button on the Help window

Exiting Minitab

To exit Minitab:

■ Choose **File > Exit**

    Minitab returns you to the Windows desktop or to the last open application.

Web site for the Book

Addison-Wesley Publishing maintains a Web site for users of this book. The address is *www.aw-bc.com*, keyword: MINITAB. On this site you will find:

• Solutions to selected end-of-tutorial problems

• A list of alternative pathways through the book that correspond to popular statistics textbooks

• Errata

0.3 | *Sample Minitab Session*

In this sample session, you learn how to perform a simple descriptive statistical analysis of the *OldFaithful.mtw* data set.

Getting Started

Before beginning this Minitab session decide where you will be storing your work. Choose either a location on your hard drive (not the Studnt14 or Data folder) or a device such as a floppy disk, a CD-ROM, or removable drive and create a folder in which to store your work.

    If you are running Minitab on a school computer you may need to ask your instructor or technical resource person where Minitab is located on the Start menu and where you should store your work.

    To open Minitab:

■ Choose **Start > Programs > MINITAB 14** (or **MINITAB 14 Student**) > **MINITAB 14** (or **MINITAB 14 Student**)

    If a First Time Alert notice appears:

■ Click the **Close** button on the alert window

Exploring an Existing Worksheet

In this section you'll explore an existing data set named *OldFaithful.mtw*. The tutorials in this book are based upon case studies and corresponding data sets, as you will see below.

CASE STUDY	GEOLOGY—DURATION OF "OLD FAITHFUL" ERUPTIONS

The "Old Faithful" geyser in Yellowstone National Park in California is one of the most popular tourist destinations in the United States. At regular intervals water, heated to over 350 degrees, erupts from the ground in spectacular fashion. As an intern with the National Park Service, you have been collecting data on these eruptions. You decide to use Minitab to analyze these data.

To open the file that contains the data:

- Click the **Maximize** button ▢ in the upper right-hand corner of the Data window so that the Data window fills the Minitab window
- Choose **File > Open Worksheet**

The Open Worksheet dialog box appears. The data file, *OldFaithful.mtw*, is located in the Studnt14 folder. If you are operating the Student version, the Studnt14 folder will be the default folder appearing in the "Look in" list box at the top of the dialog box. If you are operating the Professional version of Minitab the default folder appearing in the "Look in" list box is not Studnt14 but Data. To locate the Studnt14 folder:

- Click the **Up One Level** icon ▣ to the right of the "Look in" list box

    You should now see the Studnt14 folder icon.

- Double-click on the **Studnt14** folder
- Use the scroll bar at the bottom of the list box to scroll the list of filenames until you find *OldFaithful.mtw*
- Double-click on the icon associated with ***OldFaithful.mtw***

    At this point you may see the MINITAB message "A copy of the contents of this file will be added to the current project." If so,

- Click **OK**

    The Data window should now look as shown in Figure 0-10. Minitab displays the data in the Data window; the name of the current worksheet, *OldFaithful.mtw*, is in the title bar.

FIGURE 0-10

OldFaithful.mtw worksheet

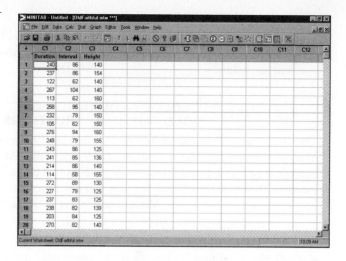

The data on Old Faithful consist of observations on 50 eruptions. The first column—named Duration—contains the time (in seconds) that the eruption lasted. The second and third columns contain, respectively, the interval (time period in minutes between eruptions) and the height (in feet) of the eruption.

Getting Information About a Data Set

As you work with data sets in the Studnt14 folder, you can learn about the variables contained in them in three ways:

1. By reading about them in Appendix A in this book, "Exploring Data Sets"
2. By looking at the description in online Help (as you did in the last section)
3. By viewing the Info window

   To view the Info window:

■ Click the **Show Info** toolbar button ⓘ

   Minitab displays the Info window—as a component of the Project Manager—to the left of the Data window, as shown in Figure 0-11.

FIGURE 0-11

The Info window

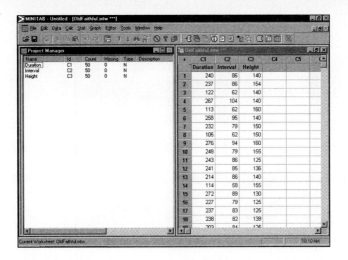

The Info window summarizes the worksheet contents; Minitab updates it automatically as the worksheet changes. The Info window for this worksheet lists the three columns in the worksheet, the name of each column, the column Id (C1, C2, and so on), the number of observations in each column (50), and the number of missing observations in each column. (In this data set there are no missing observations, but you will encounter them later in the book.) The last column in the Info window indicates the type of variable in that column. In this case all of the variables are *Numeric (N)*. Minitab also recognizes *Text (T)* variables and *Date/Time (D)* variables. You will see examples of these other variable types in subsequent tutorials. Minitab allows you to add a brief description of each variable. If you had done so, they would appear in the last column of the Info window.

To hide the Info window:

■ Maximize the Data window

The Data window fills the screen.

In your preliminary analysis you will focus on the Duration variable—how long the eruption lasts. As a first step, obtain some basic descriptive statistics for this variable.

■ Choose **Stat > Basic Statistics > Display Descriptive Statistics**

The Display Descriptive Statistics dialog box appears as in Figure 0-12.

FIGURE 0-12

*The Display Descriptive
Statistics dialog box*

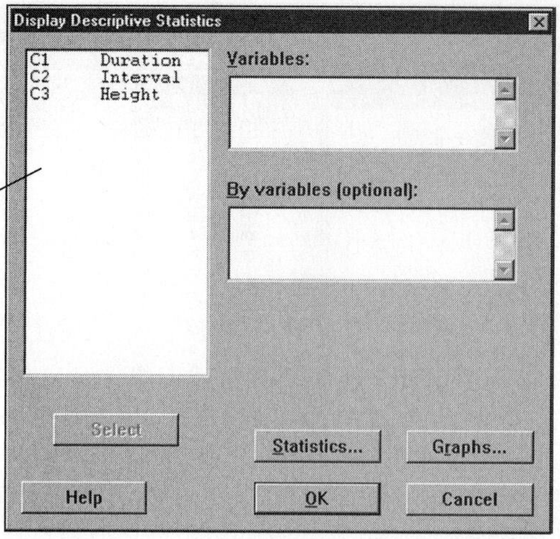

List box

- Double-click on **C1 Duration** in the list box to enter it in the "Variables" text box.
- Click **OK**

The Session window moves to the forefront and displays the output shown in Figure 0-13.

FIGURE 0-13

*Descriptive statistics for
Duration*

Descriptive Statistics: Duration

```
Variable  N  N*   Mean SE Mean StDev Minimum      Q1 Median     Q3
Duration 50  0 216.64    8.25 58.32  102.00 187.25 238.00 264.25

Variable Maximum
Duration  276.00
```

The results of the Display Descriptive Statistics command appear in the Session window. Minitab produces the number of observations (N = 50, in this case), the number of missing observations (there are N* = 0 missing observations in this case), and eight statistics related to the variable Duration, including the mean, median, and standard deviations.

▶ **Note** Minitab uses uppercase N to refer to the number of observations (missing or non-missing) in a column. To the extent that the data in a column are regarded as a sample, this practice differs from the almost universal practice of using lowercase n for the sample size (and, frequently, N for the population size). ■

The output indicates that, over the 50 eruptions you studied, the average (mean) duration was 216.64 seconds (roughly 3.6 minutes). The median duration was 238 seconds (almost 4 minutes). The fact that the median exceeds the mean suggests that the distribution of the duration of eruption is skewed to the left. There was a considerable amount of variation in the length of eruptions as indicated by the standard deviation of 58.32 seconds. The eruptions varied in duration from a low of 102 seconds to a high of 276 seconds.

Creating Graphs

To check on the shape of the distribution of eruption times, obtain a histogram of Duration.

- Choose **Graph > Histogram**

  The Histograms *pictorial gallery* appears, as shown in Figure 0-14.

FIGURE 0-14

The Histogram pictorial gallery

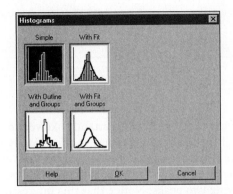

- Double-click on the picture beneath "Simple"

  The Histogram - Simple dialog box appears, as shown in Figure 0-15.

FIGURE 0-15

The Histogram - Simple dialog box

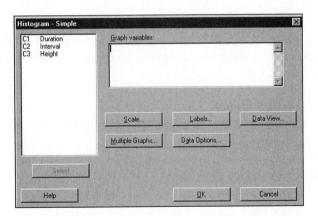

- Double-click on **C1 Duration** in the list box to enter it in the "Graph variables" text box
- Click **OK**

  Minitab displays a histogram of the 50 durations, as shown in Figure 0-16.

FIGURE 0-16

Histogram for Duration

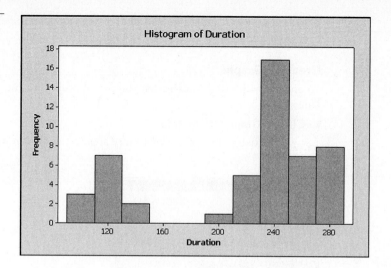

The distribution of Duration is certainly skewed to the left, but the more striking feature of the histogram is that it is quite clearly divided into two distinct components. The smaller group of durations is centered around 120 seconds. The larger group is centered around 240 seconds. This type of shape is often described as *bimodal*—having two peaks. Notice that this interesting feature of the eruption durations could not be discovered by examining the descriptive statistics. Graphs are an essential tool in understanding data. The form of the histogram of duration times emphasizes the importance of graphical displays. It is usually considered good statistical practice to obtain graphical displays before obtaining numerical summaries.

◆ **Note** You can also obtain a histogram of the 50 durations from the Display Descriptive Statistics dialog box. An example of such a *built-in* graph is given in Tutorial 3. ■

Printing a Graph

In order to include this graph in a preliminary report on the duration of eruptions, you need to print it.

- Make sure your printer is turned on
- Choose **File > Print Graph**

  Your system's Print dialog box appears.

- Click **OK**

◆ **Note** If you are using an institutional computer and the printer isn't directly attached to it, ask your instructor or technical resource person for the location of the printer. ■

You now have a copy of the graph to include in your report.

Exiting Minitab

To exit Minitab:

- Choose **File > Exit**

    A dialog box appears asking if you want to save the changes made to this project. You will learn more about saving your work in the later tutorials and do not need to save this project.

- Click the **No** button

    You have seen some of Minitab's potential. Now proceed to the more substantive tutorials, where you'll get hands-on experience using Minitab to create your own analyses. Remember to use Minitab's online Help whenever you want to learn more about commands and their options.

Review and Practice

The answers to the matching and true/false problems can be found at the end of each subsequent tutorial. The answers to the following problems can be found at the end of Tutorial 1.

Matching Problems

Match the following terms to their definitions by placing the correct letter next to the term it describes.

_____ 100

_____ 4,000

_____ 10,000

_____ Session

_____ Worksheet 1***

a. The Minitab window in which text output generally appears

b. The default maximum number of graphs that can be open in the Professional version of Minitab

c. A default name for the Minitab Data window

d. The maximum number of columns of data permitted in Minitab

e. The maximum number of cells permitted in the Student version of Minitab.

True/False Problems

Mark the following statements with a *T* or an *F*.

a. ____ Similar to electronic spreadsheet programs, Minitab worksheets can contain formulas.

b. ____ Minitab Help can be accessed only through the Help menu.

c. ____ Minitab 14 is released in a Professional version and a Student version.

d. ____ You can have only one Minitab window active at a time.

e. ____ Minitab recognizes four types of data columns: Numeric, Categorical, Text, and Date.

Working with Data

This tutorial introduces the fundamentals of working with Minitab. Much of the work will be done in the Data window. You will learn how to enter data in the Data window, navigate in this window, save and retrieve the data, and print out the data from this window.

OBJECTIVES

In this tutorial, you learn how to:

- Enter, correct, and name numeric and text data
- Save and retrieve data files
- Print data from the Data window
- Navigate the data worksheet
- Manipulate data by using expressions
- Use Minitab's Help system to find online information about a topic
- Close a worksheet
- Exit Minitab
- Enter missing value codes
- Copy and delete columns
- Insert and delete rows
- Obtain information about a data set
- Export data into Excel and import data from Excel

Opening Minitab

HEALTH MANAGEMENT— NUTRITIONAL ANALYSIS

You have just been hired as a nutritionist for a major hospital with responsibilities for planning healthy, economical meals. The first day, your supervisor, Patricia Johnson, assigns you the task of examining some data on yogurt. The data were collected by a research company that tested 17 brands of plain yogurt. The data set consists of three variables: an overall nutritional quality rating, the costs in cents per ounce, and the number of calories per eight-ounce serving. By statistically analyzing the data relating to the various brands, you hope to better understand the various yogurt products.

In this tutorial, you begin the statistical analysis by entering into Minitab the data shown in Figure 1-1. Then you will compute other measures using the cost and calorie variables.

FIGURE 1-1

Yogurt data

Row (Brand)	Nutritional Rating	Cost in Cents per Ounce	Calories per Eight-Ounce Serving
1	Excellent	13	140
2	Very Good	9	210
3	Very Good	6	180
4	Very Good	8	220
5	Good	8	170
6	Good	9	220
7	Good	11	160
8	Good	11	170
9	Good	6	210
10	Good	7	200
11	Good	7	100
12	Fair	5	210
13	Fair	9	120
14	Fair	11	100
15	Fair	11	190
16	Fair	9	120
17	Fair	8	120

Open Minitab in either of two ways:

- Double-click on the **MINITAB 14** (or **MINITAB 14 Student**) icon 📧 on the desktop

    or

- Click **Start > Programs > MINITAB 14** (or **MINITAB 14 Student**) > **MINITAB Release 14** (or **MINITAB Release 14 Student**)

    When you open the Professional version of Minitab for the first time, a First Time Alert notice appears. If you do not want to see the notice again:

- Click on the "If you do not want to see this alert anymore, check this box" check box and click on the **Close** button

    Otherwise:

- Click on the **Close** button

    A Session window appears in the upper half of the screen and a Data window appears in the lower half, as shown in Figure 1-2. A record of your Minitab session and much of your statistical analyses appear in the Session window. The Data window provides for easy data entry, modification, and viewing. In later tutorials you will work with two additional windows; the Graph and the Project Manager windows.

FIGURE 1-2

The Minitab window showing the Session and Data windows

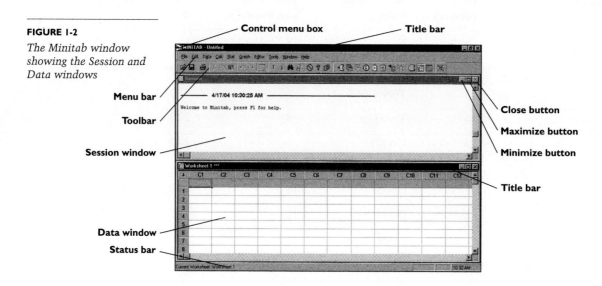

▶ **Note** Be careful to avoid opening two or more simultaneous versions of the Student version of Minitab. If you try to perform any command in such a situation, Minitab will terminate the program. ■

Before entering the yogurt data take a brief tour of the Minitab window. If you are running the Professional version of Minitab, the title bar for the Minitab window reads: Minitab - Untitled. In the Student version, the title bar will read Minitab Student - Untitled. The title bar for the Session window displays Session.

The Data window, however, is titled Worksheet 1 ***—this is its default name. When you save your current worksheet, the name you use replaces Worksheet 1. The three asterisks, ***, indicate that this is your active data set, that is, the one with which you are currently working. (The Student version of Minitab allows you to have up to five data sets open at one time.) Regardless of the version of Minitab you are running, only one worksheet can be current at a time.

The menu bar below the Minitab window title bar contains the names of Minitab's ten menus: File, Edit, Data, Calc, Stat, Graph, Editor, Tools, Window, and Help. The commands associated with each menu are organized by function. The Window menu, for example, contains commands you use to manipulate and move among windows.

The row of buttons beneath the menu bar is the toolbar. To see the function of any particular button, place your mouse cursor over the button; a tooltip appears, indicating the button's function. Single-clicking on toolbar buttons is a fast way to select commonly used operations. Most, but not all of the toolbar buttons correspond to a menu command. The Standard and Project Manager toolbars are shown in Figure 1-3. Your screen may also include another toolbar: the Worksheet toolbar.

FIGURE 1-3

The Standard and Project Manager toolbars

Print Window   Edit Last Dialog   Help   Session Window
        Undo/Redo   Current Data Window

Beneath the Data window is the status bar. This indicates the name of the current worksheet, in this case Worksheet 1. To make a window current, you can either click it, select it from the list of windows on the Window menu, or use the appropriate toolbar button. (The Window menu is handy to use when windows are not visible on screen.) You can easily switch between the Data and Session windows. For example, to go to the Session Window:

■ Click on the **Session Window** toolbar button ▦ (or choose Window > Session)

   The Session window is now active.

■ Click on the **Current Data Window** toolbar button ▦ (or choose Window > Worksheet1 ***)

   The Data window is now active.

You can invoke commands in Minitab in a variety of ways. You should follow the tutorial instructions exactly so that you will gain experience with the different ways of issuing commands. As you become familiar with them, you can use the approach that is most comfortable for you. Be aware that the contents of menus often change depending on which window is active.

◆ **Note** Instructions to choose menu commands are abbreviated in the format introduced in Tutorial 0, where the menu name (Window, for example) is followed by the menu commands you should choose, with all parts separated by > symbols, for example Window > Worksheet 1 ***. ∎

You will enter the yogurt data in the Data window. To see as many rows and columns as possible, maximize the window:

- Click the **Maximize** button ☐ on the Data window

The Data window fills the screen, as shown in Figure 1-4. When a window is maximized, its title, Untitled - [Worksheet 1 ✱✱✱], is added to the Minitab window title bar, as shown in Figure 1-4.

FIGURE 1-4

The Data window maximized

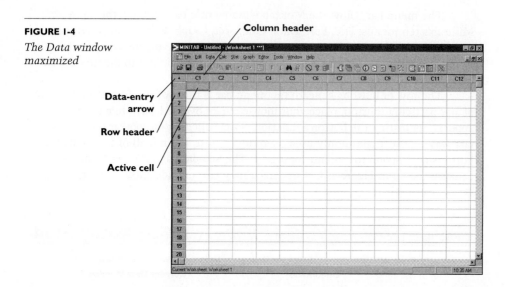

The grid, or worksheet, shown in the window consists of columns and rows. Each *column* contains the data for a specific *variable,* while each *row* contains the data for an individual *case.* The intersection of a row and column is called a *cell.* Each cell is identified by its column (C1, C2, C3, and so on) and its row (1, 2, 3, and so on). For example, the cell in the upper left-hand corner of the worksheet corresponds to column 1, row 1; its cell address is C1 row 1. When you enter a number or text into a cell, that entry becomes an *observation.*

Only one cell in the Data window is active at a given time. When you first open Minitab and open the Data window, the active cell is the *column name cell* immediately below C1. The active cell appears highlighted—that is, shaded or with a thick border—depending on your monitor. Minitab enters the information you type at the keyboard in the active cell. To make C1 row 1 the active cell:

- Click in **C1 row 1** to make it active

▶ **Note** Your screen may show fewer or more rows and columns than those shown in Figure 1-4, depending on your monitor. In the Student version of Minitab, the Data window has an upper limit of 10,000 cells. There is no comparable limit in the Professional version of Minitab. Neither version permits more than 4,000 columns. ■

1.2 *Entering Data*

You are now ready to enter the yogurt data from Figure 1-1 into the Data window. Each row will contain data from one individual case—in this example, a certain brand of yogurt. Each column will record a different variable. You will enter the yogurt's nutritional rating in column C1, its cost in cents per ounce in C2, and the calories per eight-ounce serving in C3. You do not need to have a Brand column because that column corresponds to the row number.

The value of the first case's rating is Excellent.

- Type the first rating, **Excellent**, in C1 row 1

    If you make a mistake, press (Backspace) to delete a character.

- Press (Tab) or ◄┘ (the Enter key) to accept the data you just entered

    When you press (Tab), the cursor moves one cell to the right. When you press ◄┘, it moves one cell either to the right or down, depending on the direction of the *data-entry arrow* in the upper-left corner of the Data window (see Figure 1-4). When you open Minitab this arrow points down. To change the direction of the data-entry arrow:

- Click the **data-entry arrow**

    The direction of the data-entry arrow changes. Click it again, if necessary, until the arrow is horizontal so that you can press ◄┘ to enter the yogurt data by rows:

- Click in cell **C2 row 1**, if necessary, to make it active

- Type the first cost, **13**, in C2 row 1

- Press ◄┘ to accept this entry and activate the cell to the right

- Type the first calories entry, **140**

    Now you are ready to enter the data for the second brand of yogurt in row 2:

- Press and hold down (Ctrl), then press ◄┘ to move to the beginning of the second row

▶ **Note** Directions to use a combination of two keys in sequence will be indicated by joining the names of the two keys with a "+" sign. Press (Ctrl)+◄┘ will indicate that you should press and hold down (Ctrl) as you press ◄┘. A similar notation will be used if you need to use a combination of three keys at the same time. Always press and hold the keys in the order in which they are listed. ∎

- Begin entering the next row of data by typing **Very Good** and pressing ◄┘

    Continue in this manner until you have entered into the Data window all 17 rows of data shown in Figure 1-1, except the last value, 120.

    When you have entered all but the last value correctly, deliberately enter a mistake in C3 row 17, so that you can see how to correct it:

- Type **210** and press ◄┘ to accept it

1.3 Correcting Mistakes

If you discover a typographical error after you enter data in a cell, you can easily fix it. To correct the last entry, for example:

- Double-click in the **active cell**

  When you double-click in the cell, a vertical bar, or *insertion cursor*, appears. If you type new characters Minitab inserts them at the location of the insertion cursor; any characters to the right of the insertion cursor move farther to the right.

  To delete characters:

- Use the arrow keys to move the insertion cursor to the right of the mistake (between the 1 and the 0)
- Press ⌜Backspace⌝ twice and then type the correct entry, **12**
- Press ⌸ to accept the corrected entry

  You can also press ⌜Del⌝ to delete characters to the right of the cursor.

▶ **Note** If you type a new value in a cell, but have not pressed ⌸, you can restore the previous value of the cell by pressing ⌜Esc⌝. ■

▶ **Note** It is important to know that when entering numbers greater than 999 you need not enter the comma. Commas are ignored by Minitab. So, for example, enter the number 56,789 as simply 56789. ■

1.4 Text and Numeric Data

Minitab recognizes three types of data. Columns C2 and C3 contain *numeric data* whereas C1 contains words called *text data*. When you start typing in a column, Minitab classifies a column as text if it detects any alphabetic characters. Minitab also marks the column with a T suffix. So the first column heading in the Data window is now C1-T, not C1. Later in this book you will encounter a third type of data—*date/time data*—which Minitab designates with a D.

▶ **Note** When you are entering numeric data into a column you may inadvertently enter an alphabetic character instead of a number in the first row. In this case the suffix T will automatically appear beside the column number. You can correct your mistake, but Minitab will still regard the column as a text column. You can change the data type of the column by choosing Data > Change Data Type > Text to Numeric, from the menu bar. ■

1.5 Saving Data

It is good practice to save your work frequently to guard against losing it in the event of a mistake or a power failure.

When you save your worksheet, you do not save all of your work. For example, a copy of your Session window is not saved. To save all of your work, you must save each window individually or save your work as a Minitab project. You will learn about Minitab projects in Tutorial 3.

The following initial instructions presume you save your files to an already-formatted disk in drive A. If your disk is in a drive other than A, or you are saving your work to a different location, substitute the correct drive and folder name when you save your work. If you are using your own computer, you can save your work to your hard disk.

This book also uses a file-naming convention to help you remember which version of your work you are using at any given time. Each file you save has a prefix to indicate the tutorial number. For example, you will save the yogurt data you just entered as *1YogurtData*. If you were to use this file again in Tutorial 5, for example, you would save it with a new name, *5YogurtData*, so that you can easily distinguish which file you used in which tutorial.

Now, to save the yogurt data:

- Choose **File > Save Current Worksheet As**
- Click on the "Save in" **drop-down list arrow** and select the location where you are saving your work
- Type *1YogurtData* in the "File name" text box

The Save Worksheet As dialog box should look as shown in Figure 1-5 (assuming there are no other Minitab files saved on the disk).

FIGURE 1-5

The Save Worksheet As dialog box

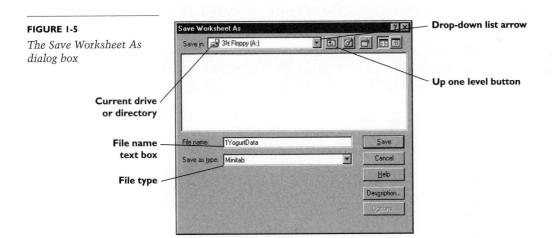

- Click **Save** (or press ⏎)

After you save your data, the status window at the foot of the screen changes from Current Worksheet: Worksheet 1 to Current Worksheet: 1YogurtData.MTW

Also, the name in square brackets in the Minitab title bar Data Window now changes from Worksheet 1 *** to 1YogurtData.mtw ***.

When you save the *1YogurtData* file, Minitab automatically adds an *mtw file extension* after the period delimiter. The three-letter suffix indicates the *file type*. Not all applications use the same file types. For example, Minitab can read files with an *mtw* file type, but other software programs, such as text editors, cannot.

▶ **Note** Notice that Minitab uses pictorial aids on many of the menu commands to illustrate the general intent of the command. ■

Naming Columns

It's easier to work with data if you assign a meaningful name to each column in your worksheet. In some output, Minitab truncates column names with more than 12 characters, so it's good practice to use names that are no longer than 12 characters. When you name a column, avoid using an apostrophe ('), a pound sign (#), or an asterisk (*). Also, do not include spaces within or on either side of a variable name. Minitab displays column names in the shaded *column name row* below the column numbers (C1, C2, C3, and so on). To name C1:

- Click in the **column name cell for C1-T** to highlight it
- Type **Rating** as the name for C1
- Press ↵ to accept the entry and move to the name cell for C2
- Type **Cost** as the name for C2
- Press ↵ to accept the entry and move to the name cell for C3
- Type **Calories** as the name for C3
- Press ↵

Your Data window should look as shown in Figure 1-6.

FIGURE 1-6

Yogurt data with columns named

▶ **Note** You can also name the columns before you enter the data. ■

1.7 *Printing from the Data Window*

The yogurt data have been entered and saved, and now you want to print out a copy of your worksheet on paper:

- Click in **C1-T**, the column heading in the upper-left-hand corner of the Data window
- Press [Shift]+[Ctrl]+[End] (recall that this means to press and hold down all three keys in order) to highlight all three columns
- Choose **File > Print Worksheet** (or click on the **Print Window** toolbar button ⬛)

    In this case, the Data Window Print Options dialog box opens.

    Whenever Minitab requires additional information, it almost always displays a *dialog box*. In a dialog box, you provide information by clicking on *option buttons* and clicking on or clearing check boxes. You also fill in *text boxes*. Whatever you type when the cursor is in the text box is entered in the box.

    Accept the defaults for the four check boxes and the option button. Also, include a title on your printout:

- Type **Yogurt Data** in the "Title" text box

    Figure 1-7 shows the completed Data Window Print Options dialog box.

FIGURE 1-7

The completed Data Window Print Options dialog box

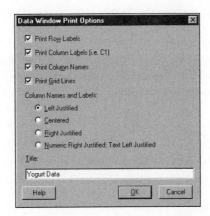

- Click **OK** to accept the print options and close the dialog box
- Click **OK** in the Print dialog box to print the option you selected

    Figure 1-8 shows the printed Data worksheet.

FIGURE 1-8

Printed Yogurt data

	C1 Rating	C2 Cost	C3 Calories
1	Excellent	13	140
2	Very Good	9	210
3	Very Good	6	180
4	Very Good	8	220
5	Good	8	170
6	Good	9	220
7	Good	11	160
8	Good	11	170
9	Good	6	210
10	Good	7	200
11	Good	7	100
12	Fair	5	210
13	Fair	9	120
14	Fair	11	100
15	Fair	11	190
16	Fair	9	120
17	Fair	8	120

1.8 Moving in the Data Window

With a small data set such as the Yogurt worksheet, it is simple to navigate through the data. With the Data window maximized, you can see all the data. Even with larger data sets it is straightforward to scroll down and/or over to find data. For large data sets, however, Minitab provides the Editor > Go To... command (not the Editor > Go To command) to navigate the data. (The ellipsis indicates that the command opens a dialog box.) To go to the 40th row in column 30, for example:

- Choose **Editor** > **Go To...** to open the Go To dialog box
- Type **30** in the "Enter column number or name" text box and press $\boxed{\text{Tab}}$
- Type **40** in the "Enter row number" text box

    The completed Go To dialog box should look as shown in Figure 1-9.

FIGURE 1-9

The completed Go To dialog box

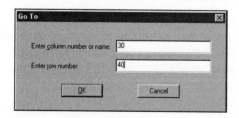

    The OK button in the Go To dialog box has a thick border around it, unlike the Cancel button. This thick border indicates that the button is highlighted.

Whenever a button is highlighted, pressing ⏎ has the same effect as clicking the right-hand button on your mouse.

- Click **OK** (or press ⏎) to implement the Editor > Go To ... command

The active cell is now C30, row 40. Now make the active cell C1, row 1 again:

- Scroll back (up and to the left) and click in the cell corresponding to C1, row 1

For some operations in Minitab you can use a *keyboard shortcut* as an alternative to the menu. Here, a keyboard shortcut is a combination of two keys that, when pressed at the same time, cause an operation to be executed. If you click on the Editor menu, for example, you will see that there is a keyboard shortcut for the Editor > Go To ... command. The shortcut is Ctrl+G. Pressing these two keys together will open the Go To dialog box. Yet another way to execute some Minitab commands is to click on the right-hand button on your mouse to access a *shortcut menu*. You will see examples of this technique later in the tutorial.

1.9 Leaving Minitab

This is a good time to take a break. Because you made changes to the data set you saved earlier by naming each of the columns, you should save the new worksheet before leaving Minitab. You have already saved your data once. Now you want to replace that saved file with this updated one and keep the same filename. To save a new version of an existing file under the same filename, choose File > Save Current Worksheet command (not the File > Save Current Worksheet As command). You will not be asked to specify a filename and location; the File > Save Current Worksheet command automatically saves the file to the previous location under the same name.

To save your worksheet again and exit Minitab:

- Choose **File > Save Current Worksheet**
- Choose **File > Exit**
- In response to the "Save changes to the project 'Untitled' before closing?" prompt, click **No**

1.10 Retrieving Data

Your data, *1YogurtData.mtw*, are stored in the folder in which your data files are located. To retrieve these data:

- Open **Minitab** and maximize the Data window (Worksheet 1 * * *), if necessary
- Choose **File > Open Worksheet**

The Open Worksheet dialog box appears.

- Navigate to the location where your data files are located.
- Double-click on ***1YogurtData.mtw*** in the "Open Worksheet" list box

You may get a notice that "A copy of the content of this file will be added to the current project." If so:

■ Click **OK**

You will learn about projects in Tutorial 3. The Data window now displays the yogurt data you saved. *1YogurtData.mtw* is the name of the current Data window.

▶ **Note** Minitab has an Open Project toolbar button but not an Open Worksheet toolbar button. ■

I.II Manipulating Data Using the Calculator Command

In developing your yogurt budget, you must compute the total cost of feeding 200 patients. You can do this with the Calculator command on the Calc menu. Calculator lets you perform many mathematical operations on columns. Minitab uses the symbols +, −, *, /, and ** to add, subtract, multiply, divide, and raise to a power, respectively. In later tutorials, you will use Calculator to use various mathematical and statistical functions.

▶ **Note** Minitab's symbol, **, to indicate raise to a power differs from the symbol, ^, favored by other software products. ■

Now, compute the total cost of feeding 200 patients for each brand of yogurt.

■ Choose **Calc > Calculator**

The Calculator dialog box appears. You intend to compute the daily cost per ounce, in dollars, of serving each brand of yogurt to 200 patients and store the results in C4. First, name the new variable Cost200:

■ Verify that your cursor is in the "Store result in variable" text box

■ Type **Cost200**

When you type in a name for a new variable, Minitab automatically assigns the entry you type as the name of the next available column (C4, in this example). If you include spaces or other special characters in a name, you will have to enter single quotation marks around the column name in some situations.

■ To move the cursor to the "Expression" text box, press ⎡Tab⎤

The cost, in dollars, of feeding 200 patients can be obtained by making C4 Cost200 equal to C2*200/100. (Recall that the original costs in C2 are in cents.) To enter this formula in the "Expression" text box either:

■ Double-click on **C2 Cost** in the list box to enter it in the "Expressions" text box

■ Click on * **200 / 100** using the calculator keypad in the dialog box

    or

■ Type **C2*200/100** (the asterisk is Minitab's symbol for multiplication)

If you typed the entries, the completed Calculator dialog box should resemble Figure 1-10.

FIGURE 1-10

The completed Calculator dialog box

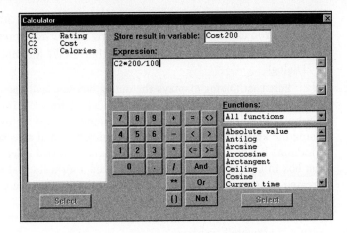

- Click **OK** to create the new variable

This formula tells Minitab to multiply each of the values in C2 by 200 and divide each of the products by 100. The Data window now contains a new column that shows the cost, in dollars, of providing each brand of yogurt to 200 patients as shown in Figure 1-11.

FIGURE 1-11

1YogurtData with the new column, Cost200

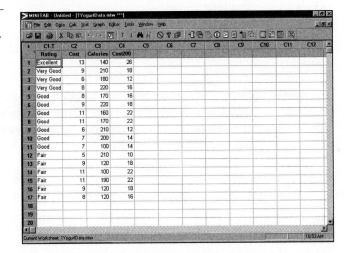

The Data window in Minitab is not a spreadsheet such as those in Microsoft Excel or IBM Smart Suite Lotus 1-2-3. In those programs, the cells may contain formulas that are updated, based upon the values in other cells. In Minitab, cells may contain values that you type or generate, but they are not formulas. For example, in the yogurt case study you used the calculator to place the values for 200*C2 in C4. If you changed the values in C2, the values in C4 would not change until you use the calculator again or use some other command to change the contents of C4.

Next, add a column that shows the cost of feeding 300, rather than 200, patients. You can do this without retyping the entire expression.

- Click on the **Edit Last Dialog** toolbar button 🔲 (or choose Edit > Edit Last Dialog)

Edit Last Dialog displays the dialog box you had open most recently.

You can create expressions in the Calculator by using any combination of clicking or typing variables from the variable list box on the left of the dialog box, using the keypad in the center of the dialog box, and selecting functions from the functions list on the right of the dialog box. You will select from the "Functions" list box in later tutorials. If you introduced a formula that was wrong, and thus created an incorrect variable in the new column, you can correct the formula by using Calc > Calculator again.

To change Cost200 to Cost300, replace the 200 with a 300 in the "Store result in variable" text box.

- Click to the right of the 2 in Cost200
- Highlight the number **2** and type **3**

The entry in the "Store result in variable" text box now reads Cost300. Next, change the 200 in the "Expression" text box to 300.

- Highlight **2** in the Expression text box and type **3**

The "Expression" text box now reads either **'Cost' * 300 / 100** (if you used the keypad) or c2*300/100 (if you typed the entry).

- Click **OK** to create the new variable, Cost300

C4 and C5 display the costs of providing each brand of yogurt to 200 and 300 patients.

▶ **Note** ⌨Tab⌨ moves the cursor from one location to another in a dialog box. If you want to return to an item that appears before your present location, either press ⌨Tab⌨ until you return to it or press ⌨Shift⌨+⌨Tab⌨ to move the cursor backwards. You can also use your mouse to quickly select any item by clicking on it. ■

1.12 Copying Columns

You initially recorded the rating for each brand in C1. Now you want to repeat this information in C6 so that it is still visible when you use columns farther to the right in the worksheet. You can use Minitab's online Help system to find out how to copy a column:

- Choose **Help > Help** (and click on the tab for the Index folder in the left-hand side of the window, if necessary)

Minitab's Index is similar to an index in a book in that you can look up a keyword to help you locate the information you need. Try entering the word "copy" as a keyword to find out more about duplicating a column:

- Type **copy** in the "Type in the keyword to find" text box

As you type, the list box below the text box scrolls to show entries that begin with the letters you type. By the time you've typed the entire word, *copy* appears, highlighted in the list box, as Figure 1-12 shows. The list box in the window shows the alphabetical list of topics closest to the keyword *copy*.

FIGURE 1-12

The Minitab Help Index

Keyword "copy"

Highlighted item

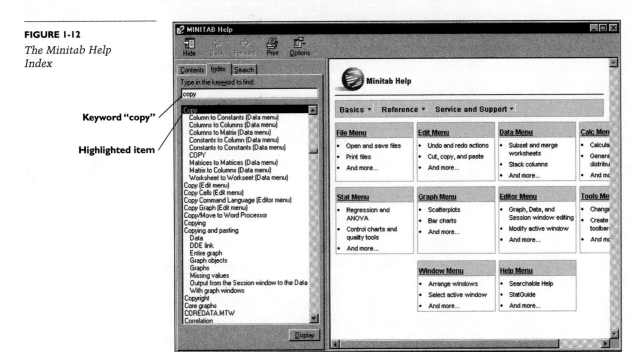

- Click on **Copy Columns to Columns (Data menu)**
- Click on the **Display** button

The right-hand Minitab Help window displays the Copy Columns to Columns entry, as shown in Figure 1-13. The Help topic describes the Data > Copy > Columns to Columns command.

When you have located a topic in Help, you can print a copy of the information in several ways: By clicking on the Print icon in the Help window, or by choosing Print from the menu that comes up when you click on the right-hand button on your mouse in the Help window.

FIGURE 1-13

The Copy Columns to Columns entry in the Minitab Help window

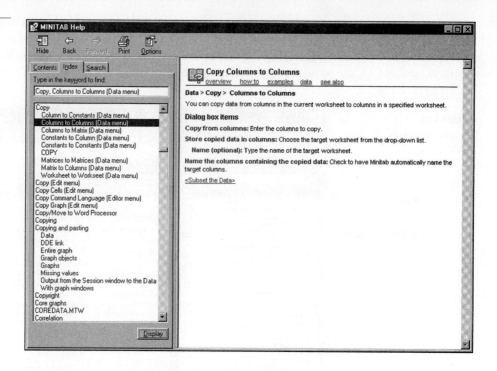

Close the Help window when you finish reading the information.

- Click the window's **Close** button ⊠ and return to the Data window

There are other ways to access online Help in Minitab. You can press F1, a function key on the keyboard, or the Help toolbar button 🔲, to display the table of contents for the Help window. A more complete introduction to Help can be found in Tutorial 0.

Next, use Minitab's Data > Copy > Columns to Columns command to copy the data from C1-T Rating to C6:

- Choose **Data > Copy > Columns to Columns** to open the Copy Columns to Columns dialog box

Similar to the Calculator dialog box, the Copy Columns to Columns dialog box indicates the currently filled columns (variables) in the Data window in a list box on the left side.

- Be sure the cursor is blinking in the "Copy from columns" text box; if not, click in the text box

- Double-click on **C1 Rating** to enter it in the "Copy from columns" text box

You can copy the selected columns to one of a variety of locations; the default location is a new worksheet.

- Click on the "Store Copied Data in Columns" **drop-down list arrow**

- Click on **In current worksheet, in columns**

- Click in the "Store Copied Data in Columns" text box and type **C6**

The completed Copy Columns to Columns dialog box should look as shown in Figure 1-14. You specified the column to which the data were to be copied but not a name for the new column. Minitab will assign a default name to columns as long as "Name the columns containing the copied data" is checked.

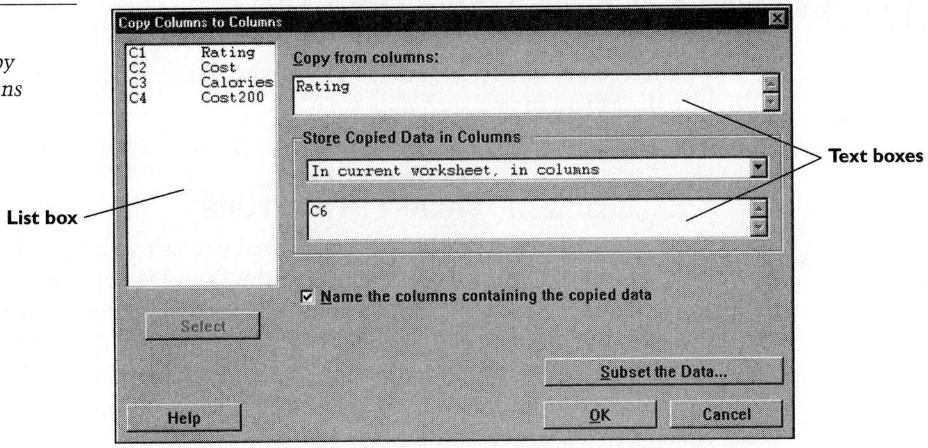

FIGURE 1-14

The completed Copy Columns to Columns dialog box

- Click **OK**

    Minitab has copied the ratings from C1-T to C6, assigning the default name Rating_1 to C6-T. You can, of course, change it to a new name. If you had specified a name for the new column (Rating 2, for instance) instead of simply C6, Minitab would have copied the data to the column immediately following the last column with data and called it Rating 2.

    This is a good time to save your work again.

- Choose **File > Save Current Worksheet**

    Minitab saves the updated *1YogurtData.mtw* file in the folder where you are saving your work.

    In preparation for the next case study:

- Choose **File > Close Worksheet** (you may also click on the **Close** button ⊠ in the upper right-hand corner of the Data window)

    The Data window will now be empty and untitled again.

    In the next case study you will be opening a stored data set that comes with Minitab. There was no need to close the worksheet *1YogurtData* before opening another data set. However, the Student version of Minitab allows a maximum of only five open worksheets at one time. So users of the Student version of Minitab are advised to close one worksheet before opening another. There is no comparable upper limit on open worksheets in the Professional version of Minitab; so users of this version need not be so concerned about closing one worksheet before opening another.

    To see the default data folder in the next case study, first exit Minitab:

- Choose **File > Exit**

Minitab will ask you if you want to save changes to the project before closing.

- Click **No**

Opening a Minitab Stored Data File

Minitab comes with a large selection of stored data sets. The following case study introduces you to one such data set.

CASE STUDY	**MANAGEMENT—**
	SALARY STRUCTURE

Tayco, a midsize corporation, recently hired you as a managerial consultant. Your task is to analyze the salary structure of the Sales Department and then point out any inequities. Brett Reid, one of your co-workers, has already gathered data on the Sales Department personnel and entered it into a Minitab file named *PayData.mtw*. Brett informs you that some employees have left Tayco because he entered the data and their records must be deleted. There also are some new employees whose records must be added.

- Open **Minitab** and maximize the Data window, if necessary
- Choose **File > Open Worksheet**

The data file, *PayData.mtw*, is located in the Studnt14 folder, the default for MINITAB Student Release 14. If you are using the Professional version of Minitab, the default folder appearing in the "Look in" box is not Studnt14, but Data. To locate the Studnt14 folder:

- Click on the **Up One Level** button 🔼 to the right of the "Look in" text box

  You should now see the Studnt14 folder icon.
- Double-click on the **Studnt14** folder

  The Open Worksheet dialog box displays a list of files similar to the one shown in Figure 1-15.

FIGURE 1-15

The Open Worksheet dialog box

Current directory

Files list box

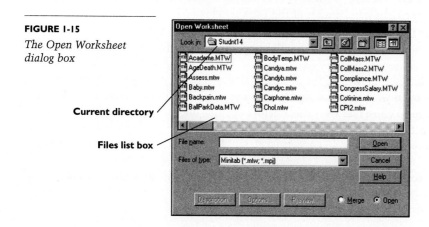

- Use the scroll bar at the bottom of the list box to scroll the list of filenames until you find *PayData.mtw*
- Double-click on the icon associated with **PayData.mtw** (or click on the icon and click **Open**)

    At this point you may see the Minitab message "A copy of the content of this file will be added to the current project." If so,
- Click **OK**

    The worksheet shown in Figure 1-16 appears on your screen.

FIGURE 1-16

The PayData.mtw worksheet

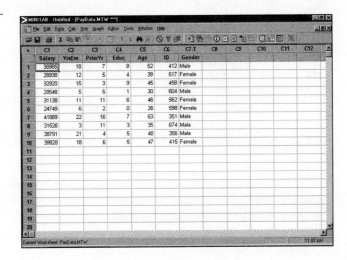

Before you modify the data, you should save the file to your disk (or another folder if you are not using a disk) under a different filename, using Save Current Worksheet As. Doing this will prevent you from accidentally overwriting this original data file with Save Current Worksheet. To save the file:

- Choose **File > Save Current Worksheet As**
- Click on the "Save in" **drop-down list arrow** and select the location where you are saving your work
- Under "File name" type the name *1PayData*
- Click **Save**

    Now, when you save your changes to the file, they will be written to *1PayData.mtw*, not *PayData.mtw*.

Getting Information About a Data Set

As you work with data sets in the Studnt14 data folder, you can learn about their variables in three ways:

1. By reading about them in Appendix A in this book, "Exploring Data"
2. By looking at the description in the online Help
3. By viewing the Info window, which is introduced in Section 1.16.

Deleting and Inserting Rows

Brett measured seven variables for each employee:

Column	Name	Description
C1	Salary	The salary of an employee in the sales department
C2	YrsEm	The number of years employed at Tayco
C3	PriorYr	The number of prior years' experience
C4	Educ	Years of education after high school
C5	Age	Current age
C6	ID	The company identification number for the employee
C7-T	Gender	Gender; Female or Male

You decide to begin updating Brett's data by removing employee 562 (row 5), who has left Tayco. You can select an entire row of data by clicking on the row number, which is called the *row header*.

- Click on the **row header** for row 5 to highlight the entire row as shown in Figure 1-17

FIGURE 1-17

The 1PayData.mtw worksheet with row 5 highlighted

Row header

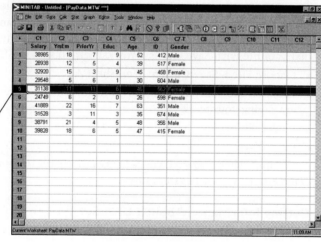

- Click on the **Cut** toolbar button

Minitab deletes row 5 and rows 6 and beyond move up one row. There are now only nine employees represented in the worksheet.

▶ **Note** To delete selected cells, rows, or columns, you could also press (Del). Other alternatives are to click the right-hand button on your mouse and choose Delete Cells, or to choose Edit > Delete Cells. The Cut command has an advantage over these delete commands. With the Cut command the deleted cells are stored in the clipboard and are available if needed. ■

If Tayco hires a replacement, you might want to insert the new employee's data into the worksheet at a particular location. Try inserting a row above row 8:

■ Click in any cell in row 8

■ Click the **right-hand button** on your mouse and choose **Insert Rows**
(or choose **Editor > Insert Rows** or click on the **Insert Row** toolbar button ▦)

Minitab inserts a blank row in this position and moves all of the rows below it down one row. Minitab displays an asterisk (*) in each numeric cell and a blank ( ) in each text cell in row 8 to indicate a missing value. Because you don't want to insert data in this row, you should delete this empty row. You can delete it as you did before, or you can use the Undo Insert option:

■ Click on the **Undo** toolbar button ↺

This operation will undo the action of the previous command, which created the new row. You could also have chosen Edit > Undo Insert, Data > Delete Rows, or Undo Insert from the right-hand mouse button menu to perform this task.

Your worksheet should now look as shown in Figure 1-18.

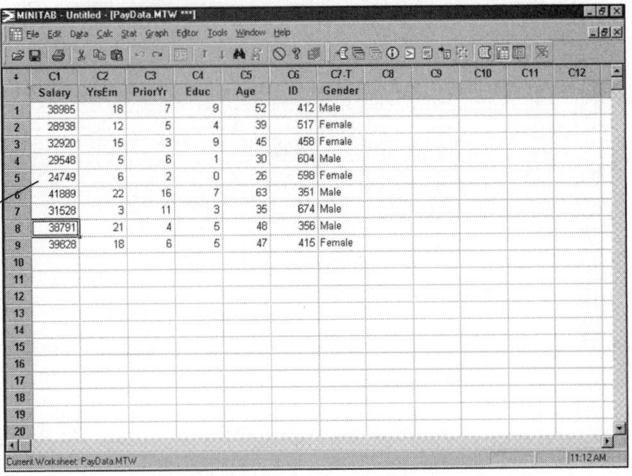

FIGURE 1-18

The 1PayData.mtw worksheet with one row deleted

Deleted row was here

▶ **Note** You may choose multiple Undo's to undo multiple editing operations. You can use the Redo toolbar button or choose Edit > Redo to restore the most recent editing operation. ■

Brett tells you that the department is hiring two new employees. You want to enter the data for these employees in rows 10 and 11. As you enter the new data, remember that if the data-entry arrow (in the upper-left corner of the screen) points to the right, then pressing ⏎ moves the cursor across the row.

- If necessary, click on the **data-entry arrow** so that it points to the right

The first new employee, whose data are to go in row 10, is a 23-year-old female who earns a salary of $28,985. She has one year of prior experience and four years' post-secondary education. Her employee number is 693. As a new employee, she has 0 years at Tayco. To enter her data:

- Click in the **C1 row 10** cell
- Type the data in row 10, starting in C1: **28985** press ⏎ **0** press ⏎ **1** press ⏎ **4** press ⏎ **23** press ⏎ **693** press ⏎ **Female**
- Press ⌈Ctrl⌉+⏎ to accept the final value and move to the beginning of row 11

▶ **Note** If you notice a mistake after you press ⏎, correct the cell entry using the techniques discussed in Section 1.3. ∎

You have less information about the second new employee. He is male, his employee number is 694, and his salary is $32,782. Place an asterisk in each cell for which you're missing numeric information.

- Type the second employee's data in row 11 (don't forget to press ⏎ between each entry): **32782 0 * * * 694 Male**
- Press ⌈Ctrl⌉+⏎ to move to the beginning of row 12

The Data window should now resemble the window shown in Figure 1-19.

FIGURE 1-19

The 1PayData.mtw worksheet with new employee information

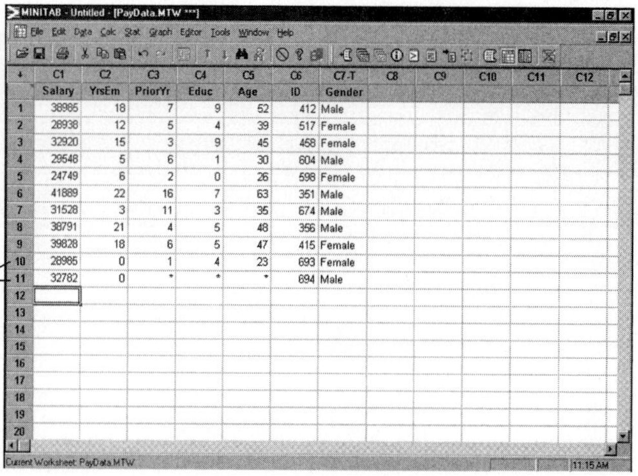

New employee information →

	C1	C2	C3	C4	C5	C6	C7-T	C8	C9	C10	C11	C12
	Salary	YrsEm	PriorYr	Educ	Age	ID	Gender					
1	38985	18	7	9	52	412	Male					
2	28938	12	5	4	39	517	Female					
3	32920	15	3	9	45	468	Female					
4	29548	5	6	1	30	604	Male					
5	24749	6	2	0	26	598	Female					
6	41889	22	16	7	63	351	Male					
7	31528	3	11	3	35	674	Male					
8	38791	21	4	5	48	356	Male					
9	39828	18	6	5	47	415	Female					
10	28985	0	1	4	23	693	Female					
11	32782	0	*	*	*	694	Male					
12												

▶ **Note** An entry of 0 (zero) is different than an entry of *, which indicates a missing numeric value. Also note that when you entered the salaries, you didn't enter commas or dollar signs, although it will do no harm. Commas and dollar signs are ignored by Minitab. ∎

▶ **Note** A blank is used to indicate a missing "value" for a text variable. ∎

1.15 Deleting Columns

Tayco recently enacted a new policy against recording age, so you need to delete the age variable. To do this:

- Highlight the entire column **C5 Age**
- Click the **right-hand button** on your mouse and choose **Delete Cells**

The contents of C5 have now disappeared and the contents of C6 and C7-T have moved left to occupy C5 and C6-T, as shown in Figure 1-20.

FIGURE 1-20

The 1PayData.mtw worksheet without the Age variable

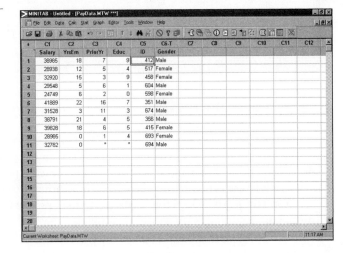

▶ **Note** There are other ways of deleting a variable. Edit > Delete Cells and Data > Erase Variables are equivalent to the above command. Data > Erase Variables deletes the contents of a column and leaves that column without any observations. ■

▶ **Note** Minitab allows you to search for a specified entry in either the Data window or the Session window. To do this, make the appropriate window active and either click the Find toolbar button or choose Editor > Find. To replace a specified entry use Editor > Replace. ■

1.16 Viewing Information About the Data Set

Minitab offers a simple way to keep track, in summary form, of the contents of each column:

- Click on the **Show Info** toolbar button ⓘ

The Project Manager window opens, as in Figure 1-21, showing summary information for each column of the current worksheet.

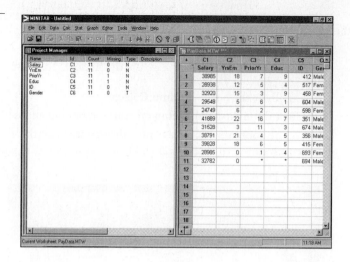

You can see that one value is missing for each of PriorYr and Educ, and that C6-T Gender is a text variable.

With a small data set such as *1PayData.mtw*, all, or nearly all, of the data and the column names can be viewed on the screen in the Data window. The information in the Project Manager window is more helpful when you are dealing with much larger data sets in which it often is difficult to keep track of the data.

You will learn about Project Manager when projects are introduced in Tutorial 3.

- Click the **Maximize** button ☐ on the Data window

After compiling the *1PayData* data you want to share it with a colleague who does not have Minitab, but does have access to Microsoft Excel. Information on how to export and import data between Minitab and Excel is given in Sections 1.17 and 1.18. If you are not interested in learning about these procedures, you should skip to the end of Section 1.18 (immediately after the two Notes) and ignore the request to exit Excel.

Exporting Data to Excel

There are several ways to export data sets from Minitab. With a small data set such as this one, a simple approach is to copy the data onto the clipboard and then paste it directly into Excel.

First, open Microsoft Excel.

- Choose **Start > Programs > Microsoft Excel**

(Excel is usually found in the Programs folder. If it is not located there you may need to use Start > Search to find it.)

A new Excel spreadsheet will open up. You should still see the Minitab icon at the bottom of the screen. To return to Minitab:

- Click on the **Minitab** icon (at the bottom of the screen)

Now copy the worksheet onto the clipboard:

- Using the column headers, highlight all six columns in the worksheet—including the column names
- Click on the **Copy** toolbar button ▣

    Now paste the data into Excel:

- Click on the **Microsoft Excel** icon (at the bottom of the screen)
- In Excel, click in the cell corresponding to **column A, row 1**

    (The location column A, row 1 is for convenience. You can put the data and names anywhere in the spreadsheet.)

- Choose **Edit > Paste**

    The Excel spreadsheet will now look as shown in Figure 1-22 with the data and column names shaded and ringed by a solid line.

FIGURE 1-22

An Excel spreadsheet containing the 1PayData data

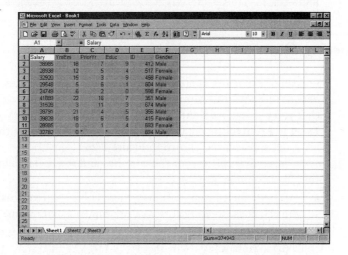

To remove the shading, click anywhere in the spreadsheet. At this point you can, of course, save the Excel file if you wish.

In the next section you will import part of this Excel spreadsheet into Minitab. To see how Minitab deals with commas in large numbers and dollar signs, you should insert these into the salaries in column A. To do this:

- Highlight the **11 numbers in column A**
- Choose **Format > Cells**
- Click on **Currency** in the "Category" list box and click **OK**

    Excel will insert commas and dollar signs into the numbers. The spreadsheet should now look as shown in Figure 1-23.

FIGURE 1-23

Salary data in Excel with commas and dollar signs

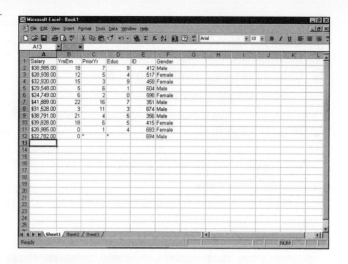

1.18 Importing Data from Excel

Instead of exporting data from Minitab to Excel, it is often necessary to import data from Excel into Minitab. You can try this process with the *1PayData* data set that you just exported into Excel and modified. Suppose you want to export from Excel to Minitab the data and column names in the first three columns (A, B, and C) of the spreadsheet, shown in Figure 1-23.

Instead of editing out the commas and the dollar signs in Excel, try copying the three columns as they are:

- Highlight the contents of these three columns (Excel uses column headers as Minitab does, so you can select the top cells containing the letters A, B, and C)
- Choose **Edit > Copy** (or use the right-hand button on your mouse to access the Copy command)
- Click on the **Minitab 14** icon (at the bottom of the screen)

  You intend to place these three columns into a new worksheet in Minitab.
- Choose **File > New**
- In the New dialog box verify that "Minitab Worksheet" is highlighted and click **OK**

  A new, empty worksheet opens. Now paste these three columns, with names, into C1, C2, and C3:
- Place the cursor in the column name cell for C1, if necessary
- Click on the **Paste** toolbar button 📋

  The new three columns copied from Excel will appear in Minitab so that your current worksheet will look as shown in Figure 1-24.

FIGURE 1-24

Three Excel columns imported into Minitab

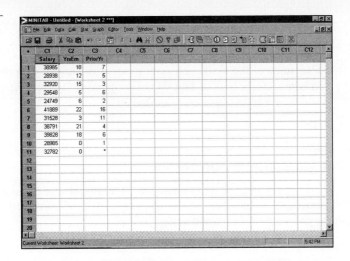

Notice that Minitab ignores the commas and the dollar signs and simply accepts the actual amounts in C1. Because you already have a worksheet with these data, close this new worksheet without saving it.

■ Choose **File > Close Worksheet**

■ At the warning "The worksheet 'Worksheet 2' will be removed from the project, 'Untitled'. This action cannot be undone. Would you like to save it in a separate file?", click **No**

This second worksheet will disappear leaving the *1PayData.mtw* worksheet current.

▶ **Note** Microsoft Excel can be used to perform some statistical analyses. However, it was not designed for this purpose as Minitab was. A growing body of experience suggests that using Excel (and relying on Excel documentation) can result in misleading and occasionally wrong output. ■

▶ **Note** In this tutorial you have used the toolbar buttons to perform the Cut, Copy, and Paste operations. You can also use the Edit menu to access these commands. A second alternative is clicking the right-hand button on your mouse and accessing them on the shortcut menu. ■

Now, save these data in the folder where you are saving your work and leave Minitab:

■ Choose **File > Save Current Worksheet**

■ Choose **File > Exit**

■ In response to the "Do you want to save the changes to the project 'Untitled' before closing?" prompt, click **No**

To exit Excel:

■ Choose **File > Exit**

- In response to the "Do you want to save the changes you made to 'Book 1'?" prompt, click **No**

Congratulations! You have finished the first tutorial. Now that you have seen some of the skills you need in order to prepare data for analysis, you are ready to take a closer look at how Minitab operates. In the next tutorial, you will learn more ways to manipulate your data.

Minitab Command Summary

This section describes the commands introduced in, or related to, this tutorial. Use Minitab's online Help for a complete explanation of all commands. Remember that the contents of Minitab menus may vary depending upon which window is active or the nature of the last command.

Minitab Menu Commands

Menu	Command	Description
File ➤	New	
	➤ Worksheet	Creates a new Minitab worksheet.
	Open Worksheet	Retrieves a previously stored worksheet in a new Data window.
	Save Current Worksheet	Replaces a previously saved file with a copy of the current worksheet. (Also saves a new worksheet the first time it is saved.)
	Save Current Worksheet As	Saves a copy of the current worksheet to a new file.
	Close Worksheet	Closes the current worksheet, removing it from the project.
	Print Worksheet	Prints the data in the worksheet. (There is a corresponding command for the Session, Graph, and History windows.)
	Exit	Exits Minitab.
Edit ➤	Redo (last command)	Restores the most recent editing operation.
	Undo (last command)	Undoes the most recent editing operation and returns the worksheet to its previous state.
	Delete Cells*	Deletes the selected cells and moves all of the remaining cells up (or all of the columns on the right to the left).
	Copy Cells*	Creates a duplicate of the selected cells.
	Cut Cells*	Creates a duplicate of the selected cells, deletes them, and moves all of the remaining cells up (or all of the remaining columns on the right to the left).
	Paste Cells*	Pastes the selected cells at the point of insertion.
	Edit Last Dialog	Opens the most recently used dialog box, retaining the previous information.
Data ➤	Delete Rows	Deletes the specified rows in the worksheet and moves all of the following rows up to fill the space.
	Erase Variables	Erases columns in the worksheet, constants, or matrices.
	Copy	
	➤ Columns to Columns	Creates a duplicate of the selected data columns.

Menu	Command	Description
Calc ➤	Calculator	Performs calculations using algebraic expressions, which may contain arithmetic operators, comparison operators, logical operators, or functions. Arguments may be columns, constants, or numbers.
Editor ➤	Find	Finds each occurrence of a specified entry.
	Replace	Finds each occurrence of a specified entry and replaces it with another.
	Go To…	Moves to any cell in the Data window.
	Insert Rows	Inserts a row above the active one and moves all the following rows down.
Window ➤	Session	Makes the Session Window active.
	Data (worksheet name)	Makes the selected Data window active and displays the selected worksheet in a spreadsheet-like format.
Help ➤	Help	Displays the Help window (and the Contents, Index, and Search folders that allow you to search for help on a single word or words).

* The Delete, Copy, Cut, and Paste commands may be used on text and graphs as well as data.

Minitab Toolbar Button Commands

Toolbar Button	Icon	Description
Print Window	🖨	Prints the contents of the active window.
Cut*	✂	Creates a duplicate of the selected cells, deletes them and moves all of the remaining cells up (or all the columns on the right to the left).
Copy*	⎙	Creates a duplicate of the selected cells.
Paste*	📋	Pastes the selected cells at the point of insertion.
Undo	↺	Undoes the most recent editing operation and returns the worksheet to its previous state.
Redo	↻	Restores the most recently used editing operation.
Edit Last Dialog	▦	Opens the the most recently used dialog box, retaining the previous information.
Find	🔍	Finds each occurrence of a specified entry.
Help	❓	Displays the Help window (and the Contents, Index, and Search folders that allow you to search for help on a single word or words).
Show Info	①	Shows summary information for each column of the current worksheet in the Project Manager window.
Session Window	▤	Makes the Session window active.
Current Data Window	▦	With one open worksheet, makes the Data window active.
Insert Row	▤	Inserts a row above the active one and moves all the following rows down.

* The Copy, Cut, and Paste commands may be used on text and graphs as well as data.

Minitab Right-Hand Mouse Button Commands

Command	Description
Redo (last command)	Restores the most recent editing operation.
Undo (last command)	Undoes the most recent editing operation and returns the worksheet to its previous state.
Delete Cells*	Deletes the selected cells and moves all of the remaining cells up (or all of the columns on the right to the left).
Copy Cells*	Creates a duplicate of the selected cells.
Cut Cells*	Creates a duplicate of the selected cells, deletes them, and moves all of the remaining cells up (or all of the columns on the right to the left).
Paste Cells*	Pastes the selected cells at the point of insertion.
Insert Rows	Inserts a row above the active one and moves all the following rows down.

* The Delete, Copy, Cut, and Paste commands may be used on text and graphs as well as data.

Review and Practice

Matching Problems

Match the following terms to their definitions by placing the correct letter next to the term it describes.

_____ Session window

_____ Text

_____ *

_____ Blank

_____ Worksheet

_____ mtw

_____ Open Worksheet

_____ Variable

_____ Cell

_____ F1

a. A series of columns in which Minitab displays data

b. Minitab's default designation for a missing text value

c. Minitab commands and output are displayed in this location

d. Minitab's default designation for a missing numeric value

e. The intersection of a column and a row in a worksheet

f. The Minitab command to retrieve a previously saved worksheet

g. The default file extension added to Minitab files created with Save Worksheet As

h. A data type that contains characters including numbers

i. Another name Minitab uses for a column of data

j. The key you press to open the Help window

True/False Problems

Mark the following statements with a *T* or an *F*.

a. ____ You can assign a name to a column by typing the name in the name row of the Data window.

b. ____ The Insert Row command inserts a row of missing values immediately below the active one.

c. ____ The Minitab symbol for multiplication is x.

d. ____ You can use any popular word processor to edit Minitab worksheet files that are saved with the mtw file type.

e. ____ For the Student version of Minitab, the default data folder for the File > Open Worksheet command is Data.

f. ____ You choose the Stop command on the File menu to leave Minitab.

g. ____ When copying columns of numeric data from Excel to Minitab there is no need to first delete the commas and the dollar signs.

h. ____ Column or variable names must be eight or fewer characters.

i. ____ When you are entering data into the worksheet, pressing ⏎ always moves you to the next cell to the right in the same row.

j. ____ The Calc > Calculator command allows you to place a formula in a cell in the active worksheet.

Practice Problems

The practice problems instruct you to save your worksheets with a filename prefixed by P, followed by the tutorial number. In this tutorial, for example, use *P1* as the prefix. If necessary, use Help or Appendix A Data Sets to review the contents of the data sets referred to in the following problems. Interpretations should use the language of the subject matter of the question. If you are using the Student version, be sure to close worksheets when you have completed a problem.

1. The data below are the SAT Math and SAT Verbal scores for the 13 freshmen in an applied statistics class

SATM	SATV
500	460
580	600
560	540
560	500
580	580
480	520
620	660
540	520
520	480
560	540
540	480
460	520
580	580

a. Enter these data into columns C1 and C2 in Minitab. Name these columns.

b. Use Calc > Calculator to create a new column (in C3), which is the total of the two scores.

c. By examining the values in C3, obtain (i) the smallest SAT total and (ii) the range of SAT totals.

2. The following data are the total snowfall (in inches) for Boston for the 21 years between 1980 and 2000:

Year	Snow	Year	Snow	Year	Snow
1980	22.3	1987	52.6	1994	14.9
1981	61.8	1988	15.5	1995	107.6
1982	32.7	1989	39.2	1996	51.9
1983	43.0	1990	19.1	1997	25.6
1984	26.6	1991	22.0	1998	36.4
1985	18.1	1992	83.9	1999	24.9
1986	42.5	1993	96.3	2000	45.9

a. Enter the 21 years (1980, 1981, ..., 2000) into C1 and name the column Year.

b. Enter these snowfall data into C2 and name that column Snow.

c. Which year saw the lowest snowfall in this period? The highest?

d. Use Calc > Calculator to create a new column named Rainfall that represents the equivalent rainfall (1" of rain equals approximately 10" of snow).

e. Save your worksheet as *P1Snow.mtw* in the folder where you are saving your work.

3. The following table summarizes the campus newspaper ratings for 13 pizza shops for two semesters.

Pizzeria	Fall Score	Score Previous Spring
A	6.7	6.73
B	5.4	7.29
C	4.9	5.77
D	4.8	6.69
E	3.9	4.33
F	3.2	*
G	3.2	*
H	3.1	5.01
I	2.8	2.58
J	2.8	6.88
K	2.6	5.08
L	2.2	7.50
M	1.5	*

a. Enter these data into C1, C4, and C5 in a new worksheet and name the columns Pizzeria, FallScore, and SpringScore, respectively. The asterisks represent missing numeric values.

b. Compute the sum of the two ratings for the 13 pizza shops and store these values in C7. Enter a suitable name for this new variable. Which shop has the highest combined rating?

c. Compute the difference (Spring – Fall) between the two ratings for each of the 13 pizza shops and store these values in C8. Enter a suitable name for this variable. Which shop has the greatest difference in ratings?

d. How did the missing values affect your sum and difference computations?

e. Erase C7 and C8.

f. Save your worksheet as *P1Pizza2.mtw* in the folder where you are saving your work.

4. An automobile dealer records details of 14 used Mercedes, which are to be advertised for sale in the city's Sunday paper. The details recorded are shown below. The variable Price is the intended asking price, and Paid is the amount that the dealer paid for the automobile.

Class	Age	Miles	Price	Paid	Condition
C	2	7000	29900	24000	Very Good
C	1	11000	33995	26750	Good
C	3	15000	29600	23850	Very Good
C	3	26000	27495	21600	Excellent
C	4	36000	22995	19200	Very Good
C	4	29000	23495	19200	Good
E	1	4000	48900	38800	Very Good
E	3	26000	39900	32650	Excellent
E	5	41000	30900	24300	Very Good
E	5	42000	32495	26300	Very Good
E	5	62000	29995	23100	Good
S	2	18000	67900	55500	Very Good
S	2	13000	61495	48150	Very Good
S	2	15000	59995	50100	Excellent

a. Enter these data into columns C1 to C6 in a new worksheet. Enter the column names used in the above table.

b. Automobile number 3 (a C class with 15,000 miles) has already been sold, so you should delete this entire row.

c. Two new cars with the following details have been purchased and are available for sale.

Class	Age	Miles	Price	Paid	Condition
S	3	35000	49900	39900	Very Good
S	5	44000	31495	24300	Excellent

Use the Editor > Go To... command to make the cell in column 1, row 14 active. Then enter these data into your worksheet.

d. Use the Calc > Calculator command to compute: (i) the difference between the price paid for the car and the price to be asked, in C7 and (ii) the percentage mark-up in price, in C8. Name these two new variables Profit and MarkUp, respectively.

e. The dealer does not intend to send the newspaper the information for the variables Paid, Condition, Profit, and MarkUp, so delete these columns from the worksheet. Why is the dealer unwilling to publish the data for these four variables?

f. Save the current worksheet as *P1Mercedes.mtw* in the folder where you are saving your work.

5. Open the *PulseA.mtw* data set. The data in this set are based upon the following experiment. All of the 92 students in an introductory statistics course recorded their own pulse rates (Pulse1). Each student then flipped a coin. If the flip resulted in heads the student ran in place for one minute; otherwise, the student did not run. All students then recorded their pulse rates for a second time (Pulse2). Those rates and related data are described in the following table.

Column	Name	Description
C1	Pulse1	Initial pulse rate, in beats per minute
C2	Pulse2	Second pulse rate, in beats per minute
C3-T	Ran	Whether or not the student ran in place; Ran or Still
C4-T	Smokes	Smoking status; Smoke or Nonsmoker
C5-T	Gender	Student's gender; Female or Male
C6	Height	Height, in inches
C7	Weight	Weight, in pounds
C8	Activity	Usual level of physical activity; Slight, Moderate, or ALot

a. Use Help to find out the exact meaning of the variables in this data set.

b. Use Calc > Calculator to compute the increase in pulse rate (Pulse2 – Pulse1) and store the results in C9. Name this new variable Increase. Some of the increases are negative. How would you interpret such values?

c. Save your worksheet as *P1PulseA.mtw* in the folder where you are saving your work. Close this worksheet.

d. Retrieve the file *P1PulseA.mtw* that you saved and closed in part c.

e. Use Calc > Calculator to compute the Body Mass Index (BMI) for each of the 92 students. Store the values in C10. When, as in this case, weight is in pounds and height is in inches, BMI is computed as:

$$\text{BMI} = 703 * \text{Weight}/\left(\text{Height}^2\right)$$

f. Very approximately, what is the range of BMI values? You will need to scroll slowly down the worksheet to scan the 92 values.

g. Use the Show Info toolbar button to obtain summary information about the worksheet.

h. Save the revised worksheet, with the same name, in the folder where you are saving your work.

6. Open the *Note02.mtw* data set, which contains information about a selection of notebook (portable) computers that were available in 2002.

a. Copy the first column (Processor) into C11 so that it is adjacent to Price.

b. Which processor is the least expensive? The most expensive?

c. Erase C11 and close the worksheet.

7. In a survey of 1503 males between the ages of 25 and 34 each individual was classified by marital status and by whether or not he was employed. The counts for each combination of categories are given below. For example, 98 single males were unemployed.

Marital Status	Employed	Unemployed
Single	493	98
Married	648	94
Divorced	135	35

a. Enter these three columns of data into C1, C2, and C3. Name the columns.

b. Use Calc > Calculator to compute (in C4) the percentage of each marital group that is unemployed. Give an appropriate name to C4. Which marital group has the highest unemployment rate?

c. Explain how the nature of the data in this question differs from the nature of the data in problems 3 and 4 above.

d. Print out a copy of your worksheet.

8. Open *TBill2.mtw*.

a. Use Help to find the exact nature of the variables in this data set.

b. Use Calc > Calculator to create a new variable equal to the difference of the daily high and the daily close in C4. Name C4 HiLessClose.

c. The values of HiLessClose should not be negative. Why? Are there any negative values in this column? If there are negative HiLessClose values, in what rows are they located?

9. Open the data set *Assess.mtw*. This data set contains a great deal of information about 81 homes.

a. Use Help to find out the exact nature of the variables in this data set.

b. Use the Show Info toolbar button to obtain summary information about the worksheet. Are there any missing values in the data set? What variables have missing values? How many homes have missing values?

c. Copy the first three variables in the data set into an Excel spreadsheet and insert commas and dollar signs into the first two variables.

d. Import these two variables from Excel into a new Minitab worksheet. How does Minitab deal with the commas and the dollar signs?

e. Compute, in C3, the difference Total$ – Land$. How would you interpret this new variable?

f. Name this new variable and then print out the contents of the new worksheet, choosing a suitable title.

On Your Own

In this project you will use secondary data sources to investigate the distribution of the number of people represented by members of the U.S. House of Representatives.

Obtain, from an almanac or from the Internet, (i) the population of each of the 50 states in the year 2000 and (ii) the number of representatives from each state in the U.S. House of Representatives in the year 2000.

Enter three columns of data into a new Minitab worksheet. The first column should be the names of the 50 states (in alphabetical order). The second column should contain the year 2000 population of the state, and the third column, the number of representatives for the state. (Remember that you need not use commas when you enter large numbers into Minitab.) Provide suitable names for each column.

Now, use the Calc > Calculator command to compute the number of people per representative for each state. Put these numbers in C4 and name this new variable. By scrolling down the worksheet, find the states that have (i) the largest number of people per representative and (ii) the smallest number of people per representative.

Finally, save your worksheet as *P1HouseRep.mtw* in the folder where you are saving your work.

Answers to Matching Problems

(c) Session Window
(h) text
(d) *
(b) Blank
(a) Worksheet
(g) mtw
(f) Open Worksheet
(i) Variable
(e) Cell
(j) F1

Answers to True/False Problems

(a) T, (b) F, (c) F, (d) F, (e) F, (f) F, (g) T, (h) F, (i) F, (j) F

Answers to Matching Problems for Tutorial 0

(b) 100
(d) 4,000
(e) 10,000
(a) Session
(c) Worksheet 1***

Answers to True/False Problems for Tutorial 0

(a) F, (b) F, (c) T, (d) T, (e) F

2

Summarizing, Transforming, and Manipulating Data

Now that you know how to enter and save data in Minitab, you are ready to organize and manipulate it to perform various statistical analyses. Many of these analyses will be done in the Session window. You also print data from this window.

In this tutorial, you learn how to:

- Summarize the data in rows (individual cases) and in columns (variables)
- Round, code, rank, and sort data
- Enable and use Session window commands
- Standardize data for different variables in order to compare them
- Obtain a subset of a worksheet
- Combine and break down columns of data
- View and print data from the Session window
- Create a text file using Notepad

CASE STUDY	EDUCATION— CLASS EVALUATION

You have just completed your first marking period as a social studies teacher. You entered your students' scores for the first three tests into a Minitab worksheet. Your principal, Ms. Taylor, wants you to prepare a report about your students' performance.

Open Minitab and open the *Marks.mtw* worksheet from the Studnt14 folder:

- Open **Minitab** and maximize the Data window
- Choose **File > Open Worksheet**
- Double-click *Marks.mtw* in the "Open Worksheet" list box (you may have to navigate to the Studnt14 folder and then scroll down to see the file)

At this point you may see the Minitab message "A copy of the content of this file will be added to the current project." If so,

- Click **OK**

The names of the 24 students are in C1 and C2. C3, C4, and C5 contain the student's test scores for Test1, Test2, and Test3, respectively.

Before analyzing the Marks data, save the data set under the name *2Marks*, an *mtw* file.

- Choose **File > Save Current Worksheet As**
- Click the "Save in" **drop-down list arrow** and select the location where you are saving your work
- Type *2Marks* in the "File name" text box
- Click **Save**

As you modify this data set, you should periodically save the modified worksheet.

You want to compute each student's average score for the three tests. You can do this in either of two ways. The first involves using Calculator on the Calc menu to add each student's test scores in the third, fourth, and fifth columns of the appropriate row and then divide each sum by 3:

- Choose **Calc > Calculator**

The Calculator dialog box appears.

- In the "Store result in variable" text box, type **Average** and then press (Tab)

Minitab assigns the name to the next empty column in the data set, in this case, C6.

Now, tell Minitab to average the three test scores. Do this by entering, in the "Expression" text box, (Test1 + Test2 + Test3) / 3. To do this:

- Click on the parentheses [()]
- Double-click on **C3 Test1**

- Click on **+**
- Double-click on **C4 Test2**
- Click on **+**
- Double-click on **C5 Test3**
- If necessary, move the cursor in the Expression text box to the right of the right-hand parenthesis
- Click on **/**
- Click on **3**
- Click on **OK**

Minitab stores in C6 the results—the average score for each student.

Notice that, in this case, you clicked rather than typed the expression you wanted to calculate.

The second way to compute each student's average score for the three tests is to use the command Row Statistics on the Calc menu. This menu selection provides an even easier way to calculate the mean (or other statistic) of some or all of the data in a particular row and then save the results in a new column. You don't even enter a formula! Try using Row Statistics:

- Choose **Calc > Row Statistics**

Row Statistics computes common statistics for each row. To compute the mean of the three scores recorded for each student in C3, C4, and C5:

- Click the "Statistic" **Mean** option button
- Press ⟨Tab⟩ to move the cursor to the "Input variables" text box

In the "Input variables" text box, you can enter the variables in any of three ways. First, you can type C3–C5 (a dash between columns designates a series of consecutive columns). Second, you can highlight and then select consecutive variables. (To select several nonconsecutive variables, hold down ⟨Ctrl⟩ while you click each one.) Or third, you can simply double-click each of the three variables:

- Double-click **C3 Test1, C4 Test2,** and **C5 Test3** in the list box on the left

"Test1 Test 2 Test3" now appears in the "Input variables" text box.

Now, tell Minitab where to display the resulting means:

- Press ⟨Tab⟩ to move to the "Store result in" text box
- Type **RowMean**

The completed Row Statistics dialog box appears as shown in Figure 2-1.

FIGURE 2-1

*The completed Row
Statistics dialog box*

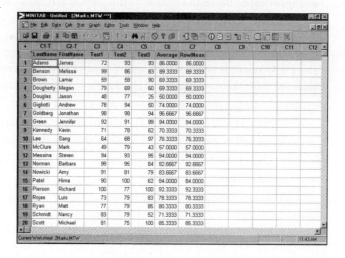

■ Click **OK**

Comparing the entries in C6 and C7 in Figure 2-2, you can see that the two methods produce the same results. You now have a test average for each student. Note that several students are performing extremely well (Jonathan Goldberg and Kathleen Sheppard, for example), while others (such as Jason Douglas and Holly Smith) are not performing well. (Again, depending on the size of your screen, you may have to scroll to see some of these data.)

FIGURE 2-2

*The Data window
showing the mean
calculated in two ways*

	C1-T	C2-T	C3	C4	C5	C6	C7	C8	C9	C10	C11	C12
	LastName	FirstName	Test1	Test2	Test3	Average	RowMean					
1	Adams	James	72	93	93	86.0000	86.0000					
2	Benson	Melissa	99	86	83	89.3333	89.3333					
3	Brown	Lamar	59	59	90	69.3333	69.3333					
4	Dougherty	Megan	79	69	60	69.3333	69.3333					
5	Douglas	Jason	48	77	25	50.0000	50.0000					
6	Gigliotti	Andrew	78	94	50	74.0000	74.0000					
7	Goldberg	Jonathan	98	98	94	96.6667	96.6667					
8	Green	Jennifer	92	91	99	94.0000	94.0000					
9	Kennedy	Kevin	71	78	62	70.3333	70.3333					
10	Lee	Sang	64	68	97	76.3333	76.3333					
11	McClure	Mark	49	79	43	57.0000	57.0000					
12	Messina	Steven	94	93	95	94.0000	94.0000					
13	Norman	Barbara	99	95	84	92.6667	92.6667					
14	Nowicki	Amy	91	81	79	83.6667	83.6667					
15	Patel	Hima	90	100	62	84.0000	84.0000					
16	Pierson	Richard	100	77	100	92.3333	92.3333					
17	Rojas	Luis	73	79	83	78.3333	78.3333					
18	Ryan	Matt	77	79	85	80.3333	80.3333					
19	Schmidt	Nancy	83	79	52	71.3333	71.3333					
20	Scott	Michael	81	75	100	85.3333	85.3333					

For additional practice, use Row Statistics to compute the median score for each student:

■ Click the **Edit Last Dialog** toolbar button to reopen the Row Statistics dialog box

■ Click the "Statistic" **Median** option button

You want to use the same input variables as you did previously:

- Press `Tab` twice to move to the "Store result in" text box
- Type **RowMedian** in the "Store result in" text box and click **OK**

C8 RowMedian, as shown in Figure 2-3, now contains the median of each student's three test scores (you may have to scroll to see this column).

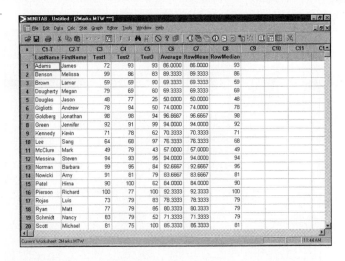

FIGURE 2-3

The Data window showing median test scores

Next, you decide to compute a weighted average of the three tests in order to reward students who improved during the first marking period:

- Choose **Calc > Calculator**
- In the "Store result in variable" text box, type **W** in front of the Average name and press `Tab`
- Modify the "Expression" text box to create the formula **(1*Test1+2*Test2+3*Test3) / 6** and click **OK**

The weighted-average expression will have a slightly different appearance depending on whether you click or type the various components. The weighted averages in C9 do reward recent performance. For example, the weighted average for Lamar Brown (whose scores were 59, 59, and 90) is more than five points greater than his unweighted average.

In your final report, you want each student's weighted average score to appear as an integer. Minitab can round the scores for you:

- Click the **Edit Last Dialog** toolbar button 🔲 to reopen the Calculator dialog box

Tell Minitab in which column to store the rounded averages:

- Type **R** in front of the WAverage name in the "Store result in variable" text box
- Highlight the entire formula in the "Expression" text box
- In the "Functions" box, scroll down until you reach the Round function, click on **Round**, and click **Select**
- Select and highlight **num digits** in the "Expression" text box and type **0**

The "num digits" placeholder is where you specify the number of decimal places to which the number should be rounded. When the value 0 is specified, the data are rounded to the nearest integer. (For example, had you wanted the scores rounded to two decimal places, you would have entered 2 instead of 0.)

The completed Calculator dialog box is shown in Figure 2-4.

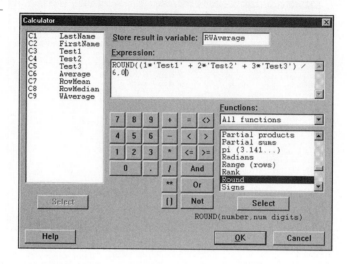

- Click **OK**

Each student's weighted average score, rounded to the nearest integer, appears in C10.

Rounding is only one of the many *functions* that Minitab provides in Calculator. You may be familiar with others, such as obtaining the absolute value, the log to the base 10, and the maximum. You will encounter a number of new and useful functions as you proceed through this book. To learn more about these functions you can examine Minitab's Help on the subject:

- Click on the **Help** toolbar button 🔲 and click on the **Search** folder's tab
- Type **calculator** in the "Type in the word(s) to search for" text box
- Click on the **List Topics** button
- Highlight **Calculator (Calc Menu)** and click **Display**
- In the right-hand side of the window, double-click on **Calculator Functions**

A list of all the available functions is shown. You can click on any of the functions to get a detailed explanation of its use.

- Click on the Help window's **Close** button 🔲

2.2 Summarizing Columns

Each row in a Minitab worksheet represents the data for an individual case, such as one particular student's test scores. When you perform a *row operation*, you calculate a *statistic* (for example, the mean for each case) and display the results in a designated column in the Data window. In contrast, Minitab's columns contain values of the same variable; for example, C3 contains all of the scores for Test1. When you want to compute statistics for a particular column or variable, you use a *column operation*.

Column operations frequently produce single numbers that appear in the Session window, not in the Data window. For example, the mean of the 24 Test1 scores will be a constant.

Explore how Minitab operates on columns (variables) by finding the mean score for Test1. Remember, this is the mean of the entire class's performance on the first test, not the mean of a single student's performance. You do this by using Column Statistics on the Calc menu:

- Choose **Calc > Column Statistics**
- Click the "Statistic" **Mean** option button
- Click in the "Input variable" text box
- Double-click on **C3 Test1** in the list box on the left to enter it in the "Input variable" text box
- Click in the "Store result in" text box
- Type **MeanTest1** as the constant name in the "Store result in" text box

The completed Column Statistics dialog box is shown in Figure 2-5.

FIGURE 2-5

The completed Column Statistics dialog box

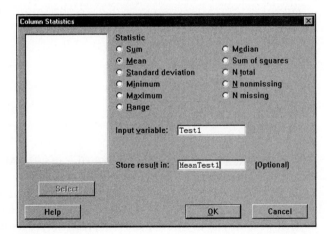

- Click **OK**

The results appear in the Session window as shown in Figure 2-6.

FIGURE 2-6

Mean of Test1

Mean of Test1

```
Mean of Test1 = 80
```

You can see the mean you requested—80—in the Session window. Your students performed relatively well on the first test.

▶ **Note** It is worth reviewing the column statistics that are available from the Column Statistics dialog box. As well as the mean, you can obtain the median of the values, the range of the values, and the standard deviation of the values. In Tutorial 3, you will learn how to obtain other statistics, such as the quartiles, that are not included under Column Statistics. ■

2.3 | Using Session Commands

Until now, you have primarily worked in the Data window and issued commands by choosing them from the Minitab menu bar. Every command you executed via the menu has a corresponding Session command that appears in the Session window using a special Session command language. At this point the Session window should be active. (If it is not, click the Session Window toolbar button or choose Window > Session.) To enable the *Session command language*:

■ Choose **Editor > Enable Commands**

The Minitab prompt, "MTB >", appears at the bottom of the Session window. To see one impact of this change, ask again for the mean of Test1:

■ Choose **Edit > Edit Last Dialog** and repeat the previous command by clicking **OK**

The output is shown in Figure 2-7.

FIGURE 2-7

Mean of Test1 with the Minitab prompt

```
MTB > Mean 'Test1' 'MeanTest1'.
```

Mean of Test1

```
Mean of Test1 = 80
```

Note that a command has appeared after the first Minitab prompt and before the second appearance of your mean output.

Minitab entered "Mean 'Test1' 'MeanTest1'" when you requested a column mean in the dialog box. You could have typed this command directly in the Session window just as Minitab did and pressed ⏎ to issue the command.

The "MTB >" prompt at the bottom of the Session window is Minitab's way of telling you that it is ready for you to type a command. To enter a command in this way, you need to know the correct words to type using Minitab's Session command language. For example, to obtain information about the worksheet, you type the Session command info:

- With your cursor at the "MTB >" prompt, type **info** and press ⏎ (Remember, always press ⏎ after typing a Session command.)

Summary information for each column of the worksheet are displayed in the Session window. (The information is comparable to that obtained with the Show Info toolbar button in the last tutorial.) Notice that Minitab saved the mean for Test1 as the named constant K1, MeanTest1.

Constants are stored as part of the worksheet, but do not appear in the worksheet's Data window. Minitab stores such values as specific names that you assign, such as MeanTest1. You can also assign locations to constants, such as K1. If you plan to use a result again later, it's useful to store it as a named constant.

Output in the Session window can be changed and edited. However, it is also easy to inadvertently change output that you didn't want to alter. To avoid this, it is recommended that newcomers to Minitab make the output *read-only*. When you make output read-only, you ensure that this display and other portions of the Session window are not altered in error. To do this:

- Choose **Editor > Output Editable** (unless the Output Editable menu command is not checked)

▶ **Note** The Output Editable and the Enable Commands menu commands are examples of "toggles". When the Session window is set to read-only, the Output Editable menu command is not checked and the Edit mode display at the bottom of the Session window displays "Read only"; when the Session window is set to be edited, the Output Editable menu command is checked. The Output Editable and the Enable Commands menu commands are available only when the Session window is active. ■

At the beginning of each of the remaining tutorials, you will be asked to enable Session commands and make the Session window read-only.

You normally view the data in your worksheet in the Data window. However, you can also display data in the Session window; this can sometimes be quicker than looking in the Data window. Suppose you want to see the names of your students and the students' rounded weighted averages. Use the Info list as a quick reference for the columns you wish to view; last name is in C1, first name is in C2, and RWAverage is in C10:

- With your cursor at the "MTB >" prompt, type **print c1 c2 c10** and press ⏎

Minitab lists the data in row and column format, along with the corresponding row numbers, as shown in Figure 2-8.

FIGURE 2-8

*Students' names and their
rounded weighted average
test scores*

```
MTB > print c1 c2 c10
```

Data Display

Row	LastName	First	RWAverage
1	Adams	James	90
2	Benson	Melissa	87
3	Brown	Lamar	75
4	Dougherty	Megan	66
5	Douglas	Jason	46
6	Gigliotti	Andrew	69
7	Goldberg	Jonathan	96
8	Green	Jennifer	95
9	Kennedy	Kevin	69
10	Lee	Sang	82
11	McClure	Mark	56
12	Messina	Steven	94
13	Norman	Barbara	90
14	Nowicki	Amy	82
15	Patel	Hima	79
16	Pierson	Richard	92
17	Rojas	Luis	80
18	Ryan	Matt	82
19	Schmidt	Nancy	66
20	Scott	Michael	89
21	Sheppard	Kathleen	98
22	Smith	Holly	43
23	Thompson	Susan	72
24	Watson	Keisha	66

As you can see, the "print" command prints to the screen, not to paper.

Notice that the Session window automatically scrolls to keep the last line of output and the current "MTB >" prompt visible. As a rule, Session window output will be displayed in the format shown in Figure 2-8 so that you can see more than one screen of output. To see the beginning of the output and other portions of the Session window, you can scroll up and down or right and left using the scroll bars. Alternatively, you can use the direction keys to move quickly through one section at a time:

- Make sure NumLock on your keyboard is off, and press (PageUp) to move to the previous screen
- Press (PageDown) to move to the next screen
- Press (Ctrl)+(Home) to move to the beginning of the Session window
- Press (Ctrl)+(End) to move to the end of the Session window

   These key shortcuts also work in the Data window.

   Here are three basic rules for typing Session commands:

1. Type the main command, followed by any arguments. If you are going to use subcommands, end the main command line with a semicolon.
2. Type any subcommands and any subcommand arguments, ending each subcommand line with a semicolon, except the last subcommand line.
3. End the last subcommand line with a period.

▶ **Note** You will develop a preference on how to issue commands: using the menus, typing them at the Session window prompt, or, where available, using a toolbar button. When the choice is between the first of these two options, this book generally directs you to use the menus, except in some cases in which it's simpler to type directly in the Session window. For an example of such a situation, compare the process of obtaining the mean for Test1 using the Calc > Column

Statistics menu command with the (simple) corresponding process using the comparable Session command:

- At the "MTB > " prompt, type **mean c3** and press ⏎

  The response appears in the Session window. ■

The display in Figure 2-8 can also be obtained using Display Data on the Data menu. Display Data corresponds to the print Session command that you used to get that figure:

- Choose **Data > Display Data**
- Double-click on **C1 LastName, C2 FirstName**, and **C10 RWAverage** in the list box to enter them in the "Columns, constants, and matrices to display" text box
- Click **OK**

  Using Display Data produces the same result as using the print command in the Session window except that Minitab issued the command a little differently in the Session window. The command Print 'LastName' 'FirstName' 'RWAverage' refers to the column names rather than their numbers. You can specify columns by either number or name when you use Session commands.

  ⬥ **Note** In the Display Data dialog box, you listed the names of three columns (LastName, FirstName, and RWAverage) by clicking them. Minitab left a space between them. You can also type the column numbers or names. In general, when listing columns, whether by number or by name, you must leave one or more spaces or type a comma between the names or numbers. Or, you may write a sequence of consecutive column numbers or names, inserting only a dash between the first and the last name or number in the sequence. For example, Minitab will interpret the list C3, C7, C9–C12, C14 as including the seven columns C3, C7, C9, C10, C11, C12, and C14. ■

  ⬥ **Note** Minitab's Help contains an alphabetical listing of all session commands (and their menu command equivalents). ■

2.4 Coding Data

The school district in which you work uses the letter grades A, B, C, D, and F. You want to code values for RWAverage that fall in the interval 90 to 100 as an A, values of 80 to 89 as a B, values of 70 to 79 as a C, values of 60 to 69 as a D, and values 59 and below as an F. You will store the letter grades corresponding to the numeric values in a new column.

To code the students' averages, you use the Code > Numeric to Text command on the Data menu:

- Choose **Data > Code > Numeric to Text** to display the Code - Numeric to Text dialog box
- Double-click on **C10 RWAverage** in the list box to enter it in the "Code data from columns" text box

- Press [Tab] and then type **Grade** in the "Into columns" text box

    To specify the intervals, you can either list all of the values in the interval separated by commas or specify the first and last values in the range separated by a colon. For example, 90, 91, 92, 93, 94, 95, 96, 97, 98, 99, 100 and 90:100 will produce the same result:

- Press [Tab] to move to the first "Original values" text box
- Type **90:100** and then press [Tab] to move to the first "New" text box

    Minitab interprets the entry 90:100 as all of the rounded weighted averages between 90 and 100, inclusive.

- Type **A** in the first "New" text box and press [Tab]
- Type **80:89**, press [Tab], type **B**, and then press [Tab] to enter data in the second "Original values" and "New" text boxes, respectively
- Type **70:79**, press [Tab], type **C**, and then press [Tab] to enter data in the third "Original values" and "New" text boxes, respectively
- Type **60:69**, press [Tab], type **D**, and then press [Tab] to enter data in the fourth "Original values" and "New" text boxes, respectively
- Type **0:59**, press [Tab], and then type **F** to enter data in the fifth "Original values" and "New" text boxes, respectively

    The completed Code - Numeric to Text dialog box should appear as shown in Figure 2-9.

FIGURE 2-9

*The completed
Code - Numeric to Text
dialog box*

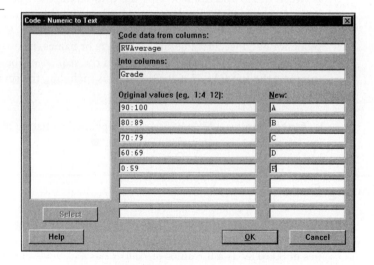

- Click **OK** to code the average scores with a letter indicating the grade
- Click on the **Current Data Window** toolbar button 🗔 to view the new column
- Scroll to the right, if necessary, to see C11-T Grade

    The resulting Data window is shown in Figure 2-10.

FIGURE 2-10

*The Data window
showing the variable
Grade*

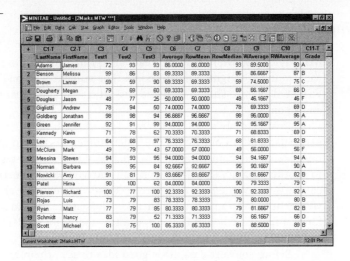

	C1-T	C2-T	C3	C4	C5	C6	C7	C8	C9	C10	C11-T	
	LastName	FirstName	Test1	Test2	Test3	Average	RowMean	RowMedian	WAverage	RWAverage	Grade	
1	Adams	James	72	93	93	86.0000	86.0000	93	89.5000	90	A	
2	Benson	Melissa	99	86	83	89.3333	89.3333	86	86.6667	87	B	
3	Brown	Lamar	59	59	90	69.3333	69.3333	59	74.5000	75	C	
4	Dougherty	Megan	79	69	60	69.3333	69.3333	69	66.1667	66	D	
5	Douglas	Jason	48	77	25	50.0000	50.0000	48	46.1667	46	F	
6	Gigliotti	Andrew	78	94	50	74.0000	74.0000	78	69.3333	69	D	
7	Goldberg	Jonathan	98	98	94	96.6667	96.6667	98	96.0000	96	A	
8	Green	Jennifer	92	91	99	94.0000	94.0000	92	95.1667	95	A	
9	Kennedy	Kevin	71	78	62	70.3333	70.3333	71	68.8333	69	D	
10	Lee	Sang	64	68	97	76.3333	76.3333	68	81.8333	82	B	
11	McClure	Mark	49	79	43	57.0000	57.0000	49	56.0000	56	F	
12	Messina	Steven	94	93	95	94.0000	94.0000	94	94.1667	94	A	
13	Norman	Barbara	99	95	84	92.6667	92.6667	95	90.1667	90	A	
14	Nowicki	Amy	91	81	79	83.6667	83.6667	81	81.6667	82	B	
15	Patel	Hima	90	100	62	84.0000	84.0000	90	79.3333	79	C	
16	Pierson	Richard	100	77	100	92.3333	92.3333	100	92.3333	92	A	
17	Rojas	Luis	73	79	83	78.3333	78.3333	79	80.0000	80	B	
18	Ryan	Matt	77	79	85	80.3333	80.3333	79	81.6667	82	B	
19	Schmidt	Nancy	83	79	52	71.3333	71.3333	79	66.1667	66	D	
20	Scott	Michael	81	75	100	85.3333	85.3333	81	88.5000	89	B	

Compare columns C10 and C11-T. Notice that each student whose rounded weighted average falls between 90 and 100, inclusive, is assigned a letter grade of A; each student whose rounded weighted average falls between 80 and 89, inclusive, is assigned a B, and so on. By scanning this column, you can quickly determine which students are doing well and which ones need help.

2.5 Ranking Data

Ms. Taylor has asked you to rank your students' average exam scores in numerical order so that she can see how each student's performance compares to the rest of the students' performances. Minitab ranks the data listed in any column by assigning a value of 1 to the lowest score, 2 to the second lowest, and so on. The top student is ranked number 24, because this is the total number of students in the class. To rank the weighted average exam scores, you use the Rank command on the Data menu.

- Choose **Data > Rank**
- Double-click **C10 RWAverage** in the list box to enter it in the "Rank data in" text box
- Type **Rank** in the "Store ranks in" text box
- Click **OK** and scroll, if necessary, to view C12 Rank

If two or more scores are tied, Minitab assigns them a rank equal to the average of the ranks of the tied scores. For example, from the scores that placed seventh and eighth in C12, you can see that Minitab assigned each the rank of 7.5 (students 6 and 9, Andrew Gigliotti and Kevin Kennedy). Verify that two students are tied for eighteenth place in your class. What are their names?

2.6 Sorting Data

Ms. Taylor also wants you to create and print a list of the students ordered by their rounded weighted mean scores. The student who had the highest rounded weighted average should be listed first, the second highest should be listed next, and so on.

If you simply sort C10 RWAverage, Minitab rearranges the entries in that particular column but leaves the others intact. Sorting in this way usually creates unexpected and confusing results, such as one student's test scores appearing next to another's name. Usually, when you sort you want to carry along all of the columns in a worksheet that corresponds to the data for an individual case. Here, you want to keep in a single row all of the data related to a given student. Note, for example, that before the sort, the test scores for student 21, Kathleen Sheppard, are 94, 97, and 100.

Before you sort data, it's a good idea to save them just in case they get mixed up during the sort process:

- Choose **File > Save Current Worksheet**

Now you're ready to sort your data based upon the values for RWAverage. You do this by using the Sort command on the Data menu:

- Choose **Data > Sort**
- Highlight all of the columns in the list box
- Click **Select** to display the columns in the "Sort column(s)" text box

Now you need to tell Minitab to use C10 RWAverage—the student's rounded weighted average on the three exams—as the sort criterion:

- Click in the first "By column" text box
- Double-click on **C10 RWAverage** in the list box to enter this column in this text box
- Click the **Descending** check box to sort the students' weighted average scores in descending order (otherwise, the weighted average scores will be sorted in ascending order—lowest score first)

Next, you need to specify where you want Minitab to store the sorted data. In this case, put them in the same location they currently occupy:

- Click the "Store stored data in" **Original column(s)** option button

Your completed Sort dialog box should look as shown in Figure 2-11.

FIGURE 2-11

The completed Sort dialog box

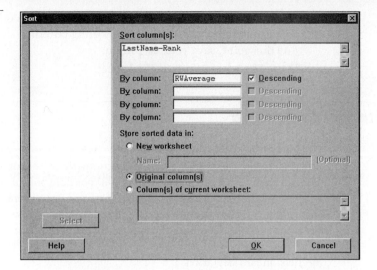

■ Click **OK**

■ Press Ctrl+Home to move to the beginning of the Data window

The students' data are ordered according to their RWAverage scores in C10, as shown in Figure 2-12. The scores for each student are carried along with them. For example, the student with the highest RWAverage score is Kathleen Sheppard, who had test scores of 94, 97, and 100.

FIGURE 2-12

The Data window after sorting

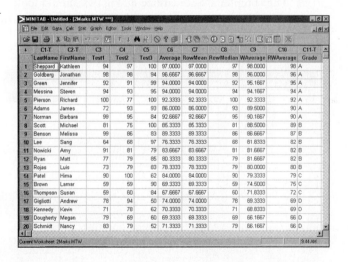

You also can sort rows by using a text variable. This is useful if you wish to place items in alphabetical order. You will do this later in this tutorial.

You now have a summary of your class's performance. Because your worksheet is sorted, you can quickly see which students are not doing well (those at the bottom of the list).

In this example, you sorted the scores on the basis of the weighted averages. Minitab can use up to four columns as sorting criteria.

At this point, save your data again:

■ Choose **File > Save Current Worksheet**

2.7 | Standardizing Data

In addition to submitting the test scores, you want to report each student's progress during the marking period by comparing the scores from the three tests. Though each test score is based upon 100 possible points, an 87 on an easy test might actually reflect less knowledge than a 75 on a more difficult test. If two scores come from two different samples, each with its own mean and standard deviation, you can compare them by *standardizing* them. You subtract from each score the appropriate mean (this is called *centering*) and then divide the difference by the standard deviation.

To compute standardized scores for three sets of test scores, you use Standardize on the Calc menu.

■ Choose **Calc > Standardize** to open the Standardize dialog box
■ Double-click on **C3 Test1, C4 Test2,** and **C5 Test3** in the list box to enter them in the "Input column(s)" text box
■ Press Tab
■ Type **Standard1 Standard2 Standard3** in the "Store results in" text box

The completed Standardize dialog box should look as shown in Figure 2-13. Notice that the option you want, "Subtract mean and divide by std. dev.", is already selected; it is the default, or preselected, option.

FIGURE 2-13

The completed Standardize dialog box

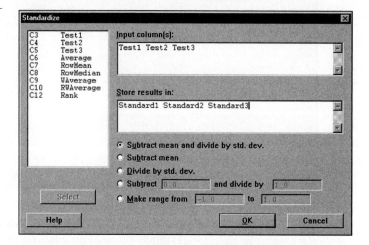

■ Click **OK** to calculate the standardized scores
■ Scroll to look at C13, C14, and C15, which contain the standardized scores

The standardized scores, the first of which are shown in Figure 2-14, allow you to easily determine how each student performed on each test, relative to the class.

FIGURE 2-14

The Data window showing standardized scores

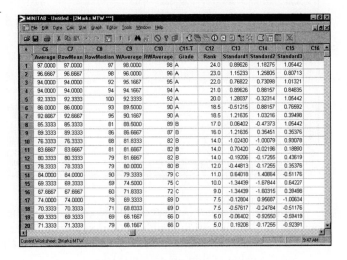

You can quickly spot trends in a student's performance. For example, relative to the class performance, the performance of Jennifer Green improved (from test 2 to test 3) while that of Barbara Norman declined.

2.8 Creating Subsets

Your older colleagues have warned you to expect complaints and questions from those students who received an F for the class. As a handy reference, you want to store the various scores for these students separately from the original worksheet. Minitab's Subset Worksheet command in the Data menu is perfect for this. It allows you to store a subset of the original data in a separate worksheet. To create a worksheet that contains the values in C1–C15, but only for those students with a grade of F:

- Choose **Data > Subset Worksheet**
- Highlight the default "Name", if necessary, and type **2MarksF**
- Click on the **Condition** button
- Double-click on **C11 Grade** to enter it in the "Condition" text box and then type = **"F"** (with the quotation marks) so that the text box contains 'Grade' = "F"
- Click **OK** twice

The new worksheet, called *2MarksF*, is shown in Figure 2-15. It contains the data for C1–C15, but only for the three F students. You will print scores for these students later.

FIGURE 2-15

*The 2MarksF
Data window*

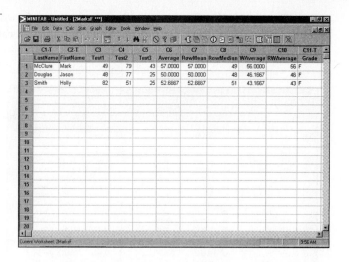

Notice that when you specified the grade of F in the "Condition" text box, you enclosed the F within double quotation marks. Whenever you write an equation that includes the value for a text variable—such as Grade = "F"—you must enclose the value within quotation marks.

Return to the *2Marks.mtw* worksheet:

■ Click on the **Current Data Window** toolbar button 🖽

▶ **Note** When multiple worksheets are open, repeatedly clicking the Current Data Window toolbar button makes each worksheet current in turn. You can also use Windows > *2Marks.mtw* to return to this worksheet. ■

▶ **Note** Minitab has a Split Worksheet command that operates in much the same way as a Subset Worksheet. It allows you to create several new worksheets based upon the unique values in a column. For example, you could create five worksheets based upon the five possible letter grades you have assigned. However, one restriction with the Student version of Minitab on the use of both Subset Worksheet and Split Worksheet is that a Minitab session can only include five worksheets. Minitab also has a Merge Worksheet command. It allows you to merge another worksheet into the active worksheet. All of these commands are on the Data menu. ■

2.9 Combining Data Using the Stack Option

Your principal, Ms. Taylor, has asked you to include statistical information based upon the three tests combined in your report. To obtain such information, you need to combine, or *stack*, the scores in Test1, Test2, and Test3 into a single column. Minitab does this with the Stack command on the Data menu. Stack creates a new column that consists of data from all of the selected columns:

■ Choose **Data > Stack > Columns**

The Stack Columns dialog box appears. To indicate the three columns you want to stack:

- Double-click on **C3 Test1, C4 Test2, and C5 Test3** in the list box to enter them in the "Stack the following columns" text box
- Click on the **Column of current worksheet** option button and press ⌨Tab
- Type **Total** in the "Store the stacked data in: Column of current worksheet" text box and press ⌨Tab

When you stack data, you can tell Minitab to create and store a subscript variable that indicates the column from which the data originated. In this case, the subscripts—Test1, Test2, and Test3 —are simply text values indicating to which test a score belongs:

- Type **Test** in the "Store subscripts in" text box

Your completed Stack Columns dialog box should look as shown in Figure 2-16. Notice that the default option, Use variable names in subscript column, is accepted.

FIGURE 2-16

The completed Stack Columns dialog box

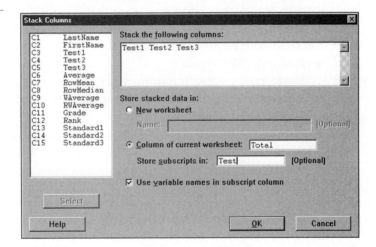

- Click **OK** to stack the three test variables in the same column
- Scroll to the right and down so that you can see the stacked scores in C16 and the subscripts in C17-T

Notice how all of the scores are stacked on top of one another in the order in which you entered the tests; in other words, all three columns, C3, C4, and C5, are listed in a single column. Note, too, that C17-T (the subscript) identifies the test from which each score originated.

You can calculate the mean of C16 and view it in the Session window:

- Choose **Calc > Column Statistics** to open the Column Statistics dialog box, which contains the selections you made the last time you used this command
- Double-click in the "Input variable" text box to highlight the variable you previously specified

- Double-click on **C16 Total** in the list box to enter it in the "Input variable" text box
- Double-click in the "Store result in" text box
- Type **TestAll** in the "Store result in" text box and click **OK**

The results, shown in the Session window, appear as shown in Figure 2-17. (Scroll down, if necessary, to see the mean of the total scores.)

FIGURE 2-17

Computing the Mean of Total score

```
MTB > Name K2 "TestAll"
MTB > Mean 'Total' 'TestAll'.
```

Mean of Total

```
Mean of Total = 78.5694
```

The figure shows that Minitab created a new constant, K2, called TestAll, which will contain the mean of all 72 (= 24*3) test scores. That mean is 78.5694. You can report this figure to Ms. Taylor.

You can return to the Data window by clicking the Current Data Window toolbar button. However, you can also use a keyboard shortcut to activate a different window:

- Press Ctrl+Tab

The Data window is now the active window.

Try activating the other Minitab windows similarly:

- Press Ctrl+Tab several times

The title bar changes as you move from one window to another, identifying which window is active.

- Continue to press Ctrl+Tab until the *2Marks.mtw* worksheet is active again

2.10 Separating Data Using the Unstack Option

Ms. Taylor has requested a separate listing of all RWAverage scores that fall within each letter grade you specified earlier with the Code command. Just as you were able to combine several columns into a single column by using Stack, you can use Minitab's Unstack command to break a single column into several new columns according to criteria you specify.

In this section, you will use the Unstack Columns command on the Data menu to unstack C10 RWAverage into five columns, C18–C22. You will put all of the scores within the range 90 to 100 (As) in the first column (C18), all scores from 80 to 89 (Bs) in the second column (C19), and so on, using the grade data that appears as the subscripts in C11-T. You will name these five new columns A, B, C, D, and F, respectively.

- Choose **Data > Unstack Columns** to open the Unstack Columns dialog box
- Double-click on **C10 RWAverage** in the list box to enter it as the variable to unstack in the "Unstack the data in" text box

- Click in the "Using subscripts in" text box and double-click on **C11 Grade** (which contains the subscripts) in the list box
- Click the "Store unstacked data" **After last column in use** option button

The completed Unstack Columns dialog box should look as shown in Figure 2-18. Note that by default "Name the columns containing the unstacked data" is checked.

FIGURE 2-18

The completed Unstack Columns dialog box

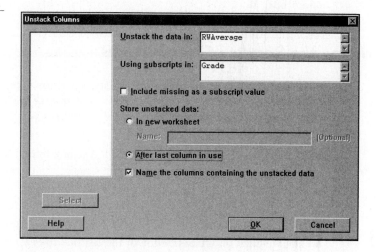

- Click **OK**
- Scroll to the right to see the new columns, C18–C22

The Data window shows the scores grouped by grade, as shown in Figure 2-19.

FIGURE 2-19

The Data window showing scores unstacked into intervals

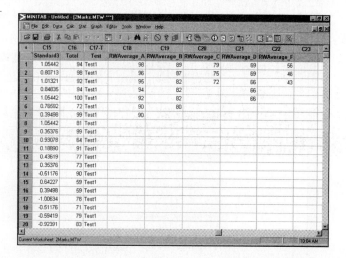

The Unstacked scores are now in the RWAverage_A, RWAverage_B, RWAverage_C, RWAverage_D, and RWAverage_F columns. You have 16 students earning grades of A, B, or C and only 8 earning Ds or Fs. Also, 2 of your 5 students with Ds are almost getting Cs. All in all, it appears your class is doing fairly well.

▶ **Note** In the Unstack Columns dialog box, you could have decided to store your unstacked data in a new worksheet for which you may provide a name. You could have also decided not to have Minitab name the unstacked columns by clearing the check from the "Name the columns containing the unstacked data" check box. ■

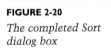

Printing the Results of Your Analysis

Now that you have summarized your class's performance, you can print your worksheet on paper for Ms. Taylor. First, re-sort your grades in alphabetical order (excluding the columns that you created with Stack and Unstack):

- Choose **Data > Sort**
- Double-click in the "Sort Column(s)" text box to highlight the previous entry, if needed
- Select **C1–C15** to enter them in the "Sort column(s)" text box (you do not want to include either the stacked or the unstacked columns you created in this sort)
- Double-click in the first "By column" text box to highlight the previous entry
- Double-click on **C1 LastName** in the list box to enter it in the "Sort by column" text box
- Click the first **Descending** check box to clear it and press Tab
- Double-click on **C2 FirstName** to enter it in the second "By column" text box (in case you have two students with the same last name)
- Click the "Store sorted data in" **Original column(s)** option button, if necessary

At this point the completed Sort dialog box should look as in Figure 2-20.

FIGURE 2-20

The completed Sort dialog box

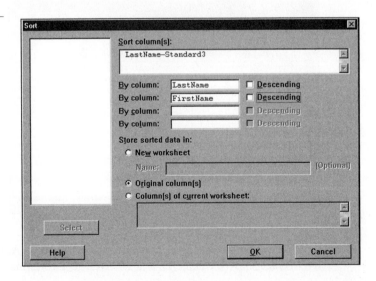

- Click **OK**

    Your worksheet is in alphabetical order again.

    Before printing the data, save the worksheet again.

- Choose **File > Save Current Worksheet**

    Now you are ready to print your grades data to paper. You could choose File > Print Window to print the entire worksheet, but Ms. Taylor probably wants to see only your students' names, weighted averages, and grades. The easiest way to select only those columns that contain these data involves using the Session window. Before moving to the Session window, verify that the columns you want to print are C1, C2, C10, and C11-T.

- Click on the **Session Window** toolbar button 🖩 to move to the Session window

- At the Minitab prompt, "MTB >", type **print c1 c2 c10 c11** and press ⏎

    Minitab displays the data you requested.

- Highlight the output by clicking just to the left of the row headings and dragging the pointer to the right of the final grade. Don't include any of the commands or prompts

    The Session window should look as shown in Figure 2-21, with the data highlighted.

FIGURE 2-21

The Session window showing the grades data highlighted

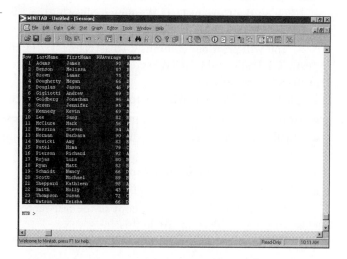

- Choose **File > Print Session Window**

    The Print dialog box lets you specify how much of the window to print. The Selection button is chosen by default, indicating that Minitab will print only the highlighted data.

- Click **OK** to print the selection you highlighted

    Figure 2-22 shows the printed grade list.

FIGURE 2-22

Printout of alphabetical grade list

Row	LastName	First	RWAverage	Grade
1	Adams	James	90	A
2	Benson	Melissa	87	B
3	Brown	Lamar	75	C
4	Dougherty	Megan	66	D
5	Douglas	Jason	46	F
6	Gigliotti	Andrew	69	D
7	Goldberg	Jonathan	96	A
8	Green	Jennifer	95	A
9	Kennedy	Kevin	69	D
10	Lee	Sang	82	B
11	McClure	Mark	56	F
12	Messina	Steven	94	A
13	Norman	Barbara	90	A
14	Nowicki	Amy	82	B
15	Patel	Hima	79	C
16	Pierson	Richard	92	A
17	Rojas	Luis	80	B
18	Ryan	Matt	82	B
19	Schmidt	Nancy	66	D
20	Scott	Michael	89	B
21	Sheppard	Kathleen	98	A
22	Smith	Holly	43	F
23	Thompson	Susan	72	C
24	Watson	Keisha	66	D

At this point, print, from the Session window, a paper copy of the names, the three test scores, and the rounded weighted averages for the three students who failed the class. To review the data for these three students:

- Click on the **Current Data Window** toolbar button 🖼 until you return to the 2MarksF worksheet

    You will print out columns C1 LastName, C2 FirstName, C3 Test1, C4 Test2, C5 Test3, and C10 RWAverage. Now return to the Session window and print out these columns:

- Click on the **Session Window** toolbar button 🖼 (or choose Window > Session)
- At the "MTB >" prompt, type print **c1–c5 c10** and press ⏎
- Highlight the output (but not the commands and prompts) in the Session window
- Choose **File > Print Session Window**
- Click **OK** to get a paper copy of the selected scores

    Now that you have a paper copy of the test scores for the F students, you can close this new worksheet. To do this:

- Click on the **Current Data Window** toolbar button 🖼
- Choose **File > Close Worksheet**
- At the prompt "The worksheet '2MarksF' will be removed from the project 'Untitled.' This action cannot be undone. Would you like to save it in a separate file?", click **No**

    If you disable the Session commands, Minitab will still provide output in the Session window, but without prompts and commands. This is helpful if you want to capture and print the results of several successive analyses. If the Session command language is not disabled, these analyses will be interrupted by prompts and commands. To disable the command language, you must be in the Session window. (If not, click on the Session Window toolbar button or use Window > Session to make it active.)

- Choose **Editor > Enable Commands** to clear it

  The next section (the last in this tutorial) illustrates the process of creating a text file featuring Minitab output, and printing such a file. If this is of interest to you, proceed to the next section. Otherwise proceed to the end of the next section to exit Minitab. You have finished this tutorial and can review the Command Summary.

2.12 Creating a Text File

Minitab printed the grades data in the Session window's default fonts without any formatting. Ultimately, you want to format the report with boldface, italics, and borders before you present it to Ms. Taylor. To do so, you first save the grades data in the Session window as an *ASCII (American Standard Code for Information Interchange) text file*. You then can open this file in any text editor or word processor.

There are several ways to create such a text file. Here you will use the copy and paste method to save the highlighted data as an ASCII file called *2Marks.txt*:

- Highlight the data for all 24 students as you did in Section 2.11, if it is not still selected
- Click on the **right-hand button** on your mouse and choose **Copy** to copy these data into the clipboard
- Choose **Tools > Notepad**

  Notepad is a Microsoft Windows accessory that lets you open and work with text files. It is a basic text editor (a miniature word processor). (In Tutorial 3, you will copy text into a more sophisticated word processor, Microsoft Word.)

- Choose **Edit > Paste** from the Notepad menu to paste the data into the Notepad

  The Notepad window displays your grades data.

  To create a title for your grades listing:

- Choose **File > Page Setup**
- Click in the "Header" text box and clear **&f** to remove the Notepad default title
- Click **OK**
- Press Ctrl+Home
- Press ⏎ twice to create two blank lines above the grades list and press ↑ twice to move back to the top
- Type *Your Name:* **Grades Report for Ms. Taylor** to create a title for your document

  The Notepad window should look as shown in Figure 2-23.

FIGURE 2-23

Grades data in Notepad

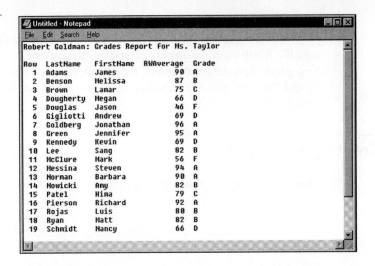

Robert Goldman: Grades Report for Ms. Taylor

Row	LastName	FirstName	RWAverage	Grade
1	Adams	James	90	A
2	Benson	Melissa	87	B
3	Brown	Lamar	75	C
4	Dougherty	Megan	66	D
5	Douglas	Jason	46	F
6	Gigliotti	Andrew	69	D
7	Goldberg	Jonathan	96	A
8	Green	Jennifer	95	A
9	Kennedy	Kevin	69	D
10	Lee	Sang	82	B
11	McClure	Mark	56	F
12	Messina	Steven	94	A
13	Norman	Barbara	90	A
14	Nowicki	Amy	82	B
15	Patel	Hima	79	C
16	Pierson	Richard	92	A
17	Rojas	Luis	80	B
18	Ryan	Matt	82	B
19	Schmidt	Nancy	66	D

▶ **Note** You could have saved all or part of the Session window with a File > Save Session Window As command. Minitab automatically adds the extension *txt*, which is a common extension for text files that makes them easy to find. ■

You can only do limited formatting in Notepad. However, you can open the file in a more sophisticated word processor and work with it there. For now, print the file with your name on it:

- Choose **File > Print,** and click the **Print** button when prompted
- Choose **File > Exit** to exit Notepad
- Click **Yes** to save the *2Marks* file in the location where you are saving your work

   You are returned to Minitab.

   You now have two paper copies of your grades report (one with a title) and a file that you can format in your word processor. At the end of the next marking period, it should be much easier for you to organize your grades. Now, save the file.

- Choose **File > Save Current Worksheet**

   When you save a worksheet, you save all of the data—the columns and the constants associated with the data set. The contents of the Session window, the Info window, and graphs you created, are not saved with the worksheet. These may be saved, however, as part of a *project*. In the next tutorial, you will learn to use and save projects. In addition, you will learn how to produce graphical displays, as well as more sophisticated numerical summaries, of your data.

- Choose **File > Exit**

- In response to the prompt "Save changes to the project 'Untitled' before closing?" click **No**

   Congratulations! You have completed Tutorial 2. You learned a variety of important methods for handling data. These methods include coding, ranking and sorting, creating subsets, and stacking and unstacking columns of data. Although these methods are not themselves data analysis tools, they will prove extremely valuable when you do analyze data in later tutorials.

Minitab Command Summary

This section describes the menu commands introduced in, or related to, this tutorial. Use Minitab's online Help for a complete explanation of all commands. Remember that the contents of Minitab menus may vary depending upon which window is active or the nature of the last command.

Minitab Menu Commands

Menu	Command	Description
File ➤	Print Session Window	Prints the selected contents of the active window.
Data ➤	Subset Worksheet	Creates a separate worksheet that is a subset of the original worksheet.
	Split Worksheet	Creates several new worksheets based upon the unique values in a column.
	Merge Worksheet	Merges another worksheet into the active worksheet.
	Unstack Columns	Copies and unstacks the contents of one column to form two or more columns. The number of columns equals the number of values in the subscript column.
	Stack	
	➤ Columns	Copies and stacks the contents of two or more columns to form a new column. The number of cells in the stacked column equals the total number of cells in the columns selected for stacking.
	Sort	Sorts selected columns of the worksheet using up to four columns as the sorting criteria.
	Rank	Ranks the data in a column and stores the ranks (usually in another column).
	Code	
	➤ Numeric to Text	Creates a new column whose text values are based upon the values in another column. There are similar commands for coding from Text to Numeric, from Numeric to Numeric, and so on.
	Display Data	Displays one or more columns of data or constants in the Session window.
Calc ➤	Calculator	Performs calculations using algebraic expressions, which may contain arithmetic operators, comparison operators, logical operators, or functions. Arguments may be columns, constants, or numbers.
	Column Statistics	Calculates various statistics for the selected column, displaying the results and, optionally, storing each result in a constant.
	Row Statistics	Computes a specified statistic for each row based upon a set of columns.
	Standardize	Creates new values by subtracting the mean and dividing by the standard deviation.
Editor ➤	Enable Commands	Hides or shows Session commands and the MTB > prompt in the Session window.
	Output Editable	Controls whether or not the Session window can be edited.
Tools ➤	Notepad	Opens a basic text editor that you can use to create simple documents.
Window ➤	Session	Moves you to the Session window, which contains a record of all commands issued in the current session and the resulting output. This is where you type Session commands.

Minitab Session Commands

Session Command	Description
info	Displays worksheet information, such as the columns and stored constants, in the Session window.
print	Displays one or more columns of data (or constants) in the Session window.

Minitab Toolbar Button Command

Toolbar Button	Icon	Description
Session Window		Moves you to the Session window, which contains a record of all commands issued in the current session and the resulting output. This is where you type Session commands.

Review and Practice

Matching

Match the following terms to their definitions by placing the correct letter next to the term it describes.

_____ Display Data

_____ Enable Commands

_____ Functions

_____ Ascending

_____ Stack

_____ Unstack

_____ Descending

_____ Subscripts

_____ print

_____ MTB >

a. The menu command that creates the MTB > prompt in the Session window

b. The Minitab command that creates a new column by placing several columns on top of each other

c. An optional indicator variable created when Minitab stacks several columns into one column

d. The Minitab command that creates several columns of data by splitting a given column

e. The default order of data created with the Sort command

f. The optional order for data created with the Sort command

g. A Minitab menu command that prints one or many columns in the Session window

h. A Minitab Session command that prints one or many columns in the Session window

i. Standard, preprogrammed numeric operations stored in Minitab

j. Minitab prompt

True/False Problems

Mark the following statements with a *T* or an *F*.

a. _____ Minitab will interpret C1–C5 as all of the columns C1, C2, C3, C4, and C5.

b. _____ The Round function rounds only to the nearest whole number.

c. _____ The Session command info display includes all columns that contain data in the active worksheet, the column names (if any), and the number of observations in the column.

d. _____ The Sort command automatically sorts all columns in the worksheet.

e. _____ You use the Display Data command to print directly to paper.

f. _____ The first quartile is one of the statistics available in the Column Statistics dialog box.

g. _____ The default operation of Standardize subtracts the mean of a variable and then divides by its standard deviation.

h. _____ The Subset Worksheet command creates several new worksheets based upon the unique values in a column.

i. _____ All of the statistics available in the Row Statistics dialog box are also available in the Column Statistics dialog box.

j. _____ The Code command interprets 10:20 as the numbers between 10 and 20, inclusive.

Practice Problems

The practice problems instruct you to save your worksheets with a filename prefixed by P, followed by the tutorial number. In this tutorial, for example, use *P2* as the prefix. If necessary, use Help or Appendix A Data Sets, to review the contents of the data sets referred to in the following problems. Interpretations should use the language of the subject matter of the question. If you are using the Student version, be sure to close worksheets when you have completed a problem.

1. Open *Textbooks.mtw*. You are interested in analyzing the prices of the textbooks.

   a. Use Minitab's Help to determine the contents of this file.

   b. Compute the larger of the list price and the price at Amazon.com and the smaller of the two, (i) using the Calc > Calculator command, and (ii) using the Row Statistics command. Store the results in C13, C14, C15, and C16. Provide suitable names for these four variables. Verify that the two methods produce the same results. Scroll down to compare the larger and the smaller prices. Do you see a different pattern for the General Education texts and the Business/Economics texts? Explain briefly.

   c. Use the Calc > Column Statistics command twice, to obtain the mean list price of the 36 textbooks and the mean price at Amazon.com. Similarly, obtain the median price for the two sources. Which source has the higher mean price? The higher median price?

   d. If necessary, use the Editor > Enable Commands to enable the MTB > prompt. Issue commands to obtain the same answers you got in part c.

2. Open *ExamScores.mtw*. The contents of this file are similar to the contents of *Marks.mtw*, the file used in Tutorial 2. Note that the scores on the final exam are out of 200. This is why these scores are divided by 2 in the various formulas below.

   a. Obtain the standardized scores for each of the first three exams. Name these variables. Identify any student with a standardized score on Exam2 that is greater than that on Exam1 and a standardized score on Exam3 greater than that on Exam2?

b. Compute the average of the scores on the first three exams, (i) using the Calc > Calculator command, and (ii) using the Row Statistics command. Store the results in C9 and C10, respectively. Name these variables Mean1 and Mean2. Verify that the two methods produce the same means.

c. Compute course scores for the students (in C11 CourseScore) by using the formula:

CourseScore = (Mean1 + Final/2)/2

d. Construct another new variable (in C12 RCourseScore) equal to the course scores rounded to the nearest whole number. Sort all of this worksheet's variables by the variable containing the rounded course scores. Which student obtained the highest rounded course score? Which student obtained the lowest rounded course score? Do you consider either of these students to be strikingly different from the other students? Explain briefly.

e. A colleague suggests an alternative scheme for computing course scores. This scheme involves computing the student's best score (Best) and second best score (SecBest) on the first three exams and then using the following formula

CourseScore2 = (3*Final/2 + 2*Best + SecBest)/6

i. Use the Max option under Row Statistics to compute, in C13, the student's Best score over the first three exams.

ii. Similarly, use the Min option to compute, in C14, the student's worst score over the first three exams.

iii. Use the Sum option under Row Statistics to compute, in C15, the student's total score over the first three exams.

iv. Use the Calc > Calculator command to compute in C16 SecBest the student's second best score. It will be the student's total score minus the sum of their best and worst scores.

v. Finally compute C17 CourseScore2 using the above formula. Does this scheme give significantly different course scores than those you computed in CourseScore?

3. Open *MBASurvey.mtw*. Note that five columns contain a 1 or a 0 depending on whether the student has a particular credit card. You are interested in analyzing these columns.

a. Use the Row Statistics command to create (in C11) the sum of the entries in these five columns. Explain the meaning of the values in C11.

b. Compute the mean and the median number of credit cards over the 28 students.

c. Place the ranks associated with C11 in C12. What is the smallest rank? How do you account for this value?

d. Compute the mean of the values in AmEx. Name this value PropAmEx. What does this value tell you about the extent to which the students in this class carry the American Express card?

4. Open *AgeDeath.mtw*. In this problem you will investigate the age distribution of the 151 individuals.

a. Use the Code command to create a new text variable (in C6-T AgeGroup), which takes the text value Young if the person was 49 years old or younger, MiddleAged if the person was between 50 and 69, Senior if the person was between 70 and 84, or Old if the person was between 85 and 105 at the time of death.

b. Unstack the ages into four new columns C7–C10, using the AgeGroups as the subscripts.

c. What is the median age in each AgeGroup? Use the N Total option under Column Statistics to determine the number of persons in each age group. Which is the most numerous group? Does your answer make sense?

5. Open *Pizza2.mtw*, which contains the data you worked with in the Tutorial 1 Practice Problems.

a. Compute the median score for the FallScore column and for the SpringScore column. In which semester was the median score higher? By how much was it higher?

b. Rank the items in the FallScore column and place the results in C7. Name this variable FSRank1.

c. Note that one way the ranks you computed in C7 differ from the ranks given in C3 is that 1 does not indicate the greatest value. Transform C7 to conform to the newspaper ranks by subtracting each number from 14. Place the results in C8. Name this variable FSRank2.

d. What is another difference between FallRank and FSRank2? Which variable is most informative? Explain briefly.

e. Use the Session command print to place the worksheet into the Session window.

f. Use the Row Statistics command to create a variable that is the sum of FallScore and SpringScore? Name this variable ScoreSum. Why might the information in this variable be of greater value than the information in either FallScore or SpringScore?

g. Use Subset Worksheet to create a new worksheet named *Pizza2Loc.mtw* that is based only upon those pizzerias that are local.

h. Answer part a for this new subset of pizzerias. How do your results differ from those in part a?

6. Open *Radlev.mtw*.

a. Use the Session command info to determine the information that has been saved in this data set. Which columns are used? What names have been assigned to them? How many cases does the data set contain? Are there any missing data?

b. Scroll down the worksheet looking for missing values in column 3 and column 4. How does Minitab represent missing values in these two columns? Why does it use a different code for missing values in these two columns?

c. Obtain the maximum age and the minimum age of the houses in this study.

d. Code the variable Age into a new column, C6, so that new homes are those that are less than 2 years old. Assign newer homes a code value of 0 and older homes a code of 1. Name C6 AgeCode.

e. Save the worksheet as *P2Radlev.mtw* in the location where you are saving your work.

f. Sort the data set using the new variable AgeCode. Scroll down the worksheet. How has your sorting affected the data set?

g. Delete the rows that correspond to older homes. You don't need to save your changes.

7. Open *Survey.mtw*.

   a. Note that the numeric variable Hand takes the value 1 if the student is left-handed, 2 if right-handed, and 3 if ambidextrous. Code the variable Hand into a new, text variable, C12-T Handed. Use suitable text values for Handed.

   You will explore the ages of the students in the three groups in two ways; using Unstack and using Subsets.

   b. Unstack the ages of the 100 students into C13, C14, and C15 using Handed as the subscripts. Compute the mean age for each of these three columns.

   c. Use the Subset Worksheet command to create three new worksheets, one for each text value of the variable Handed. Again, compute the mean age for each of the three groups.

   d. Explain which of these two processes you prefer and why.

8. Open *PulseA.mtw*, which contains the data you worked with in the Tutorial 1 Practice Problems.

   a. Compute the mean, median, and standard deviation of the initial pulse rates (in Pulse1) and the weights.

   b. Compute the increase in pulse rate (C2-C1) and store the results in Increase. Unstack the increases into two columns, C10 and C11, based upon whether or not the student ran in place.

   c. Compute the mean, median, and standard deviations of the increases in pulse rate for those who did and those who did not run in place. Write a brief summary of your output, indicating why the different results for the two groups make sense.

   d. Unstack the weights by gender and standardize each of the two new variables. Explain what information is gained by the standardized weights relative to the original weights.

9. Open *Murderu.mtw*.

   a. Obtain the median murder rate for each region. Write a brief account of how the median murder rates vary by region.

   b. Stack the fours sets of murder rates into a single column (C13 MurderRate) retaining the regions as the subscripts in C14 Region. What is the median murder rate over the 50 states?

   c. Open *Murders.mtw*. How do the variables in this data set differ from the variables you created in part b?

   d. Return to the *Murderu* data set. Obtain a paper copy of your modified data set.

10. Open *Salary02.mtw*.

   a. If necessary, use the Editor > Enable Commands to enable the MTB > prompt. Issue commands to obtain the mean 2002 salary for each of the three professorial ranks (ignore the two instructors). Explain how average salary changes with rank.

   b. Stack the three columns of 2002 salaries into C17 2002Salary (again, ignore the salaries for instructors). Retain the ranks as the subscripts (in C18 Ranks). What is the mean 2002 salary for the 58 professors at the college?

c. For 2003, each of the 58 professors will be receiving a 2.5 percent salary increase and then, in addition, a $1000 bonus. Create a new column (C19 2003Salary), which contains the 2003 salaries. Round these 2003 salaries to the nearest dollar.

d. At the MTB > prompt, print out the contents of the three columns you created in parts b and c. Highlight this display and obtain a paper copy of the three columns (with names).

e. Paste these same three columns into Notepad and save the data as *P2Salary03.txt*.

11. Open *MonthlySnow.mtw*. Use Minitab's Help to determine the contents of this file.

a. What is the maximum seasonal snowfall over this 111-year period? What is the month and year during which this amount of snowfall occurred?

b. Code the numeric Month variable to a new text variable called Season. Do this by assigning Winter for January, February, and March; Spring for April, May, and June; Summer for July, August, and September; and Fall for October, November, and December. Unstack the snowfall amounts into four new columns corresponding to the four seasons. Obtain the maximum monthly snowfall amounts for each of the four seasons. Write a brief account of your findings.

c. Display the coded worksheet in the Session window.

d. Place a copy of this output into Notepad and save it as *P2MonthlySnow.txt*.

On Your Own

The United States government frequently classifies the 50 states and the District of Columbia into four regions: Northeast, North Central, South, and West. Determine how each is classified. In addition, obtain a statistic that interests you for each state and the district. You might choose, for instance, the unemployment rate or the infant mortality rate. Enter the following three variables into Minitab: the name of the state or district, the name of its region, and the value of its statistic. Unstack these variables by region. Obtain the mean, median, and standard deviations of your interesting statistic. Comment on any similarities and differences for this statistic among the four regions.

Answers to Matching Problems

(g) Display Data
(a) 1
(i) Functions
(e) Ascending
(b) Stack
(d) Unstack
(f) Descending
(c) Subscripts
(h) print
(j) MTB >

Answers to True/False Problems

(a) T, (b) F, (c) T, (d) F, (e) F, (f) F, (g) T, (h) F, (i) T, (j) T

Data Analysis for One Variable

In this tutorial you will begin the process of analyzing data by learning how Minitab can be used to explore and summarize data for a single variable, both graphically and numerically. With Minitab you can produce professional-quality graphs for both *qualitative* and *quantitative* data. You can also quickly obtain a variety of commonly used data summaries, such as percentages, means, and standard deviations.

OBJECTIVES

In this tutorial, you learn how to:

- Save your data, session, and graphs in a project
- Summarize qualitative data
- Create a bar chart and a pie chart (from both raw and summarized data)
- Obtain data tips for a graph
- Name, print, edit, and close a graph
- Navigate the Project Manager
- Save a graph outside a project
- Use the Microsoft calculator
- Create a histogram, a stem-and-leaf display, a dotplot, an individual value plot, and a boxplot
- Produce descriptive statistics
- Produce descriptive statistics with accompanying graphs
- Use StatGuide and the Glossary
- Create your own descriptive statistic
- Copy text output to ReportPad and to Microsoft Word
- Open an existing project

3.1 | *Creating a Project*

In Tutorials 1 and 2, you saved a copy of your worksheet. Minitab provides a convenient method for saving, not only your data, but a complete record of your entire session, including commands, output, and all open data sets. You need only tell Minitab to store all this information in a *project*. A project also contains the session's graphs and History window. Projects are particularly helpful when you wish to leave Minitab before finishing an assignment. When you save your work in a project, you can return to it at the exact point at which you stopped. In the tutorials from this point on, you will save all of your work in projects.

CASE STUDY	PUBLIC HEALTH— INFANT NUTRITION

The research unit of a healthcare group where you work has been conducting a study of the factors that appear to be associated with a new mother's decision to breast-feed her infant. Sixty-eight low-income, pregnant women who attended a clinic affiliated with the group are the subjects. Your initial task is to summarize the data collected on these women and their newborn children.

Begin by opening Minitab:

- Open **Minitab** and maximize the **Session window**
- If necessary, **Enable Commands** and clear **Output Editable** from the Editor menu

Minitab automatically opens a new, untitled project each time you open the program. You can work in this project or open a different one. In this tutorial you will work with this project.

The data on the new mothers are in the file *Infants.mtw*.

- Choose **File > Open Worksheet**
- Double-click *Infants.mtw* in the "Open Worksheet" list box (you may have to navigate to the Studnt14 folder and then scroll down to see the file)

At this point you may see the Minitab message "A copy of the contents of this file will be added to the current project." If so,

- Click **OK**

The worksheet opens, displaying information about the new mothers and their infants. Refer to Help or Appendix A Data Sets for complete information on the nine variables in the worksheet.

When you open a worksheet in a project, Minitab makes a copy of it and adds it to the project. The original remains unchanged (unless you use the File > Save Current Worksheet command). The copy of the worksheet and any changes made to it will be stored with the new project file. Copies of a single worksheet file can be stored in several different projects.

This is a good time to name and save the project that you will work on in this tutorial:

- Click the **Save Project** toolbar button 🖫 and, in the location where you are saving your work, save this project as *T3.mpj*

You can also use the command File > Save Project to save a project.

Notice that Minitab saves a project in a file with an *mpj* suffix and that the name of the project precedes the name of the current worksheet in the Minitab title bar.

▶ **Note** Minitab has a toolbar button for saving a project but, oddly, does not have a toolbar button for saving a worksheet. ■

Minitab allows the creator of a data set to attach a description of the data to the file. To see a description for *Infants*:

- Place your cursor over the red triangle located beneath the data-entry arrow in the upper-left corner of the worksheet.

A pop-up window will open containing the information about the data set. Notice that the authors of this book created the worksheet on March 31, 2003 and you are invited to contact them for further information. In practice, the worksheet description may be used to provide the kind of background information regarding the data set that you can find in Help and in Appendix A Data Sets toward the end of the book.

▶ **Note** You can also obtain this description of the data set by choosing the sequence Editor > Worksheet > Description. ■

Just as you can attach a description to a data set, you also can attach a description to the project you are working on:

- Choose **File > Project Description**
- Type *your name* as the "Creator" of the project and type *the current date* as the "**Date(s)**"
- Enter the following under "Comments": **This project, T3, contains both numerical and graphical summaries of the Infants data set.**
- Click **OK**

Save the project again with the attached description:

- Click the **Save Project** toolbar button 🖫

When analyzing data it is important to distinguish between quantitative and qualitative variables. Qualitative (or categorical) variables fall into different categories. Quantitative variables take numerical values for which operations such as averaging are appropriate.

These two types of variables need not correspond to Minitab's distinction between numeric and text data. For example, it is quite common to enter data for qualitative variables into Minitab as numbers (1 = male, 2 = female, for example). Minitab will happily compute the average gender, regardless of the fact that the result is meaningless. This book will use the terms *qualitative variables* and *quantitative variables*, not the terms *categorical variables* and *numerical variables*.

▶ **Note** You will notice that in this tutorial the bulleted actions are more concise than those in the previous tutorials. ■

Summarizing Qualitative Variables

Three of the nine variables, Ethnic, Smoke, and BreastFed in *Infants.mtw* are qualitative. In this section you will summarize the data in one of these three columns, the column with the variable Smoke. Summarizing the data for a qualitative variable is straightforward and involves computing the counts with which each text value (category) occurs and the corresponding percentages.

- Choose **Stat > Tables > Tally Individual Variables**
- Double-click **C3 Smoke** as the "Variables"
- Click the "Display" **Percents** check box to obtain percentages as well as the default Counts

The completed Tally Individual Variables dialog box should look as shown in Figure 3-1.

FIGURE 3-1

The completed Tally Individual Variables dialog box

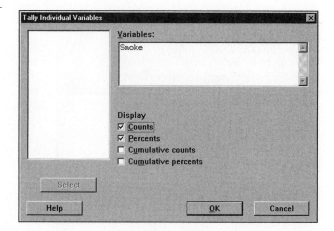

- Click **OK**

The output appears in the Session window as shown in Figure 3-2.

FIGURE 3-2

Tally output for the variable Smoke

```
MTB > Tally 'Smoke';
SUBC>   Counts;
SUBC>   Percents.
```

Tally for Discrete Variables: Smoke

Smoke	Count	Percent
HeavySmoker	10	14.71
LightSmoker	9	13.24
NonSmoker	49	72.06
N=	68	

Minitab refers to *counts* rather than frequencies. Of the 68 mothers, 10 are heavy smokers, 9 light smokers, and the remainder are non-smokers. The corresponding percentages are 14.71%, 13.24%, and 72.06%.

Notice that Minitab lists the three text values for smoking in alphabetical order by default, with heavy smokers, then light smokers, and then non-smokers. A more standard ordering would be non-smokers, light smokers, and heavy smokers. You can stipulate this ordering by using the Editor > Column > Value Order command. First, return to the worksheet.

■ Click the **Current Data Window** toolbar button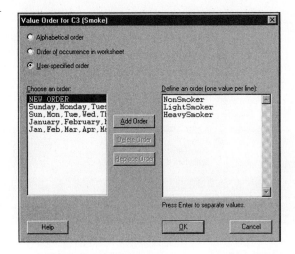

■ Click anywhere in the column **C3-T Smoke**

■ Choose **Editor > Column > Value Order**

■ Click the **User-specified order** option button

When you click this button, the text values in the right-hand text box appear in the order that they appear in the worksheet as shown in Figure 3-3. (You can see this by looking at the entries in C3-T.) By coincidence, this order is exactly the order you want. (If you wanted a different order, you would edit the entries in the right-hand text box.)

FIGURE 3-3

*The completed Value
Order for C3 (Smoke)
dialog box*

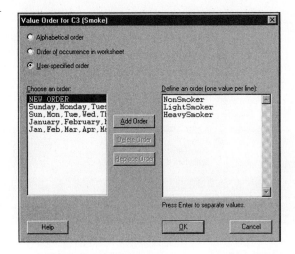

■ Click **OK**

Now, obtain a second tally of C3-T Smoke:

■ Choose **Stat > Tables > Tally Individual Variables**

■ Verify that Smoke is listed in the "Variables" text box and that "Counts" and "Percents" are checked, and click **OK**

The output should now show the same counts and percentages as before, but in the new order: NonSmoker, LightSmoker, and HeavySmoker.

Minitab can also provide *cumulative counts* and *cumulative percents*.

■ Click the **Edit Last Dialog** toolbar button

■ Click the "Display" **Cumulative counts** and **Cumulative percents** check boxes, and click **OK**

The resulting output is shown in Figure 3-4. Fifty-eight (58) of the mothers are either non-smokers or light smokers. These two groups make up just over 85% of the total.

FIGURE 3-4

Tally output for the variable Smoke with cumulative counts and percents

```
MTB > Tally 'Smoke';
SUBC>   Counts;
SUBC>   CumCounts;
SUBC>   Percents;
SUBC>   CumPercents.
```

Tally for Discrete Variables: Smoke

Smoke	Count	CumCnt	Percent	CumPct
NonSmoker	49	49	72.06	72.06
LightSmoker	9	58	13.24	85.29
HeavySmoker	10	68	14.71	100.00
N=	68			

The Tally command can be used with text data, as it was above, or with numeric data. However, in the latter case it is most useful when there are only a small number of different values.

3.3 Creating Bar Charts

You want to graphically represent the percentage associated with each smoking group. Minitab offers two ways to do this: a bar chart and a pie chart. Begin by obtaining a bar chart:

- Choose **Graph > Bar Chart**

    The Bar Chart pictorial gallery appears.

- Double-click on the picture beneath "Simple"
- Double-click on **C3 Smoke** as the "Categorical variables"
- Click **Bar Chart Options**, click the **Show Y as Percent** check box, and click **OK**
- Click **Labels**, type **Bar Chart of Smoke** as the "Title", and click **OK**

    The completed Bar Chart - Counts of unique values, Simple dialog box should look as shown in Figure 3.5.

FIGURE 3-5

The completed Bar Chart - Counts of unique values, Simple dialog box

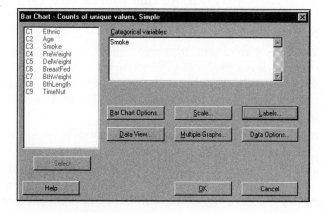

- Click **OK**

    The resulting Bar Chart appears in its own Graph window as shown in Figure 3-6.

FIGURE 3-6

Bar Chart of Smoke

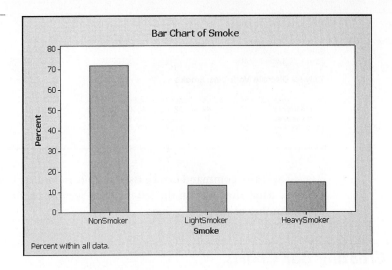

The chart shows the large percentage of the mothers who are non-smokers and the much smaller and similar percentages that are light and heavy smokers. Notice that the order of the smoking groups is as you stipulated earlier: NonSmoker, LightSmoker, and HeavySmoker.

▶ **Note** If you had not selected the "Show Y as Percent" check box above, Minitab would have shown the actual counts on the vertical (Y) axis of the Bar Chart rather than percentages. ■

Move your cursor to any point inside the bar for non-smokers. After a second, a pop-up window will open. Minitab refers to such windows that provide information about some aspect of the data as *data tips*. This data tip contains the result "Bar: Value = 72.0588". This value is the percentage of mothers who are non-smokers. You can do the same with the other two bars.

Naming Graphs

When you plan to generate a considerable number of graphs—as you will in this tutorial—it is a good idea to name each one as you create it. Naming a graph is not the same as giving it a title. The title is internal to the graph, while the name is external to it and can be referenced by Minitab. You can use the Project Manager to name and delete graphs.

■ Click the **Project Manager** toolbar button 🖼 (or use Window > Project Manager)

    The Project Manager double window opens. In the right-hand window you can see a graph icon 📈 followed by the default name Chart of Smoke. Now, modify the name to *Your Initials*: Bar Chart of Smoking Status.

■ Hold your cursor over the graph icon 📈, click the **right-hand mouse button**, and choose **Edit Title**

- In the Title box, change the name to *Your Initials*: **Bar Chart of Smoking Status** and click outside the box to complete the change

▶ **Note** In the interests of brevity you will be encouraged to provide brief titles and names for graphs. This is fine when you are producing graphs for exploration. However, longer, more informative titles are appropriate when producing graphs for presentation. Here, an example, might be "Bar Chart of Smoking Status Showing Over 70% of Mothers are Non-Smokers." ■

The Project Manager

Before returning to the worksheet, it's worth examining the Project Manager window. It should look similar to Figure 3-7.

FIGURE 3-7

Project Manager window showing the Session folder

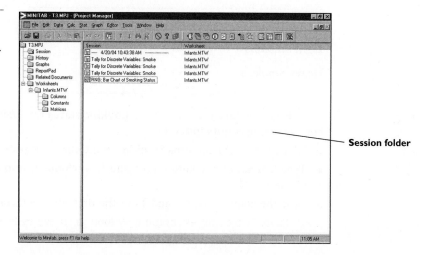

As the name suggests, the Project Manager is where you can manage all aspects of your work in Minitab. In the left-hand side of the window are folder icons that allow you to view and manipulate the various parts of your project. These are:

- The Session folder, which is open in the right-hand side of the window, contains a list of all the output in the Session window and a list of all the graphs you have produced.
- The History folder contains a record of all the session commands you have issued.
- The Graphs folder contains your graphs, which you manage (Open, Save, and Copy to ReportPad, for instance).
- The ReportPad folder is a simple word processor that you will use later in this tutorial.
- The Related Documents folder can contain such items as a URL related to your project.
- The Worksheets folder contains information about all the worksheets that have been opened and left open because the project was created.

Now return to the worksheet.

■ Click the **Current Data Window** toolbar button 🔲

Bar Graphs Using Summarized Data

The worksheet Infants contains *raw data*—that is data as they were recorded on each individual (mother/child, in this case) in the study. Virtually all worksheets contain raw data. The bar chart in Figure 3-6 was based upon the raw data for the column Smoke. Minitab can also create many graphs from *summarized data*. To see how this works, obtain a bar chart from data you saw in an article. In a study of 180 mothers similar in some respects to the group for whom you have data, the following summarized data were published:

Smoking Status	Number of Mothers
Non-Smokers	101
Light Smokers	46
Heavy Smokers	33

Begin by entering these three smoking status text values into C11 and the corresponding counts into C12:

■ Name these two columns **Smoking** and **Counts**, respectively

■ Type **NonSmoker**, **LightSmoker**, and **HeavySmoker** into the first three cells in Column C11

■ Type the counts **101**, **46**, and **33** in the first three cells of C12

These two columns should now look as shown in Figure 3-8.

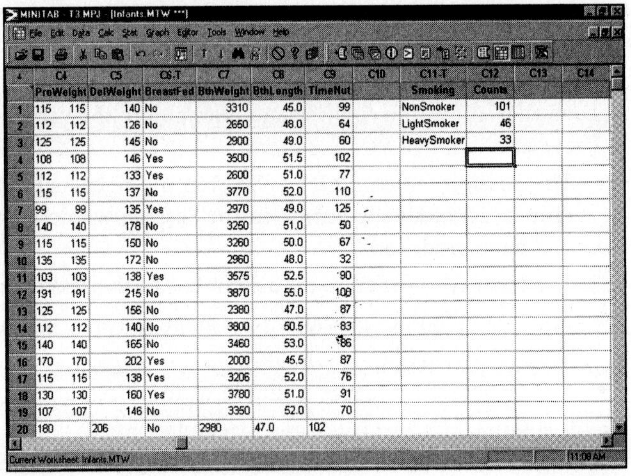

FIGURE 3-8

Summarized data in C11-T and C12

Now, to get the bar graph:

- Click the **Edit Last Dialog** toolbar button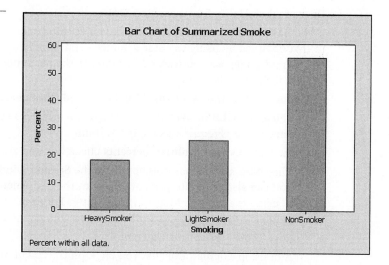
- Double-click **C11 Smoking** as the "Categorical variables"
- Click **Data Options** and then click on the **Frequency** tab
- Double-click **C12 Counts** as the "Frequency variable(s)" and click **OK**
- Click **Labels** and type **Bar Chart of Summarized Smoke** as the "Title"
- Click **OK** twice

The resulting bar chart is shown in Figure 3-9. The percentage of mothers that are non-smokers is smaller than in your study (about 55%), but this percentage is still much larger than that for light smokers (about 25%) and heavy smokers (about 20%).

FIGURE 3-9

Bar chart based upon summarized data

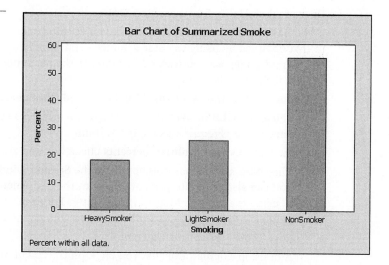

Notice that the ordering of the smoking groups in the bar chart in Figure 3-9 is the default alphabetic ordering. If you wish, you can use the Editor > Column > Value Order command on column C11-T as you did with column C3-T earlier.

Saving Graphs

The graphs that you have produced and will produce later in this tutorial are being added to the Minitab project *T3.mpj*. Just as you can save worksheets as independent files, you can also save graphs for use in other projects. You can to save this bar chart with the summarized data because it might be useful for a workshop that you are presenting soon.

To save this bar chart in a graph file:

- Choose **File > Save Graph As**
- Click the "Save in" **drop-down list arrow** and select the location where you are saving your work
- Type **BarChartSumSmoke.mgf** as the "File name"

- Click **Save**

Minitab saves the graph as an *mgf* (Minitab Graphics Format) file type. This graph remains a part of the current project, but is also stored as a separate graph file for use outside the project.

- Use the Project Manager toolbar button ▣ and rename the default name of this graph to *Your Initials*: **Bar Chart of Summarized Smoking Status** then return to the Session window

▶ **Note** Just as you opened and printed a text file earlier, you can open and print the contents of a graph file. The command File > Open Graph is used to open a saved graph. To print an open graph, make sure the graph is showing in the active window and use File > Print Graph. ◼

Percentages Using Summarized Data

If you have access to only summarized data, Minitab can construct a table similar to that obtained by using the Stat > Tables > Tally Individual Values command. For example, suppose you wanted to compute the percentages of the smoking status for the 180 mothers:

- Choose **Stat** > **Tables** > **Cross Tabulation and Chi-Square**
- Double-click **C11 Smoke** as the "Categorical variables For rows" and **C12 Counts** as the "Frequencies are in" column
- Click the "Display" **Column percents** check box and click **OK**

The count and percentages appear in the Session window.

You can also compute percentages from the frequencies listed above (and in C12) by using the built-in Microsoft calculator. For example, suppose you wanted to compute the percentage of the 180 mothers who are non-smokers:

- Choose **Tools** > **Microsoft Calculator**

A desktop calculator opens. You click on the calculator keypad just as you would press the keys on an ordinary calculator or use the numbers on your keyboard. To compute $100*101/180$:

- On the calculator keypad, click on **1 0 0 * 1 0 1 / 1 8 0 =**

The rounded answer (56.11) is the percentage of the 180 mothers who are non-smokers. To exit the calculator:

- Click the **Close** button ☒

▶ **Note** A less intuitive approach to this problem would be to use the Calc > Calculator command. ◼

Creating Pie Charts

A second display that can be used with qualitative data is the *pie chart*. This is perhaps the Minitab display that you will be most familiar with. To obtain a pie chart of the variable Smoke, for example:

- Choose **Graph > Pie Chart**
- Click in the "Categorical variables" text box and double-click **C3 Smoke**
- Click **Labels** and type **Pie Chart of Smoking Status** as the "Title"
- Click **OK** twice

The resulting pie chart is shown in Figure 3-10. The legend identifies which segment of the chart corresponds to which smoking group. The segment corresponding to the non-smokers does appear to be roughly five times the area associated with either the light or the heavy smokers.

FIGURE 3-10

Pie chart of Smoke

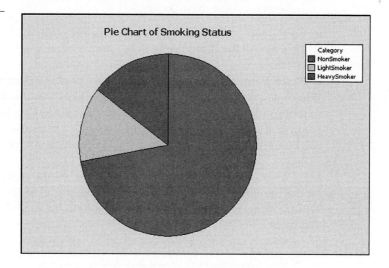

Notice that the individual segments in the pie chart are not labeled. You can easily remedy this:

- Click the **Edit Last Dialog** toolbar button ▦
- Click **Labels** and then click on the **Slice Labels** tab
- Click the **Category name** and **Percent** check boxes
- Click **OK** twice

The new pie chart, shown in Figure 3-11, includes the names of each segment and the corresponding percentages.

FIGURE 3-11

*Pie chart of Smoke with
category names and
category percentages*

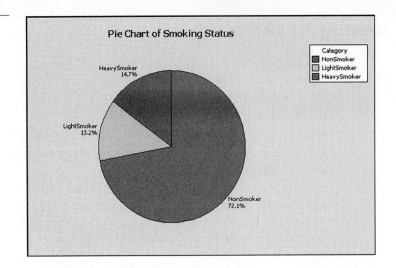

- Go to Project Manager and rename this last graph in the list to *Your Initials*: **Pie Chart of Smoking Status** then return to the Session window

▶ **Note** Both bar charts and pie charts are used to represent qualitative data. Which, do you think, most accurately represents the distribution of the three smoking groups? There is some evidence that readers are better able to accurately perceive differences in heights than angles. As a consequence, bar graphs are probably more suitable than pie charts for comparing the relative counts of groups. ■

You may have noticed a similarity between the appearance of the dialog box for the bar chart and the pie chart. In fact, nearly all of the graph dialog boxes in Minitab share the same five dialog box options. To view them, open the Bar Chart dialog box:

- Choose **Graph > Bar Chart** and double-click *Simple*

The common graph dialog boxes are:

- The *Scale* dialog box, which allows you to control the appearance of the axes and to add gridlines and reference lines.
- The *Labels* dialog box, which allows you add a title, footnotes, and data labels.
- The *Data View* dialog box, which can be used to shape all aspects of the appearance of the graph.
- The *Multiple Graphs* dialog box, which controls the organization of multiple versions of the same type of graph.
- The *Data Options* dialog box, which allows the user to specify the data that will be used in the graph.

To leave the dialog box:

- Click **Cancel**

Quantitative Variables: Creating Histograms

The variables Age, PreWeight, DelWeight, BthWeight, BthLength, and TimeNut are quantitative. Minitab offers a number of graphs designed to display quantitative data. In the sections below you will examine histograms, stem-and-leaf displays, dotplots, individual value plots, and boxplots.

One of the key components of the care given to pregnant women at the study clinic is nutritional counseling. The column C9 TimeNut records the amount of time that each woman spent with a nutritionist during her pregnancy. You will examine some graphical displays of these 68 times, beginning with a histogram.

- Choose **Graph > Histogram** and double-click *Simple*

  The Histogram dialog box resembles those for bar charts and pie charts.

- Double-click **C9 TimeNut** to enter this variable as the "Graph variables"

- Click **Labels,** type **Histogram of Time with Nutritionist** as the "Title", and click **OK** twice

  The resulting histogram appears as shown in Figure 3-12.

FIGURE 3-12

Histogram for TimeNut

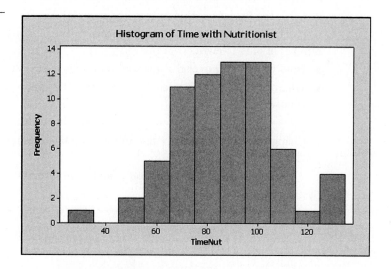

The histogram is centered at about 80 minutes though the actual times seem to vary from a low of about 30 minutes to a high of around 130 minutes. The overall shape is very slightly skewed to the left though it is not too far from symmetric. The histogram divides the data into *intervals* or *bins*. The horizontal axis lists the midpoint of every second bin. By default, the height of each bar is the number (frequency) of times that fall in that bin.

Move your cursor to any point inside the bar with midpoint 70. A data tip provides the information "Bar: Value = 11, Bin = 65, 75". This should be interpreted as indicating that this bin runs from 65 minutes up to, but not including, 75 minutes, and 11 of the 68 times fall in this bin. If you place your cursor inside the next bar, with midpoint 80, the data tip will read "Bar: Value = 12, Bin = 75, 85". Again, interpret this as indicating that 12 times fall in the interval from 75 minutes up to, but not including, 85 minutes. It is Minitab's practice to include the upper bound only on the highest bin, in this case "Bar: Value = 4, Bin = 125, 135". The largest of the 68 times is 128 minutes; had this value been 135 minutes it would have been included in this bin.

- Go to Project Manager and rename the graph to *Your Initials*: **Histogram of Time Spent with Nutritionist** then return to the Session window

▶ **Note** This is the last time you will be asked to name your graph, but you are urged to continue this practice. ∎

Suppose you prefer percentages rather than frequencies on the vertical (Y) axis and that you would like to see how the appearance of the histogram changes with just eight bins rather than the 11 in the histogram in Figure 3-12. This last change can be made only by editing the completed graph. To change the vertical axis to percentages rather than frequencies:

- Click the **Edit Last Dialog** toolbar button 🔲
- Click **Scale** and then click on the **Y-Scale Type** tab
- Click the **Percent** option button and click **OK** twice

The new histogram appears as shown in the Graph window. To change the number of bins to eight:

- Hold your cursor over the horizontal (X) axis of the plot
- When the data tip indicates "X Scale" click the **right-hand mouse button** and choose **Edit X Scale**
- Click on the **Binning** tab and click the **Number of intervals** option button
- Change the default value 11 to **8** in the "Number of intervals" text box and click **OK**

The histogram with percentages and eight bins should look like Figure 3-13.

FIGURE 3-13

Histogram for TimeNut with percentages and eight bins

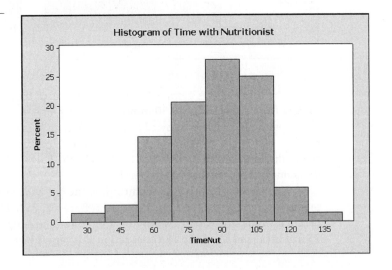

The new histogram looks a little less ragged than the original, but there is sufficient definition to be able to see the near symmetry of the graph.

If you move your cursor into the bar centered at 105 the data tip will inform you that the bin value = 25, indicating that 25% of the 68 times lie in the bin running from 97.5 minutes up to, but not including, 112.5 minutes.

▶ **Note** Clicking on the Binning tab also allows you to change the histogram interval type from midpoint to *cutpoint* (or *endpoint*). ■

▶ **Note** One of the (few) deficits in Minitab is a command to construct a frequency distribution for a numeric variable by constructing intervals and providing the frequency with which values fall in each interval. You can, however, use data tips on each bin of a histogram to construct such a frequency display for yourself. Or you can place the frequency of each bin of a histogram by clicking Labels, clicking the Data Labels tab, and clicking the Use y-value labels check box. ■

Assume that you prefer this histogram to the earlier one and so will close that one. You can close a graph that is not currently active in two ways:

1. Delete the graph in the Project Manager
2. Delete the graph in the Graphs folder

In this case use the Graphs folder.

- Click the **Show Graphs Folder** toolbar button 🖻

The names of all the graphs you have created thus far in this tutorial are shown in the left-hand window.

- Hold your cursor over the graph name *Your Initials*: **Histogram of Time Spent with Nutritionist**

- Click the **right-hand mouse button** and choose **Close**

- At the prompt "The graph '*Your Initials*: Histogram of Time Spent with Nutritionist' will be removed from the project. This action cannot be undone. Would you like to save it in a separate file?", click **No**

- Click the **Session Window** toolbar button 🖻.

3.6 Creating Stem-and-Leaf Displays

One of the drawbacks to a histogram is that the display does not enable you to see the actual data. For example, you know that 25% of the times are in the interval from 97.5 up to, but not including, 112.5 minutes, but you cannot tell from the display what the times are. A display that helps in this regard while retaining much of the same features of a histogram is a *stem-and-leaf display*, sometimes called a *stemplot*.

To construct a stem-and-leaf display of the 68 times in C9 TimeNut:

- Choose **Graph > Stem-and-Leaf**
- Double-click **C9 TimeNut** as the "Graph variables" and click **OK**

Figure 3-14 shows the resulting stem-and-leaf display in the Session window. You may need to scroll to see the entire display.

```
MTB > Stem-and-Leaf 'TimeNut'.

Stem-and-Leaf Display: TimeNut

Stem-and-leaf of TimeNut   N  = 68
Leaf Unit = 1.0

    1    3   2
    2    4   8
    4    5   07
   14    6   0034556779
   23    7   011226779
  (15)   8   012234445556778
   30    9   000123668899
   18   10   00022225555
    7   11   005
    4   12   5568
```

Depths — (arrow to left column)   **Leaves** — (arrow to right column)

Stems — (arrow to center column)

If you can imagine turning the display 90 degrees you will have a graph similar to the histogram of the same variable. Notice that this display differs from the others in that it is constructed from typed numbers and appears, not in its own graph window, but in the Session window. The stem-and-leaf display is an example of a *character graph*. The earlier graphs, which appear in their own graph window, are called *professional graphs*.

For now, ignore the numbers in the left column of the display. The numbers in the center column represent the *stems*, or leftmost digits, of the data values. The column on the right contains the leaves. Because Minitab reports a leaf unit of 1 (Leaf Unit = 1.0), each leaf represents the one's digit of a data value and each stem represents the ten's digit of the data value. For example, the last line of the stem-and-leaf display shown in Figure 3-14 indicates a stem of 12 followed by leaves of 5, 5, 6, and 8. This stem and these four leaves represent the data values 125, 125, 126, and 128.

The first left-hand column in the stem-and-leaf display indicates the *depths*, which are used to show cumulative frequencies. Starting at the top, the depths indicate the number of leaves that lie in the given row (or stem) plus any previous rows. For example, the 4 in the third line of this column in Figure 3-14 indicates that there are four leaves (32, 48, 50, and 57) in the first three rows (or stems). If the line containing the median has any entries, the number in the left column represents the number of leaves on that line and appears in parentheses. The median for this data set is 85.5, which would fall between two values in the fifth line from the bottom of the display. The number 15 in parentheses is the number of leaves in this stem. The 18 in the third line from the bottom indicates that there are a total of 18 leaves in the three lines at the bottom of the display.

Minitab uses the term *increment* to refer to the difference in value between any two lines of the display. In the display in Figure 3-14, Minitab uses an increment of 10. You can see this by noting that the smallest value possible for each line of the display is, respectively, 30 minutes, 40 minutes, 50 minutes, and so on. In each case, the difference is 10 minutes. To see the effect of changing the increment to 5:

- Choose **Edit > Edit Last Dialog** and press `Tab` three times
- Type **5** in the "Increment" text box and click **OK**

Your new display should look like the one shown in Figure 3-15.

FIGURE 3-15

Stem-and-leaf display for the times spent with a nutritionist, with an increment of 5

```
MTB > Stem-and-Leaf 'TimeNut';
SUBC>   Increment 5.
```

Stem-and-Leaf Display: TimeNut

```
Stem-and-leaf of TimeNut  N  = 68
Leaf Unit = 1.0

 1    3   2
 1    3
 1    4
 2    4   8
 3    5   0
 4    5   7
 8    6   0034
14    6   556779
19    7   01122
23    7   6779
31    8   01223444
(7)   8   5556778
30    9   000123
24    9   668899
18   10   0002222
11   10   5555
 7   11   00
 5   11   5
 4   12
 4   12   5568
```

This is an example of *split stems* where each stem appears twice. Leaves 0, 1, 2, 3, and 4 go on the upper stem and leaves 5, 6, 7, 8, and 9 on the lower stem. This creates a greatly elongated display. Split stems are not very helpful in this case, but they are important in situations in which there would otherwise be only a small number of stems.

Now obtain a stem-and-leaf display of the infant birth weights.

- Click the **Edit Last Dialog** toolbar button

You want to examine a different variable than before, so you want to clear the previous settings that appear in the Stem-and-Leaf dialog box and restore them to the default settings. To quickly restore the default settings:

- Press `F3`

The "Graph variables" text box is now empty and the "Increment" setting is cleared.

To create the stem-and-leaf display of the infant birth weights:

- Double-click **C7 BthWeight** as "Graph variables"

The completed Stem-and-Leaf dialog box is shown in Figure 3-16. Because a stem-and-leaf display is a character graph the various options available with the earlier graphs (Scale, Labels, and so on) are not available here.

FIGURE 3-16

The completed Stem-and-Leaf dialog box

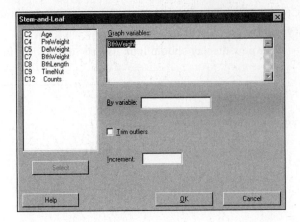

- Click **OK**

  The stem-and-leaf display for the infant birth weights is shown in Figure 3-17.

FIGURE 3-17

Stem-and-leaf display for infant birth weights

```
MTB > Stem-and-Leaf 'BthWeight'.
```

Stem-and-Leaf Display: BthWeight

```
Stem-and-leaf of BthWeight   N  = 68
Leaf Unit = 100

    1     2  0
    3     2  33
    4     2  4
    9     2  66677
   21     2  888889999999
   30     3  000000111
  (15)    3  222222333333333
   23     3  44444555
   15     3  66677777
    7     3  88889
    2     4  0
    1     4
    1     4  4
```

The distribution of birth weight is close to symmetric, although there may be an outlier in the right tail. In this example the leaf unit is 100 grams. So, the largest birth weight, for instance, is recorded as 4400 grams. (In fact, if you sort the 68 birth weights you will see that the largest is 4458 grams. This suggests that Minitab is not rounding numbers to the nearest 100 but is simply dropping the last two digits.) This stemplot also uses split stems; each of the stems 2 and 3 appear five times; once with the leaves 0 and 1, once with the leaves 2 and 3, and so on. Stem 4 appears three times.

When you are deciding whether to use any of these stem-and-leaf displays in a presentation, consider how much opportunity you will have to interpret them. As you can see, they require some careful explanation. However, if your audience is familiar with them, such plots can provide an excellent picture of your data.

Because stem-and-leaf displays are not professional graphs they will not be named in the Project Manager.

This is a good time to save your project, *T3.mpj*.

■ Click the **Save Project** toolbar button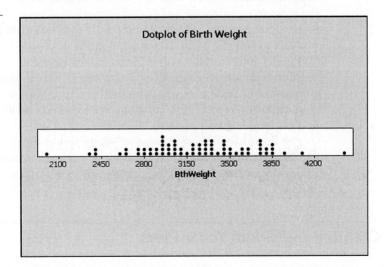

The stem-and-leaf displays as well as all the other graphs that you have created are saved when you save the projects, along with the rest of the output in the Session window.

▶ **Note** If you want to save your current project under a different name, choose File > Save Project As. ■

3.7 Creating Dotplots

Next, you will look at a dotplot representation of the data in C7 BthWeight. This will give you an opportunity to compare the relative advantages of a dotplot and a stem-and-leaf display. You then will be in a better position to decide which graph types you want to include in your presentation.

■ Choose **Graph > Dotplot** and double-click *Simple*

■ Double-click **C7 BthWeight** as the "Graph variables"

■ Click **Labels** and type **Dotplot of Birth Weight** as the "Title"

■ Click **OK** twice

The dotplot appears as shown in Figure 3-18.

FIGURE 3-18

Dotplot of infant birth weights

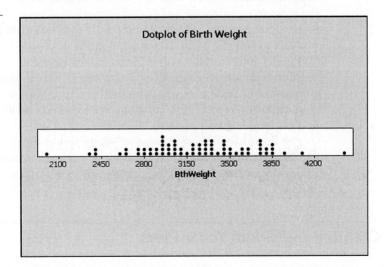

A *dotplot* usually displays each data value as a dot above a horizontal axis so that that you can see approximately where each value falls. You can see the outliers at either end of the plot that were not included in the stem-and-leaf display of these same data. The distribution of birth weights is almost symmetric, centered at approximately 3200 grams. The weights vary from approximately 2000 grams to approximately 4400 grams.

Note When there are a large number of observations Minitab will construct a dotplot in which each dot represents two or more observations. ∎

Hold your cursor over the dot representing the second largest birth weight (immediately to the left of data label 4200). A data tip shows the result "Symbol, Row 36, bin = 4075, 4125". The symbol refers to the dot your mouse is pointed at. The dot represents the birth weight for the infant in row 36. The notation "Bin = 4075, 4125" alerts you to the fact that Minitab actually creates bins or intervals for a dotplot, though many more than it does with histograms. All values (birth weights) falling in the same bin are stacked one above another. If you examine the stem-and-leaf display of these same data in Figure 3-17, you can see that the second largest birth weight is, to three significant digits, 4080 grams.

Now, return to the Session window.

■ Click the **Session Window** toolbar button 🖼

When the Session commands are enabled, each time you make a menu selection, Minitab displays the corresponding command in the Session window. Look at the last two commands that appear in the Session window. The first is a command that instructs Minitab to construct a dotplot of the birth weight data. The second is a subcommand, as indicated by SUBC>. A subcommand provides additional information about the preceding command, in this case the command to add a title. They represent the options you select in dialog boxes.

Where there is a choice between using menus or typed commands, most users find that choosing from the Minitab menus and selections in dialog boxes is the most straightforward and error-free method of issuing commands. However, more advanced users of Minitab, especially those who are proficient typists, may find that typing the Minitab command in the Session window is more efficient. (Tutorial 2 and Minitab's online Help contain more information about the Minitab Session command language.) These tutorials, however, primarily use the menu commands. Regardless of which method you use, Minitab will provide a record of your activities by printing commands in the Session window as long as the Session commands are enabled.

A right-hand mouse command, where it is available, is generally easier to use than a menu command or a Session window command. A toolbar button command, where it is available, is generally easier than any of the other three options.

3.8 Creating Individual Value Plots

A display similar to the dotplot is the *individual value plot*. Examine such a plot of the infant birth weights.

■ Choose **Graph > Individual Value Plot** and double-click *Simple*
■ Double-click **C7 BthWeight** as the "Graph variables"
■ Click **Labels** and type **Individual Value Plot of Birth Weight** as the "Title"
■ Click **OK** twice

The individual value plot should appear as shown in Figure 3-19.

FIGURE 3-19

Individual value plot of infant birth weights

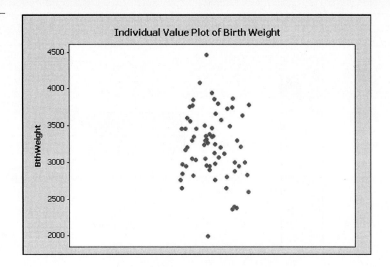

The plot bears some resemblance to a vertical dotplot except that the values are not stacked in piles but are randomly scattered.

▶ **Note** Each time you construct an individual value plot you will obtain a different random scatter of the same values, unless you set a base value immediately before you generate the plot. You can do this in Minitab by choosing the Calc > Set Base command. More information on this command can be found in Tutorial 6. ■

It is not easy to get an idea of the shape of the distribution of birth weights though you can see the outliers at the top and bottom of the plot. Move your cursor over the dot representing the second largest birth weight. You are informed again that this dot represents the birth weight in row 36, but on this plot the actual value for the birth weight is given (4080 grams). The scattering of the points allows for each point on the display to represent exact values.

3.9 Creating Boxplots

A *boxplot* (or *box-and-whisker plot*) provides you with a rather skeletal view of a data set. It highlights useful statistical information, such as quartiles and outliers. To obtain a boxplot of times spent with a nutritionist in C9:

- Choose **Graph > Boxplot** and double-click *Simple*
- Double-click **C9 TimeNut** as the "Graph variables"
- Click **Labels** and type **Boxplot of Time with Nutritionist** as the "Title"
- Click **OK** twice

   The Boxplot appears as shown in Figure 3-20.

FIGURE 3-20

Boxplot of time spent with nutritionist

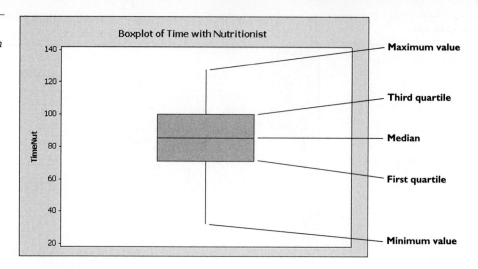

A boxplot uses five numbers to describe a set of data:

1. Maximum value
2. Third quartile (75th percentile)
3. Median (50th percentile)
4. First quartile (25th percentile)
5. Minimum value

Collectively, these five numbers are known as the *five-number summary*. Minitab constructs a rectangle (box) between the first and the third quartiles and displays a horizontal line at the location of the median. This box encloses the middle half of the data. The *whiskers* that extend in either direction indicate the non-outlying data. If there are no outlying values, the whiskers extend to the smallest and largest values in the data set. If there are outlying values, Minitab identifies them with asterisks (*).

Place the cursor over any part of the boxplot and you will be informed that the first quartile of times is Q1 = 71.25 minutes, the median is 85.5 minutes, and the third quartile is Q3 = 100 minutes. The data tip also provides the value for the interquartile range (IQR = Q3 – Q1 = 28.75 minutes) and the value of the sample size (68 women).

Obtain a boxplot of the infant birth weights to compare with your other displays of these data. You wonder whether it might not be better to transpose the axes, however, so that the plot appears horizontally across the page rather than vertically as shown in Figure 3-20.

- Click the **Edit Last Dialog** toolbar button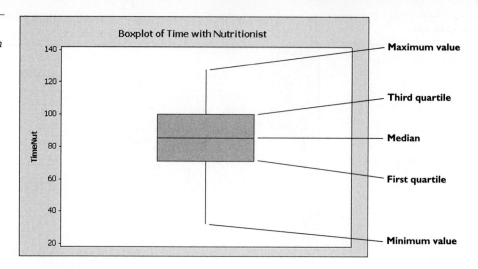
- Double-click **C7 BthWeight** as the "Graph variables"
- Click **Labels**, type **Boxplot of Birth Weight** as the "Title", and click **OK**
- Click **Scale** and click the **Transpose value and category scales** check box
- Click **OK** twice

The boxplot of infant birth weights is shown horizontally in Figure 3-21.

FIGURE 3-21

Boxplot for infant birth weight

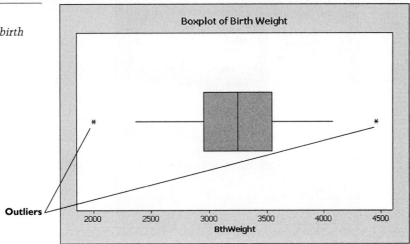

The distribution is almost exactly symmetric with the two outlying weights that you have seen before. Minitab defines outliers as values that either are less than Q1 – 1.5(Q3 – Q1) or exceed Q3 + 1.5(Q3 – Q1). If there are outliers greater than Q3 + 1.5(Q3 – Q1), Minitab will extend the upper whisker to the largest non-outlier. If there are outliers lower than Q1 – 1.5(Q3 – Q1), Minitab will extend the lower whisker to the smallest non-outlier. Outliers are indicated by asterisks.

A boxplot does not show the level of detail as a histogram, a stem-and-leaf display, or a dotplot. However, in Tutorial 4 you will see that these graphs are excellent choices when you want to compare different distributions.

3.10 | *Quantitative Variables: Summarizing Data Numerically*

In Sections 3.5 through 3.9 you investigated different ways of summarizing data for quantitative variables in the form of displays. Now you will use Minitab to find numerical summaries for these variables.

In Tutorial 2 you used the command Calc > Column Statistics to obtain the mean value for a column of test scores. You can use this technique to compute many common statistics such as the mean, median, standard deviation, range, and so on. However, it is a little cumbersome to use if you want to obtain multiple statistics for a column of data because the Calc > Column Statistics allows you to obtain only one statistic at a time. Moreover, there are some statistics, such as quartiles, which this command does not provide.

Minitab provides a command that solves both of these problems: Stat > Basic Statistics > Display Descriptive Statistics. You are going to start by looking at descriptive statistics for the quantitative variable, time spent with a nutritionist.

- Choose **Stat > Basic Statistics > Display Descriptive Statistics**
- Double-click **C9 TimeNut** as the "Variables"

The completed Display Descriptive Statistics dialog box should resemble Figure 3-22.

The completed Display Descriptive Statistics dialog box

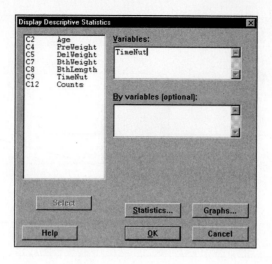

■ Click **OK**

Minitab automatically displays in the Session window the statistics you requested, as shown in Figure 3-23.

FIGURE 3-23

Descriptive Statistics for time spent with a nutritionist

```
MTB > Describe 'TimeNut';
SUBC>   Mean;
SUBC>   SEMean;
SUBC>   StDeviation;
SUBC>   QOne;
SUBC>   Median;
SUBC>   QThree;
SUBC>   Minimum;
SUBC>   Maximum;
SUBC>   N;
SUBC>   NMissing.
```

Descriptive Statistics: TimeNut

Variable	N	N*	Mean	SE Mean	StDev	Minimum	Q1	Median	Q3
TimeNut	68	0	86.18	2.35	19.36	32.00	71.25	85.50	100.00

Variable	Maximum
TimeNut	128.00

By default, the Stat > Basic Statistics > Display Descriptive Statistics command produces a variety of measures of central tendency and variability for each variable you select. Here is a list of the default statistics (you can also see them listed as part of the command):

• N, the number of cases for that column

• N*, the number of cases with missing values in that column (there are none for TimeNut)

• Mean, the mean value

• SE Mean, the standard error of the mean

• StDev, the standard deviation

• Minimum, the smallest value in the column

- Q1, the first quartile
- Median, the median value in the column
- Q3, the third quartile
- Maximum, the largest value in the column

You have probably studied most of these statistics at some point, but some, such as SE Mean, may be unfamiliar to you. SE Mean is actually the standard deviation divided by the square root of the number of non-missing values. In samples, this is a measure of the expected variation in the value for the sample mean over repeated samples. If you are not familiar with this statistic, you will learn more about it when you cover statistical inference. Notice that the last five statistics in the list above form the five-number summary referred to earlier when they were used to construct a boxplot.

▶ **Note** Minitab uses uppercase N to refer to the number of cases (missing or non-missing) in a column. To the extent that the data in a column are regarded as a sample, this practice differs from the almost universal practice of using lowercase n for the sample size (and, frequently, N for the population size). ■

▶ **Note** The Stat > Basic Statistics > Display Descriptive Statistics command produces statistics based upon a sample, not statistics based upon a population. ■

The statistics produced in Figure 3-23 are the default statistics, but are not the only ones that can be obtained with the Stat > Basic Statistics > Display Descriptive Statistics command. You can explore the options available.

- Click the **Edit Last Dialog** toolbar button 🔲
- Click **Statistics** to open the Descriptive Statistics - Statistics dialog box shown in Figure 3-24.

FIGURE 3-24

The Descriptive Statistics - Statistics dialog box

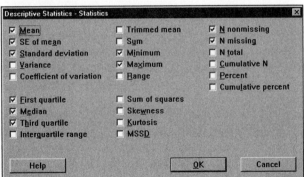

The default options are checked, but many more are available. You wonder how the *Trimmed mean* differs from the mean and so decide to include the Trimmed mean and omit the SE of Mean:

■ Clear the SE of Mean check box, click the **Trimmed mean** check box to select that statistic, and click **OK**

It is nearly always helpful to look at a graph as you study descriptive statistics and in earlier sections you looked at various graphical displays. Minitab makes it particularly simple to obtain graphs with descriptive statistics by including *built-in graphs* with the Stat > Basic Statistics > Display Descriptive Statistics command. To see the built-in histogram:

■ Click **Graphs** and click the **Histogram of data** check box
■ Click **OK** twice

The histogram of the 68 times is exactly the same as the one you obtained earlier (shown in Figure 3-12) except for the title. To see the text output:

■ Click the **Session Window** toolbar button 🗔

The new list of descriptive statistics is shown in Figure 3-25.

FIGURE 3-25

New Descriptive Statistics for time spent with a nutritionist

```
MTB > Describe 'TimeNut';
SUBC>   Mean;
SUBC>   StDeviation;
SUBC>   QOne;
SUBC>   Median;
SUBC>   QThree;
SUBC>   TRMean;
SUBC>   Minimum;
SUBC>   Maximum;
SUBC>   N;
SUBC>   NMissing;
SUBC>   GHist.
```

Descriptive Statistics: TimeNut

Variable	N	N*	Mean	TrMean	StDev	Minimum	Q1	Median	Q3	Maximum
TimeNut	68	0	86.18	86.31	19.36	32.00	71.25	85.50	100.00	128.00

Notice that Minitab prints the statistics in the order in which they were listed in the dialog box. This order cannot be changed. The trimmed mean time (86.31 minutes) differs only slightly from the value for the mean time (86.18) minutes. You decide to use Minitab's StatGuide to see how Minitab computes this statistic.

▶ **Note** Sometimes you need to be able to save the descriptive statistics, for example, for use in a later calculation. In this case, you can use the Stat > Basic Statistics > Store Descriptive Statistics command instead of the Stat > Basic Statistics > Display Descriptive Statistics command. The two commands offer the same extensive list of statistics from which to select. ■

Using StatGuide and the Glossary

Minitab has a resource, StatGuide, that is designed specifically to provide guidance in interpreting the results of analyses performed in Minitab. You will use it to find out more information about the first quartile. StatGuide can be accessed in a number of ways but the two most efficient ways are to:

1. Click the StatGuide toolbar button ▣, or
2. Hold your cursor over the Session window output that you would like guidance on, click the right-hand mouse button, and choose StatGuide.

- Click anywhere on the Descriptive Statistics output
- Click the **StatGuide** toolbar button ▣

   StatGuide opens two windows as shown in Figure 3-26. The MiniGuide window on the left-hand side contains a list of the various descriptive statistics. The right-hand StatGuide window provides an overview of the purpose and uses of descriptive statistics.

FIGURE 3-26

The initial MiniGuide and StatGuide windows

MiniGuide

StatGuide

- Click on **First and third quartile (Q1 and Q3)** in the MiniGuide window
   In the MiniGuide window Minitab provides an example of the default Descriptive Statistics output with the value of the first quartile highlighted. In the StatGuide window Minitab provides an explanation of the first quartile. The window and the explanation are shown in Figure 3-27. The first quartile is the highest values for the lowest 25% of the observations. You can use the Print button on the StatGuide menu bar to print a copy of the explanation. You can find a detailed explanation of how to calculate Q1 by clicking on Calculating quartiles in the "In Depth" section of the StatGuide window.

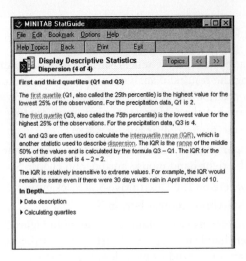

Notice that the MiniGuide does not list the trimmed mean among the Topics. However, another Minitab resource that can provide guidance in this case is the Glossary. To see what the Glossary has to say about the trimmed mean:

- Close both the MiniGuide and StatGuide Help Windows
- Click the **Help** toolbar button ![help icon]
- In the right-hand window, click **Reference** and then click **Glossary**
    The Glossary window appears on the right.
- Either scroll down in the Glossary window or click the **T** button and scroll down, and then click **Trimmed mean (TRMEAN)**
    A brief explanation of the Trimmed Mean is provided.
- Close the Help window to return to the Session window

3.12 Constructing Other Descriptive Statistics

As you have seen, Minitab can produce a large number of descriptive statistics. However, there are some that it does not produce, such as the *Mean Absolute Deviation (MAD)*, which is the average of the absolute values of the deviations of the observations from the mean. This measure of variation is commonly used in business forecasting. It can be obtained using other Minitab functions. You will calculate the MAD for the 68 times in C9 TimeNut.

The first step is to create a new variable containing the absolute values of the differences (MAD) between the values for TimeNut and the mean for C9 TimeNut.

- Choose **Calc > Calculator**
- Type **AbsDiff** as the "Store results in variable"
- Enter, by typing or clicking, **MEAN (ABSO ('TimeNut' – MEAN ('TimeNut')))** as the "Expression" and click **OK**

Note If you are clicking, you will need to click the "Absolute value" function "ABSO" will appear in the text box. ■

- Use the **Current Data Window** toolbar button ⊞ to return to the Data window to see that the column C13 AbsDiff contains the MAD value

    To print the MAD value in the Session window:

- Choose **Data > Display Data**
- Choose **C13 AbsDiff** as the "Columns, constants, and matrices to display" column, and click **OK**

    The Session window displays the value of the MAD statistic—15.2111 minutes—as shown in Figure 3-28. This indicates that the 68 values in TimeNut vary from their mean, 86.18 minutes, by an average of 15.2111 minutes.

FIGURE 3-28

The value for the MAD statistic

```
MTB > Let 'AbsDiff' = MEAN(ABSO('TimeNut' - MEAN('TimeNut')))
MTB > Print 'AbsDiff'.
```

Data Display

```
AbsDiff
   15.2111
```

As a measure of variability, the MAD is comparable to the standard deviation. For the variable TimeNut, the value for MAD is less than the standard deviation (19.36 minutes, from Figure 3-25). Because the standard deviation is a key component of so many statistical methods, it is much more widely known and used than the MAD.

The next section explains how to copy text into Minitab's ReportPad and into Microsoft Word. If you are not interested in learning about these operations, go on to Section 3.14.

3.13 *Copying Text Output into ReportPad and Microsoft Word*

In the last tutorial you learned how to copy text output into the word processor Notepad. Here you will copy output into Minitab's ReportPad and into Microsoft Word.

The ReportPad folder in the Project Manager allows you to create and edit reports consisting of both text output that you have obtained in the Session window and graphs. In this section you will copy the descriptive statistics for the times spent with a nutritionist into ReportPad.

- If the Session window is not already active, make it active

    You should be able to see the descriptive statistics for TimeNut. If not, scroll back until you see them.

- Highlight the descriptive statistics for TimeNut beginning with the title, Descriptive Statistics: TimeNut

    At this point you could copy this highlighted material into the clipboard and then paste it into ReportPad. However, Minitab offers a more convenient option.

- Click the **right-hand mouse button** and choose **Append Selected Lines to Report**
- Click the **Show ReportPad** toolbar button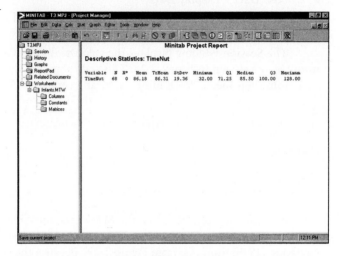

Your Descriptive Statistics output is shown in the ReportPad under the heading Minitab Project Report, as shown in Figure 3-29. You can edit this output, add commentary, and save the contents as *rtf* (Rich Text Format) that can be opened in other word processors.

FIGURE 3-29

The contents of the ReportPad

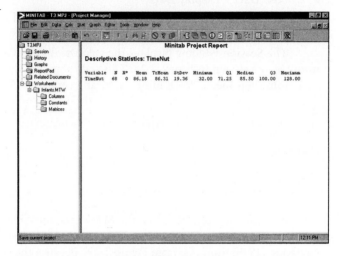

Practice saving the current contents of the ReportPad to an *rtf* file.

- Hold your cursor over the Report Pad folder, click the **right-hand mouse button**, and choose **Save Report As**
- Type **3DescStatTimes** as the "File name"
- Click **Save**

You can also use File > Save Report As to perform this operation. The file you saved, *3DescStatTimes.rtf*, can be opened in most word processors, including Microsoft Word.

▶ **Note** You can also save the contents as a Web page. ■

▶ **Note** You can also append an active graph into Report Pad by clicking the right-hand mouse button and choosing Append Graph to Report. ■

- Click the **Minimize** button to minimize Project Manager and make the Session window active

Your descriptive Statistics output should still be highlighted. Many Minitab users prefer to copy text output directly from the Session window into a word processor, such as Microsoft Word or Corel WordPerfect. Next, you will practice doing this with Microsoft Word.

In preparation for this you should have both Minitab and Microsoft Word open. Use the Start menu (usually located at the lower left-hand corner of your screen) to open Microsoft Word:

- Choose **Start > Programs > Microsoft Word** (if Microsoft Word is not in the Programs list, you will have to search for it)

  You can toggle between Microsoft Word and Minitab by using the Alt + Tab combination.

- Press Alt + Tab to return to the Session window in Minitab
- If the descriptive statistics for TimeNut are not highlighted, highlight this output
- Click the **Copy** toolbar button 🖺
- Press Alt + Tab to return to Microsoft Word

  In Microsoft Word:

- Choose **Edit > Paste**

  Your output should now appear in the open Word document. You can save this Word file if you wish, but exit Word to return to Minitab.

3.14   *Saving and Reopening a Project*

To understand the function of a project more fully, you will save the project *T3.mpj*, leave Minitab, return, and then open the project:

- Click the **Save Project** toolbar button 🖫
- Choose **File > Exit**

  You have now saved the entire contents of the Session window for Tutorial 3 along with the worksheet and the graphs you produced.

  ▶ **Note** If you forget to save your project, Minitab will ask, at the File > Exit command: "Save changes to this project before closing?" To save your project, click Yes. ■

  Now return to Minitab.

- Open **Minitab**

  You can open a project by using the Open Project toolbar button or with the command File > Open Project.

- Click the **Open Project** toolbar button
- Click the "Look in" **drop-down list arrow** and select the location where you are saving your work
- Click *T3.mpj* in the Project list box

  Minitab allows you preview the contents of a project. To preview the contents of *T3.mpj*:

- Click **Preview**

  The list of items available in the project T3 includes all of the graphs you created and did not close and the worksheet *Infants.mtw*. Other windows, such as the Session window and the Project Manager, are not shown (but they were saved!).

> **Note** You can also preview the contents of a worksheet when you open it. This can be particularly helpful when you need to open a large data set that you are unfamiliar with or a data set in a format other than Minitab. ∎

- Click **OK** to return to the Open Project dialog box
- Click **Open** to open *T3.mpj*

The graphs and the various windows open in rapid succession, and you are returned to exactly the point you were at when you exited Minitab. The Session window that was open when you left is still open. In the Windows menu you will find all of the graphs that you named and didn't close during this tutorial. Scroll back in the Session window and you will discover all of the stem-and-leaf displays that you created, as well as all of the commands you issued. Further, all of the settings you created at the beginning of the tutorial are still in effect. For example, the Session command language is enabled.

Now you can exit Minitab again:

- Choose **File > Exit**

Well done! You've finished Tutorial 3. You have reviewed the various ways that Minitab can provide graphical and numerical summaries for both qualitative (or categorical) and quantitative variables. In the next tutorial, you will explore how Minitab can be used to compare groups.

MINITAB *at work*

EDUCATION

Medical education experts overseeing the training of young physicians are concerned that doctors do not write prescriptions accurately enough. The Department of Medical Education at a Pittsburgh research hospital decided to test this concern by conducting a study of prescriptions written by residents studying to become doctors.

They collected copies of prescriptions written by residents over a four-month period and then recorded and categorized error types, hoping to find patterns that could be corrected. Error categories included omissions, ambiguities, and mistakes. (Contrary to popular belief, unreadable handwriting is rarely a problem, though the use of Latin terms, abbreviations, and unfamiliar drug names make it seem so to patients.)

The educators used Minitab's > Stat > Table command to determine error rates for each category. When they studied Minitab's displays, they found that most errors occurred when residents did not provide enough information to the pharmacist.

Experience seemed to be a good teacher; third-year residents made half as many errors as first-year residents. Studies of this kind use Minitab to show medical educators how to teach residents to be better communicators—and, at the same time, better doctors.

Minitab Command Summary

This section describes the commands introduced in, or related to, this tutorial. Use Minitab's online Help for a complete explanation of all commands.

Minitab Menu Commands

Menu	Command	Description
File ➤	Open Project	Opens the named project.
	Save Project	Saves the current project.
	Save Project As	Saves a project with the name to be specified.
	Project Description	Shows or creates a description of a project.
	Open Graph	Opens a professional graph that has been saved in a graph file.
	Save Graph As	Saves a professional graph (when the Graph window is active).
	Save Report As	Saves the contents of the ReportPad (when the ReportPad is the active window).
	Print Graph	Prints a professional graph (when the Graph window is active).
Stat ➤	Basic Statistics	
	➤ Display Descriptive Statistics	Provides a variety of numerical measures, including ones of central tendency and variability for one or more columns. Various built-in graph options are available.
	➤ Store Descriptive Statistics	Stores a selected number of numerical measures including ones of central tendency and variability for one or more columns. The measures are stored in the first row of successive columns.
	Tables	
	➤ Tally Individual Values	Calculates a count (or frequency) of each value of a variable or variables.
	➤ Cross Tabulation and Chi-Square	Calculates percentages from summarized count data.
Graph ➤	Histogram	
	Simple	Produces a histogram for each variable indicated.
	Dotplot	
	Simple	Produces a dotplot for each variable indicated.
	Stem-and-Leaf	Produces a stem-and-leaf display for each variable indicated.
	Boxplot	
	Simple	Produces a boxplot for each variable indicated.
	Individual Value Plot	
	Simple	Produces an individual value plot for each variable indicated.
	Bar Chart	
	Simple	Produces many kinds of charts, including bar, line, symbol, and area charts for each variable indicated.
	Pie Chart	Produces a pie chart for each variable indicated.

Menu	Command	Description
Editor ➤	Worksheet	
	➤ Description	Shows or creates a description of a worksheet (when the worksheet is the active window).
	Column	
	➤ Value Order	Specifies the order in which the text values of a text variable appear in output (when the worksheet is the active window).
Tools ➤	Microsoft Calculator	Opens the Microsoft Calculator.
Window ➤	Project Manager	Opens the Project Manager.

Minitab Toolbar Button Commands

Toolbar Button	Icon	Description
Open Project	🖼	Opens the named project.
Save Project	🖼	Saves the current project.
StatGuide	🖼	Opens StatGuide.
Show Worksheets Folder	🖼	Shows the contents of the Worksheets folder in the Project Manager.
Show Graphs Folder	🖼	Shows the contents of the Graphs folder in the Project Manager.
Show ReportPad	🖼	Opens the ReportPad.
Show Related Documents	🖼	Shows the contents of the Related Documents folder in the Project Manager.
Project Manager	🖼	Opens the Project Manager.

Minitab Right-Hand Mouse Button Commands

Command	Description
Edit Title	Changes the name of a graph in the Project Manager.
Edit X Scale	Changes the number of bins in a histogram.
Close	Removes a graph from the Project Manager.
Append Selected Lines to Report	Places the selected text in the ReportPad.
Append Graph to Report	Places the active graph in the ReportPad.
Save Report As	Saves the contents of the ReportPad (when the ReportPad is the active window).

Review and Practice

Matching

Match the following terms to their definitions by placing the correct letter next to the term it describes.

_____ MAD	a. A graph that is based upon just five numbers
_____ Trimmed mean	b. A file that preserves worksheets, Session window contents, and graphs
_____ Tally Individual Variables	c. A graphical description of the data in which some digits of the original data are evident
_____ Boxplot	

_____ Count
_____ Stem-and-Leaf display
_____ Project
_____ Labels
_____ Dotplot
_____ StatGuide

d. A resource that explains of some of the tools found in Minitab

e. A statistic that measures the central tendency of a set of data

f. A command to determine the counts for the data in a column

g. A statistic that measures spread around the mean

h. The number of times each distinct value occurs

i. A graph in which every observation is usually represented by a point

j. An option within a dialog box that allows you to enter titles on graphs

True/False Problems

Mark the following statements with a *T* or an *F*.

a. ____ Stem-and-leaf displays are viewed in the Session window.

b. ____ Minitab lets you annotate most graphs with titles.

c. ____ Save Graph As is a command on the Windows menu.

d. ____ Minitab projects include all open worksheets and the contents of the Session window, but not the graphs created; these have to be saved separate from the project.

e. ____ The order in which statistics are displayed in the Display Descriptive Statistics output cannot be changed.

f. ____ Histograms and dotplots are graphs designed to be used with qualitative data.

g. ____ The Enable Commands command is available only when the Data window is active.

h. ____ The Tally Individual Variables command can be used only with text data.

i. ____ ReportPad provides explanations for the statistical tools available in Minitab.

j. ____ Professional graphs can be saved as part of a Minitab project and outside of a Minitab project.

Practice Problems

The practice problems instruct you to save your worksheets with a filename prefixed by P, followed by the tutorial number. In this tutorial, for example, use *P3* as the prefix. If necessary, use Help or Appendix A Data Sets, to review the contents of the data sets referred to in the following problems. Interpretations should use the language of the subject matter of the question. If you are using the Student version, be sure to close worksheets when you have completed a problem.

1. Open *PulseA.mtw*.

   a. The variable Activity has a missing value in row 54. Use the command Data > Delete Rows to eliminate this row. Use the Stat > Tables > Tally Individual Variables command to obtain the percentage distribution of Activity. Which activity level has the smallest count? What percentage of the total class is in this group?

   b. Produce two bar charts of the variable Activity, one with counts on the vertical axis and one with percentages on the vertical axis. Why might you prefer the latter chart?

   c. Repeat part b, using pie charts rather than bar charts.

   d. Refer to the percentage charts in parts b and c. Which of the two graphs gives you the more accurate visual impression of the percentage distribution of Activity?

   e. Open the Project Manager. What are the default names for the four displays you created in this question? What is the drawback to Minitab's system of default naming of displays? Rename the two displays that show percentages.

   f. Close the two displays showing counts rather than percentages.

2. Open *Backpain.mtw*. This problem deals with the most appropriate way to view percentages.

    a. Use the Stat > Tables > Tally Individual Variables command to find the percentage of patients that are female.

    b. Produce (i) a bar chart and (ii) a pie chart for the variable Gender. In both cases ensure that the graph shows percentages. Use data tips on both graphs to obtain the percentage of patients that are female. How do these percents compare (with regard to the number of decimal places) with your answer in part a?

3. Open *Compliance.mtw*.

    a. Obtain a tally of the variable Educ that includes counts and percentages. Upon what basis has Minitab ordered the five education levels? Use Editor > Column > Value Order to verify that the five education levels are correctly ordered. Obtain the tally again to verify that the ordering is correct.

    b. Obtain another tally of the variable Educ, but now include both cumulative counts and cumulative percentages. Interpret the values 14, 36.36, and 42.42 in the output.

    c. Obtain a tally of the variable Employ. In this case include only counts and percentages. Explain the values 7 and 21.21, and the notation "* = 1" in the output. Why are cumulative counts and percentages not appropriate in this case?

    d. Highlight the output in the Session window from parts b and c. Append this material to the ReportPad as you did in Section 3.13. Also in ReportPad, but preceding the output, write a brief report summarizing the two tallies.

    e. Repeat part d, except copy the highlighted output and paste it into Microsoft Word and write the summary in Microsoft Word.

4. An instructor teaching an intermediate statistics course collects the following summarized data on the students in her class.

Major	Number of Students with This Major	Number of Math Courses	Number of Students Taking This Number of Math Courses
Economics	9	1	1
Mathematics	6	2	0
Math/Econ	5	3	4
Psychology	4	4	7
Comp Sci	2	5	9
		6	3
		7	1
		8	1

    a. Enter these data into columns C1–C4 on a new Minitab worksheet. Provide appropriate names for each column.

    b. Obtain a bar chart showing the distribution of majors. Upon what basis does Minitab order the majors? Use Bar Chart - Options to have Minitab order the bars in decreasing order of occurrence. Obtain that bar chart and use Bar Chart - Options again to change back to the default order.

c. Use Editor > Column > Value Order to change the order that the majors appear in output to Math/Econ, Mathematics, Economics, Comp Sci, and Psychology. Obtain the bar chart again to make sure that the change has occurred.

d. Obtain a histogram showing the distribution of the number of math courses per student. (*Hint:* Choose the number of math courses as the Graph variable. Click Data Options and then on the Frequency tab. Select the number of students as the Frequency variable.)

e. Investigate which other plots for quantitative data allow you to enter summarized data. Explain which of the plots you prefer for these data.

5. Open *BallparkData.mtw*.

a. Use Calc > Calculator to create a new variable that is the percentage that average attendance is of capacity. Name this new variable %Cap.

b. Obtain an individual value plot of %Cap. (If you want to obtain the same random scatter as in the solutions, use the Calc > Set Base command with 52 as the base.) Move your cursor over the one outlier in the plot to find the value for %Cap and the corresponding row. Go to this row in the worksheet and determine which team is the outlier.

c. Obtain a histogram of ParkBlt. What is the most striking feature of the display?

d. Display a data tip on each of the two smallest bars of the histogram to obtain the bin interval for each. What are these intervals?

e. How many Major League teams play in ballparks that were built before 1930? Which teams are they?

f. Save this graph as a graph file.

g. Obtain a boxplot of ParkBlt. Is the feature of these data that you noted in the histogram equally obvious in this boxplot? Explain briefly.

6. Open *OldFaithful.mtw*.

a. Produce a stem-and-leaf display of the duration of eruptions. What is most striking about the distribution of the durations? What is the meaning of the notation "Leaf Unit = 10"? What is the range of durations? Interpret (in the context of durations) the values (9) and 23 in the left-most column of the plot.

b. Obtain the descriptive statistics for Duration, selecting the range as an option. Does the range of durations match your answer in part a? Explain briefly.

c. One of the default statistics that you produced in part b is SE Mean (the standard error of the mean). Use StatGuide, as you did in Section 3.11, to investigate the meaning of this statistic.

7. Open *MBASurvey.mtw*. This problem shows how a histogram can also be constructed with cutpoint intervals.

a. Create a histogram of the column Cash. Obtain a data tip on the first bar. What is the bin interval? Does this interval make sense in this context? Explain briefly.

b. Hold your cursor over the horizontal (X) axis of the plot, click the right-hand mouse button, and choose Edit X Scale. Click on the Binning tab and click the Cutpoint option button. Click OK. In what way has this sequence of commands changed the histogram? Explain how the change has improved the histogram.

8. Open *CollMass.mtw*.

   a. In this problem you will focus on the variable %Accept. Obtain a histogram, a dotplot, and an individual value plot of the data in this column. (If you want to obtain the same random scatter as in the solutions, use the Calc > Set Base command with 81 as the base.) Which plot gives you the best impression of the shape of the distribution of %Accept? How would you characterize this shape? Given the nature of this variable, does the shape make sense? Would you expect the mean or the median value for %Accept to be the larger? Explain briefly.

   b. You are interested in using data tips to compute the exact range of values for %Accept. With which of the three plots is this possible? Compute this range. Which college has the lowest %Accept? Does the answer surprise you?

   c. Obtain the default descriptive statistics for this variable and include a built-in individual value plot. Were you correct in your choice of which of the mean or median was greater? How, if at all, does this individual value plot differ from the one you obtained in part a?

   d. Use the Microsoft Calculator to compute how many standard deviations the lowest and the highest values for %Accept are from the mean value.

9. Open *PulseA.mtw*.

   a. Produce a dotplot of Weight. Name the dotplot Weight Dotplot. The dotplot is bimodal, with one peak at around 150 pounds and a smaller peak at about 125 pounds. Can you explain this feature of this plot? Rename this plot in the Project Manager.

   b. Produce a stem-and-leaf display of Weight. How much does the lightest person in the class weigh? How much does the heaviest person in the class weigh? Interpret (in the context of weights) the values 11 and 16 in the left-hand column of the display. Explain the notice at the top of the output "Leaf Unit = 1.0." Can you save this plot in a graph file outside of Minitab? Explain briefly.

   c. Use the Calc > Calculator command to place the increases in pulse rates (Pulse2 – Pulse1) in a new column. Increase. Unstack the contents of Increase into two new columns, and based upon whether or not the student ran in place. Obtain the default descriptive statistics for the increases in pulse rate for those who ran in place.

   d. Highlight the output in the Session window from parts b and c. Append this material to the ReportPad as you did in Section 3.13. Also in ReportPad, but preceding the output, write a brief report summarizing the mean, standard deviation, minimum, and maximum increases for the students who ran in place.

   e. Repeat part d, except now copy the highlighted output and paste it into Microsoft Word, and write the summary in Word.

   f. Close this worksheet without saving it.

10. Open *Backpain.mtw*.

    a. Produce a dotplot of the variable LostDays. What is the most striking feature of the plot? Given this shape, explain which of the mean or the median you would expect to be greater.

    b. Use the Stat > Basic Statistics > Display Descriptive Statistics command to obtain the mean, median, and trimmed mean. Do the values for the mean and the median support your answer in part a? Is the trimmed mean smaller or greater than the mean? Why do you think this is so?

c. Obtain a boxplot of the ages of the patients. Move your cursor over the plot to determine the quartiles and the interquartile range of the ages. Which plot components are not revealed by data tips? What two different commands might you use to obtain them?

11. Open *Height.mtw*. In this problem you will verify Minitab's calculation of the variance of the sample of heights. The variance is the square of the standard deviation and is computed by (i) summing the square of the difference between each value (height) and the mean height, and (ii) dividing this sum by one fewer than the number of non-missing values.

a. Obtain the descriptive statistics for Heights. Be sure to select the variance as one of the statistics. Record the values for the number of heights, the mean height, and the variance of height.

b. Use the Calc > Calculator command to compute a new column of values consisting of the squared differences between the heights and the mean height. Name this column SqDiff.

c. Use the Calc > Column Statistics command to obtain the sum of the values in SqDiff.

d. As the final step, use the Microsoft Calculator to divide the sum of the squared differences by one fewer than the number of heights. Does your answer agree with the value Minitab gave you in part a?

e. Use the Calc > Calculator command to compute a new column consisting of the variance by using one expression. Name this column Variance. Does your answer agree with the value Minitab gave you in parts a and d?

On Your Own

There is an enormous amount of data available on the Internet. In this project you are invited to visit a particularly rich source of data on the Internet—the Data and Story Library (DASL) at the Carnegie-Mellon University. This library contains a large number of data sets, each accompanied by a story that provides background on the data. The library's URL is http://lib.stat.cmu.edu/DASL/.

When you get to the homepage, scroll down and read the overview of the site. The data sets in the library are classified by the subject matter of the data and the type of analysis (methods) that can be performed on the data. If you click List all methods, for example, you will see several topics that have been covered in this tutorial, including boxplots, histograms, means, and medians.

Select a data set that interests you, read the "story" that goes with it, and examine the data. The data are always in a simple enough form so that if you highlight and copy the data to the clipboard, it can be pasted right into a Minitab worksheet.

Try copying a data set from DASL into Minitab and practice using as many of the tools covered in this tutorial as you can.

Answers to Matching Problems

(g) MAD
(e) Trimmed mean
(f) Tally Individual Variables
(a) Boxplot
(h) Count
(c) Stem-and-Leaf display
(b) Project
(j) Labels
(i) Dotplot
(d) StatGuide

Answers to True/False Problems

(a) T, (b) T, (c) F, (d) F, (e) T, (f) F, (g) F, (h) F, (i) F, (j) T

4

Data Analysis: Comparing Groups

In this tutorial, you will construct displays (graphs and tables) to compare groups based upon two variables. You will examine situations in which both variables are qualitative and in which one is qualitative and the other quantitative. Also in this tutorial, you will learn how to export graphs. Two case studies, one familiar and one new, will be used to illustrate these displays.

OBJECTIVES

In this tutorial, you learn how to:

- Construct contingency tables that can be used to explore the relationship between two qualitative variables
- Construct cluster and stack bar charts that are used to illustrate the relationship between two qualitative variables
- Use multiple dotplots, individual value plots, and boxplots to compare groups
- Describe data for one variable separately for each category of another variable
- Construct a chart that displays summary statistics for a quantitative variable
- Export a Minitab graph to ReportPad or Microsoft Word

4.1 Contingency Tables

CASE STUDY | **PUBLIC HEALTH—INFANT NUTRITION (CONTINUED)**

The research unit of a health-care group where you work has been conducting a study of the factors that appear to be associated with a new mother's decision to breast-feed her infant. Sixty-eight low-income, pregnant women who attended a clinic affiliated with the group are the subjects. Your initial task is to summarize the data collected on these women and their newborn children.

In Tutorial 3, you examined the smoking status of the 68 women. In this tutorial, you will use displays to investigate how smoking status is related to ethnicity. In the previous tutorial you also found considerable variation in the birth weights of the newborn children. Now you will use displays and summary measures to investigate how birth weight is related to smoking status.

Begin by opening Minitab:

■ Open **Minitab** and maximize the Session window

■ If necessary, **Enable Commands** and clear Output Editable from the Editor menu

Recall that Minitab automatically opens a new untitled project each time you open the program. You can work in this project or open a different one. In this and subsequent tutorials you will name and work in the new project.

The data on the 68 women are in the file *Infants.mtw*. First, open this worksheet.

■ Open the worksheet **Infants.mtw** (from the Studnt14 folder)

■ If necessary, click the **Current Data Window** toolbar button 🗔

Now, save the project.

■ Click the **Save Project** toolbar button 🗔 and in the location where you are saving your work, save this project as *T4.mpj*

First, investigate the relationship between smoking status and ethnicity. Because both of these variables are qualitative, you explore the relationship between them by obtaining a two-way table of counts called a *contingency table* or, sometimes, a *cross-tabulation*. Before obtaining such a table use the Editor > Column > Value Order command to verify that the ordering of the text values will be: non-smokers, light smokers, and then heavy smokers.

■ Click anywhere in the column **C3-T Smoke**

■ Choose **Editor > Column > Value Order**

■ Click the **User-specified order** option button

By coincidence, the order of the text values in the text box is exactly the one you want.

■ Click **OK**

To obtain the contingency table:

- Choose **Stat > Tables > Cross Tabulation and Chi-Square**
- Double-click **C3 Smoke** as the "Categorical variables: For rows" and **C1 Ethnic** as the "Categorical variables: For columns"

The completed Cross Tabulation and Chi-Square dialog box is as shown in Figure 4-1.

FIGURE 4-1

The completed Cross Tabulation and Chi-Square dialog box

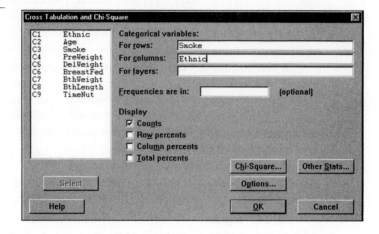

- Click **OK**

The resulting table is shown in Figure 4-2. Each cell contains the count as indicated by Minitab in its legend at the bottom of the output.

FIGURE 4-2

Default contingency table of smoking status and ethnicity

```
MTB > XTABS 'Smoke' 'Ethnic';
SUBC>    Layout 1 1;
SUBC>    Counts;
SUBC>    DMissing 'Smoke' 'Ethnic'.
```

Tabulated statistics: Smoke, Ethnic

```
Rows: Smoke    Columns: Ethnic

              Black  Hispanic  White  All

NonSmoker       18       11      20    49
LightSmoker      4        0       5     9
HeavySmoker      3        1       6    10
All             25       12      31    68


Cell Contents:      Count
```

In addition to providing the counts of each smoking status (49 non-smokers, 9 light smokers, and 10 heavy smokers) that you obtained from a tally in Tutorial 3, the contingency table indicates the counts of each smoking status for each ethnic group. For instance, there are 18 non-smoking black women, 11 non-smoking Hispanic women, and 20 non-smoking white women. For each ethnic group, what are the percentages of the 68 women in the study who are non-smokers for each ethnic group? Also, which ethnic group has the highest proportion of non-smokers?

To answer queries such as these obtain row, column, and total percentages for the contingency table:

- Click the **Edit Last Dialog** toolbar button ▣
- Click the "Display" **Row percents, Column percents**, and **Total percents** check boxes (Counts should already be checked)
- Click **OK**

FIGURE 4-3

Percentages of women in various categories

```
MTB > XTABS 'Smoke' 'Ethnic';
SUBC>   Layout 1 1;
SUBC>   Counts;
SUBC>   RowPercents;
SUBC>   ColPercents;
SUBC>   TotPercents;
SUBC>   DMissing 'Smoke' 'Ethnic'.
```

Tabulated statistics: Smoke, Ethnic

```
Rows: Smoke   Columns: Ethnic

               Black   Hispanic   White     All      Count
NonSmoker         18         11      20      49
               36.73      22.45   40.82  100.00       % of Row
               72.00      91.67   64.52   72.06
               26.47      16.18   29.41   72.06       % of Column

LightSmoker        4          0       5       9       % of Total
               44.44       0.00   55.56  100.00
               16.00       0.00   16.13   13.24
                5.88       0.00    7.35   13.24

HeavySmoker        3          1       6      10
               30.00      10.00   60.00  100.00
               12.00       8.33   19.35   14.71
                4.41       1.47    8.82   14.71

All               25         12      31      68
               36.76      17.65   45.59  100.00
              100.00     100.00  100.00  100.00
               36.76      17.65   45.59  100.00

Cell Contents:      Count
                    % of Row
                    % of Column
                    % of Total
```

The legend at the bottom of the output provides an explanation of each cell's contents. The first row contains the cell's count, the second row this count as a row percentage, the third row this count as a column percentage, and the fourth row this count as a total percentage. You can now determine that 26.47% of the women in the study are non-smoking blacks by looking at the "% of Total" line in the first cell (18 / 68*100%). You also observe that 16.18% of the women are non-smoking Hispanics and that 29.41% of the women are non-smoking whites. By examining the "% of Columns" lines you determine the smoking status percentage of women for each ethnicity. In particular, you observe that Hispanic women have the highest percentage of non-smokers at 91.67%. This is greater than 72.00% for black women and 64.52% for the white women. You can construct you construct a contingency table with only the counts and these column percentages.

- Click the **Edit Last Dialog** toolbar button ▣
- Clear the "Display" Row percents and Total percents check boxes
- Click **OK**

```
MTB > XTABS 'Smoke' 'Ethnic';
SUBC>   Layout 1 1;
SUBC>   Counts;
SUBC>   ColPercents;
SUBC>   DMissing 'Smoke' 'Ethnic'.
```

Tabulated statistics: Smoke, Ethnic

Rows: Smoke   Columns: Ethnic

	Black	Hispanic	White	All
NonSmoker	18	11	20	49
	72.00	91.67	64.52	72.06
LightSmoker	4	0	5	9
	16.00	0.00	16.13	13.24
HeavySmoker	3	1	6	10
	12.00	8.33	19.35	14.71
All	25	12	31	68
	100.00	100.00	100.00	100.00

Cell Contents: Count
                   % of Column

From this contingency table you observe that there are different patterns of smoking among the three ethnic groups—although, for each ethnic group the non-smokers are the *modal* (most frequently occurring) category. In Tutorial 12 you will have an opportunity to test if these patterns are more different than would be expected by chance.

▶ **Note** If you wanted your contingency table with ethnicity as the row variable and smoking status as the column variable, then you need only switch these variables in "Categorical variables: For rows" and "Categorical variables: For columns" text boxes. ■

▶ **Note** You can add a third variable to a contingency table by selecting it in the "For layers" text box (see Figure 4-1). For instance, if you selected BreastFed as the third variable Minitab would create two contingency tables similar to those in Figure 4-2, one for those mothers who did not breast-feed their infant (No), and one for those who did (Yes). ■

4.2 Cluster and Stack Bar Charts

In Tutorial 3, you constructed a chart showing the percentage distribution of the variable Smoke. The chart showed that approximately 70% of the women were non-smokers (NonSmoker), with almost equal percentages of light and heavy smokers (LightSmoker and HeavySmoker). Here you will construct two bar charts to represent the relationship between the variables Smoke and Ethnic that you tabulated in the previous section. To generate the first bar chart you define Ethnic as a *cluster variable*. This enables the resulting bar chart to display three bars for each ethnic group, one for each smoker category:

- Choose **Graph > Bar Chart** and double-click *Cluster*
- Double-click **C1 Ethnic** and then **C3 Smoke** as the "Categorical variables (2-4, outermost first)"
- Click **Bar Chart Options**, click the **Show Y as Percent** check box, and click **OK**
- Click **Labels**, type **Cluster Bar Chart of Smoke by Ethnic** as the "Title", and click **OK**

At this point, the completed dialog box should look as shown in in Figure 4-5.

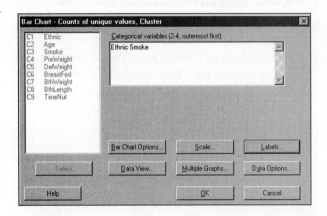

In this dialog box, Minitab is using the term "outermost" as the cluster variable, in this case, Ethnic.

- Click **OK**

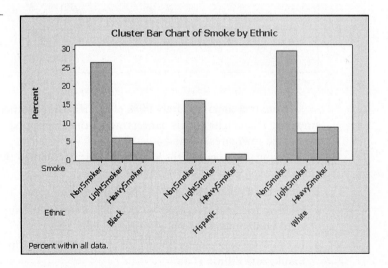

The *side-by-side bar charts* of Smoke are clustered by Ethnic. You can see that most Hispanics do not smoke (approximately 15% of the 68 women are non-smoking Hispanics), there are no Hispanic light smokers, and a few Hispanics are heavy smokers (approximately 1% of the women). Different patterns are present for blacks and whites.

Your graph would be more informative if the percentages corresponded to the column percentages from the last section. That is, for each ethnic group the sum of the percentages totaled 100%.

- Click the **Edit Last Dialog** toolbar button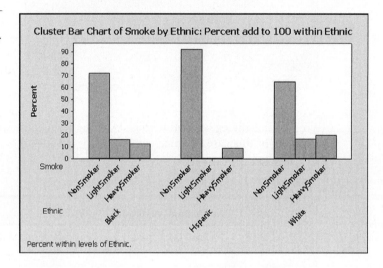
- Click **Bar Chart Options**, click the "Take Percent and/or Accumulate" **Within categories at level 1 (outermost)** option button, and click **OK**
- Click **Labels**, type **Cluster Bar Chart of Smoke by Ethnic: Percents add to 100 within Ethnic** as the "Title," and click **OK** twice

The resulting graph is shown in Figure 4-7.

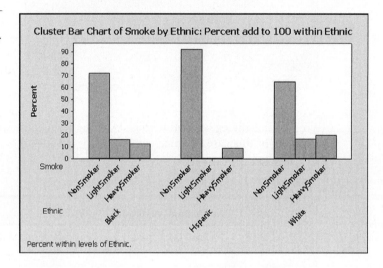

Note that approximately 90% of the Hispanic women are non-smokers. The height of this bar represents the column percentage 91.67 that you calculated in Figures 4-3 and 4-4.

Name this graph in the same way that you named the smoking status bar chart that you created in Tutorial 3.

- Click the **Project Manager** toolbar button
- Hold your cursor over the last icon in the right-hand window, click the **right-hand mouse button**, and choose **Edit Title**
- In the Title box, change the name to *Your Initials*: **Cluster Bar Chart of the Smoke and Ethnic Data**
- Click the **Session Window** toolbar button

A second way to represent the distribution of Ethnic for each smoker status is to stack the bars for each Ethnic category rather than show them side by side.

To create this chart:

- Choose **Graph > Bar Chart** and double-click *Stack*
- Double-click **C1 Ethnic** and then **C3 Smoke** as the "Categorical variables (2-4, outermost first)"
- Click **Bar Chart Options**, click the **Show Y as Percent** check box, click the "Take Percent and/or Accumulate" **Within categories at level 1 (outermost)** option button, and click **OK**
- Click **Labels**, type **Stack Bar Chart of Smoke by Ethnic** as the "Title", and click **OK** twice

The resulting chart is shown in Figure 4-8.

FIGURE 4-8

Stack bar chart for Smoke by Ethnic

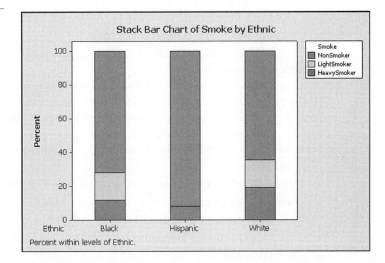

The chart in Figure 4-8 shows exactly the same results as that in Figure 4-7, but it is harder to read. For instance, it is difficult to compare the stacked bars above the base bar. The reader also needs to reference the default legend. Most statisticians prefer cluster bar charts over stack bar charts. You can close this current chart:

- Click the Graph window's **Close** button
- At the prompt "The graph 'Chart of Ethnic, Smoke' will be removed from the project. This action cannot be undone. Would you like to save it in a separate file?", click **No**

Before beginning the next case study, save the project *T4* again.

- Click the **Save Project** toolbar button

There is no need to close the Infants data.

In the next case study you will compare numerical summaries of one or more quantitative variables for categories of a qualitative variable.

4.3 Comparing Dotplots

| CASE STUDY | SPORTS—
BASEBALL STADIUMS |

You've been asked to write an article about Major League Baseball ballparks and their effects on the success of the teams that play in them. You've gathered some basic data about the capacity of each of the 30 parks and when each was built, as well as the average attendance at each park in 2001 and the team performance in 2001. You will begin the article by comparing the American and National Leagues on some of these variables.

The data on baseball stadiums are in the file *BallparkData.mtw*.

■ Open the Worksheet ***BallparkData.mtw***

Refer to Help or Appendix A Data Sets for complete information on the six variables in this worksheet.

In Tutorial 3, you obtained five different graphs for a quantitative variable: a histogram, a stem-and-leaf display, a dotplot, an individual value plot, and a boxplot. Minitab allows you to compare subgroups by showing any one of these graphs separately for each subgroup. For instance, you can compare a dotplot of attendance (C5 Attend) for the American League to a dotplot of attendance for the National League:

■ Choose **Graph > Dotplot** and double-click ***One Y, With Groups***

■ Double-click **C5 Attend** as the "Graph Variables," click Tab, and double-click **C2 League** as the "Categorical variables for grouping (1-4, outermost first)"

■ Click **Labels** and type **Dotplots of Attendance for Each League** as the "Title"

■ Click **OK** twice

The resulting dotplot is shown in Figure 4-9.

FIGURE 4-9

Dotplot for attendance by league

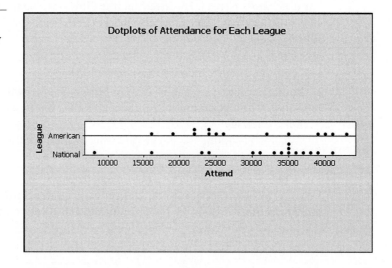

The dotplots help you to easily compare the two sets of attendance. You can see the distributions of attendance for each league. There is a greater range of attendance for the National League.

◆ **Note** With bar charts, Minitab uses the term *clusters* to refer to the groups being compared (Ethnic groups, in Section 4.3). In the case of dotplots (and, individual value plots and boxplots) Minitab uses the term *groups* ("With Groups", "… for grouping", and so on). ■

4.4 Comparing Individual Value Plots

The procedure for comparing individual value plots of attendance for each league is similar to that for comparing dotplots.

- Choose **Graph > Individual Value Plot** and double-click *One Y*, *With Groups*
- Double-click **C5 Attend** as the "Graph Variables" and **C2 League** as the "Categorical variables for grouping (1-4, outermost first)"
- Click **Labels** and type **Individual Value Plots of Attendance for Each League** as the "Title"
- Click **OK** twice

The resulting individual value plots are shown in Figure 4-10.

FIGURE 4-10

Individual value plots for attendance by league

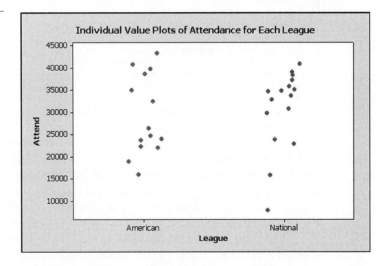

◆ **Note** As mentioned in the last tutorial, each time you construct an individual value plot you will obtain a different random scatter of the same values, unless you set a base value immediately before you generate the plot. ■

These plots present another way to easily compare the two sets of attendance. You observe the same characteristics of attendance for each league that you observed with the dotplots.

4.5 Comparing Boxplots

You construct boxplots to compare attendance for each league in a similar fashion.

- Choose **Graph > Boxplot** and double-click ***One Y, With Groups***
- Double-click **C5 Attend** as the "Graph Variables" and **C2 League** as the "Categorical variables for grouping (1-4, outermost first)"
- Click **Labels** and type **Boxplots of Attendance for Each League** as the "Title"
- Click **OK** twice

The resulting side-by-side boxplots are shown in Figure 4-11.

FIGURE 4-11

Side-by-side boxplots for attendance by league

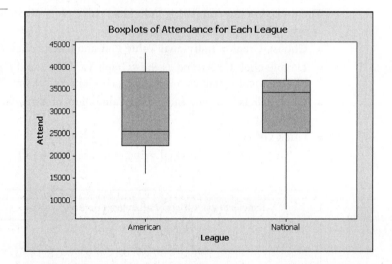

These plots provide more useful information for comparing attendance in each league than the dotplots or individual value plots. You can see that the attendance for the American League is slightly right-skewed based upon the location of its median, while the attendance for the National League is left-skewed based upon the location of its median and the lengths of its whiskers. You also observe that there are no outlying attendance figures for teams in either league. (Recall that in boxplots outliers are represented by asterisks beyond the whiskers.) As you can see, boxplots are an effective way to compare entire distributions.

When you filled in the Boxplot - One Y, With Groups dialog box, you selected Attend as the graph variable and League as the categorical variable for grouping. When multiple boxplots are compared in this way, the graph variable is always a quantitative variable. Usually, but not always, the categorical variable for grouping will be a qualitative variable, similar to League. The grouping variable can be quantitative, though it will typically take only a small number of numerical values.

Recall that each boxplot is based upon a five-number summary. (Minimum, Q1, Median or Q2, Q3, and the Maximum). You can obtain the value of three of these numbers (Q1, Median, and Q3) for each boxplot by obtaining a data tip on the boxplot.

■ Hold the cursor over any part of the boxplot for the American League

The data tip reveals that, for the American League "Q1 = 22234.5, Median = 25523.5, Q3 = 38960", and the "IQRange = 16725.5". Recall that all these values refer to numbers of people (attendance). Three of the five numbers in the five-number summary are given, but not the minimum or the maximum.

In the next section obtain these same three numbers in a different way as well as the minimum and maximum attendance.

4.6 Describing Subgroups

In the last tutorial you obtained descriptive statistics for one quantitative variable. You can obtain similar statistics for all of the quantitative variables in *BallparkData.mtw*.

■ Choose **Stat > Basic Statistics > Display Descriptive Statistics**

■ Double-click **C3 ParkBlt, C4 Capacity, C5 Attend**, and **C6 WinPct** as the "Variables"

■ Click **OK**

Minitab automatically displays the statistics you requested in the Session window as shown in Figure 4-12.

FIGURE 4-12

Default descriptive statistics for four quantitative variables in BallparkData.mtw

```
MTB > Describe 'ParkBlt' 'Capacity' 'Attend' 'WinPct';
SUBC>    Mean;
SUBC>    SEMean;
SUBC>    StDeviation;
SUBC>    QOne;
SUBC>    Median;
SUBC>    QThree;
SUBC>    Minimum;
SUBC>    Maximum;
SUBC>    N;
SUBC>    NMissing.
```

Descriptive Statistics: ParkBlt, Capacity, Attend, WinPct

Variable	N	N*	Mean	SE Mean	StDev	Minimum	Q1	Median	Q3
ParkBlt	30	0	1977.2	4.47	24.5	1912.0	1966.0	1984.5	1995.3
Capacity	30	0	47451	1402	7680	33871	41243	46782	50466
Attend	30	0	30062	1621	8880	7935	23479	32616	37534
WinPct	30	0	0.5001	0.0147	0.0805	0.3830	0.4200	0.5075	0.5575

Variable	Maximum
ParkBlt	2001.0
Capacity	66307
Attend	43362
WinPct	0.7160

The averages for each of the four variables are 1977.2 for the year that a park was built, 47,451 for the capacity of a park, 30,052 for the attendance at a park, and .5001 for the winning percentage. The standard deviations of these same variables are 24.5, 7680, 8880, and 0.0805, respectively. How do these values differ if you look at the American League and the National League separately? You are particularly interested in any difference in the attendance statistics. It is simple to obtain a description of Attend for each league and to obtain boxplots of Attend for each league.

- Click the **Edit Last Dialog** toolbar button 🔲
- Press F3 to clear previous entries
- Double-click **C5 Attend** as the "Variables"
- Double-click **C2 League** as the "By variable"
- Click **Graphs** and click the **Boxplot of data** check box
- Click **OK** twice

The built-in boxplots of Attend by League appear. The display is identical to that obtained in the previous section, except for its title.

To see the numerical output in this case, return to the Session window:

- Click the **Session Window** toolbar button 🔲

The output is shown in Figure 4-13. You may need to scroll up to see all of the output.

FIGURE 4-13

Descriptive statistics for Attend by League

```
MTB > Describe  'Attend';
SUBC>    By 'League';
SUBC>    Mean;
SUBC>    SEMean;
SUBC>    StDeviation;
SUBC>    QOne;
SUBC>    Median;
SUBC>    QThree;
SUBC>    Minimum;
SUBC>    Maximum;
SUBC>    N;
SUBC>    NMissing;
SUBC>    GBoxplot.
```

Descriptive Statistics: Attend

Variable	League	N	N*	Mean	SE Mean	StDev	Minimum	Q1	Median
Attend	American	14	0	29150	2394	8958	16026	22235	25524
	National	16	0	30861	2256	9026	7935	25277	34240

Variable	League	Q3	Maximum
Attend	American	38960	43362
	National	36907	40877

The output consists of all of the summary statistics referred to previously but for each league separately. The average attendance in the National League (30861) exceeds that in the American League by about 1,700. There is slightly more variability in the National League attendance (StDev = 9026) than in the American League (StDev = 8958). You may also obtain the minimum and maximum attendance of each league from this output. (Recall that you were not able to obtain these statistics by opening a data tip on the boxplots.)

In general, a description of a first variable by a second variable consists of a separate set of summary statistics for the first variable for the group defined by each value of the second variable. Usually the second variable, in this case, League, is either qualitative or quantitative with just a small number of numeric values.

▶ **Note** Recall that sometimes you need to be able to save the descriptive statistics, for example, for use in a later calculation. In this case, you can use the Stat > Basic Statistics > Store Descriptive Statistics command instead of the Stat > Basic Statistics > Display Descriptive Statistics command. ■

Another approach for describing subgroups is to choose the command Stat > Tables > Descriptive Statistics. For example, to obtain only the mean attendance and standard deviation of attendance for each league by this method:

- Choose **Stat > Tables > Descriptive Statistics**
- Double-click **C2 League** as the "Categorical variables: For rows"
- Click the "Display summaries for" **Associated Variables** button
- Double-click **C5 Attend** as the "Associated variables"
- Click the "Display" **Means** and **Standard deviations** check boxes
- Click **OK**

The completed dialog box is as shown in Figure 4-14.

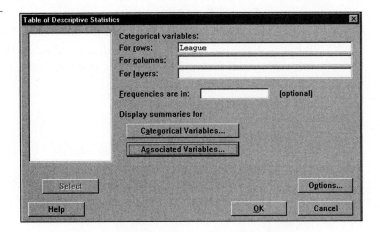

- Click **OK**

The resulting table is shown in Figure 4-15.

```
MTB > Table 'League';
SUBC>    Layout 1 0;
SUBC>    DMissing 'League';
SUBC>    Means 'Attend';
SUBC>    StDev 'Attend';
SUBC>    Counts.
```

Tabulated statistics: League

Rows: League

	Attend Mean	Attend StDev	Count
American	29150	8958	14
National	30861	9026	16
All	30062	8880	30

The table contains the means and standard deviations for attendance for each league. It also contains these statistics for both leagues combined in the All row. These statistics for the two leagues combined were not present in the output generated by the Stat > Basic Statistics > Display Descriptive Statistics command with the By variables option. Having these All statistics is a distinct advantage of this method.

At this point, save your project again.

- In the location where you are saving your work, save the work you have done in this tutorial in the project *T4.mpj*

4.7 Using Charts to Display Descriptive Statistics

Minitab's Graph > Bar Chart command can be used to produce many kinds of displays of summary statistics, such as sums, means, maximums, and minimums. To explore this capability, chart the mean attendance for the American and National Leagues.

- Choose **Graph > Bar Chart**
- Click the "Bars represent" **drop-down list arrow** and click **A function of a variable**

The dialog box has changed to reflect the displays that can be constructed with this option. The revised dialog box is shown in Figure 4-16.

FIGURE 4-16

The revised Bar Charts pictorial gallery

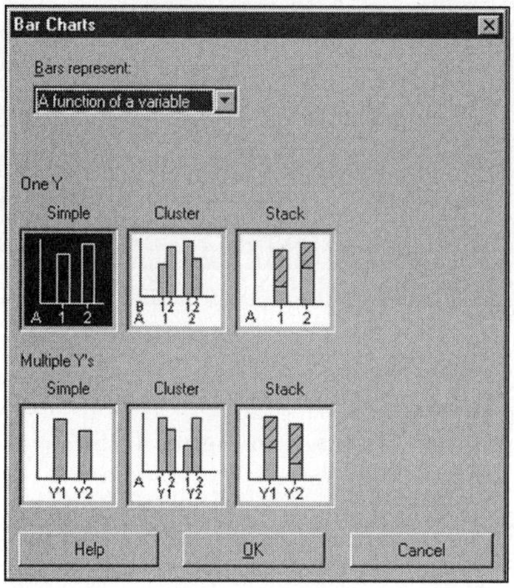

- Double-click ***One Y, Simple***

The mean is the default function so you do not need to change it.

- Double-click **C5 Attend** as the "Graph variables" and **C2 League** as the "Categorical variables"
- Click **Labels**, type **Mean Attendance Bar Chart for Each League** as the "Title", and click **OK**

The completed dialog box is shown in Figure 4-17.

FIGURE 4-17

*The completed Bar Chart -
A function of a variable,
One Y, Simple dialog box*

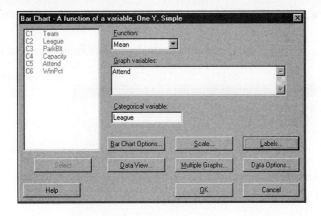

- Click **OK**

    The new graph is shown in Figure 4-18.

FIGURE 4-18

*Mean attendance bar
chart for each league*

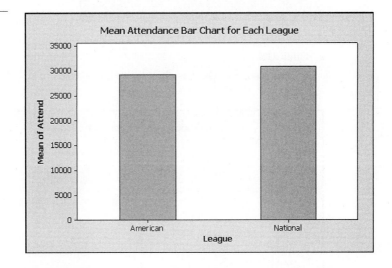

Unlike the previous bar charts in which the heights of the bars represented counts or percentages, the bars in the chart shown in Figure 4-18 represent the mean attendance for each league. The bars show that the mean attendance is slightly larger for the National League. It offers still another way to display this information.

The Y axis of this graph goes from 0 to 35000. Because it starts at the origin it is ideal for comparing the ratio of the heights between the two bars. If you wish to emphasize the difference between, rather than the ratio of, the heights of the two bars, you can edit the Y axis to go from 29000 to 31000.

- Hold your cursor over the Y axis, click the **right-hand mouse button**, and choose **Edit Y Scale** (or double-click on the Y axis)
- Clear the "Scale Range Auto" Minimum and Maximum check boxes

■ Type **29000** as the "Scale Range: Minimum" and **31000** as the "Scale Range: Maximum"

The completed Edit Scale dialog box is as shown in Figure 4-19.

FIGURE 4-19

The completed Edit Scale dialog box for Y Axis

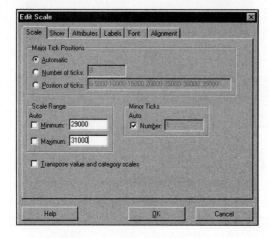

■ Click **OK**

The resulting chart is shown in Figure 4-20.

FIGURE 4-20

Mean attendance bar chart emphasizing differences between leagues

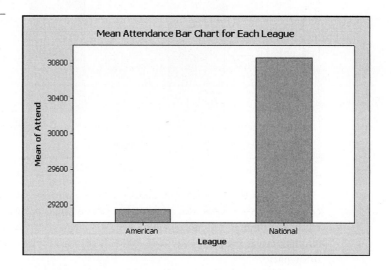

This display shows that the average attendance in the National League is approximately 1,600 more than the average attendance in the American League.

Now save the project *T4* again:

■ Click the **Save Project** toolbar button 🖫

In the next tutorial, you revisit this case study in the context of examining the relationship between two quantitative variables. The next section of this tutorial illustrates procedures for exporting graphs from Minitab. If this is of no interest to you, then you have finished this tutorial. Exit Minitab and proceed to the Command Summary.

4.8 Exporting Graphs

Suppose you wanted to incorporate the last chart into a report. In this section you will use two methods for doing this: exporting the graph into the ReportPad and exporting it into Microsoft Word. First, make your chart more suitable for presentation by introducing a more informative title, providing a better Y-axis label, and including the source of the chart's data.

- Activate the last display by clicking anywhere on the bar chart
- Hold your cursor over the title, click the **right-hand mouse button**, and choose **Edit Title: Mean Attendance Bar Chart for Each League** (or double-click on the Title)
- Type **Greater Mean Attendance for the National League** as the "Text"

The completed Edit Title dialog box is shown in Figure 4-21.

FIGURE 4-21

The completed Edit Title dialog box

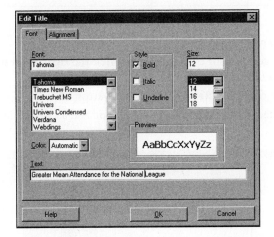

- Click **OK**

    In a similar fashion, change Attend to Attendance in the Y-axis label.
- Hold your cursor over the Y-axis label, click the **right-hand mouse button**, and choose **Edit Y Axis Label**
- Type **Mean of Attendance** as the "Text"
- Click **OK**

    Footnotes to graphs can be used to supply information, such as the source of the data or some special feature of the data. To add a footnote to indicate the source of the chart's data:

- Hold your cursor over the graph, click the **right-hand mouse button**, and choose **Add > Footnote**
- Type **Source: *BallparkData.mtw*** as the "Footnote"
- Click **OK**

    The revised graph is shown in Figure 4-22.

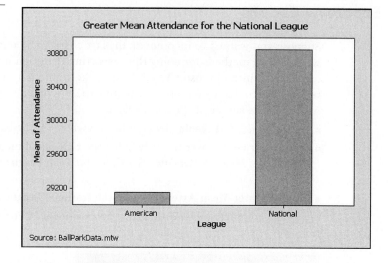

FIGURE 4-22

Revised mean attendance bar chart emphasizing differences between leagues

To place this chart into the ReportPad:

- Hold your cursor over the graph, click the **right-hand mouse button**, and choose **Append Graph to Report**
- Click the **Show ReportPad** toolbar button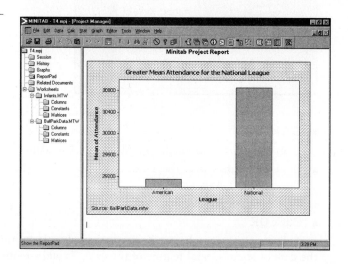

    The graph is now part of the ReportPad as shown in Figure 4-23.

FIGURE 4-23

Revised mean attendance bar chart for each league in ReportPad

It is often necessary to export a graph directly into a word-processing package. To move the last chart into Microsoft Word, for example:

- Click the **Show Graphs Folder** toolbar button 🖻 to return to the graph itself.
- Hold your cursor over the graph, click the **right-hand mouse button**, and choose **Copy Graph**

The graph is now in the Clipboard so that you can paste it into Microsoft Word or another word-processing package.

For the last time, save your project:

- Click the **Save Project** toolbar button 🖫

Now you may exit Minitab:

- Choose **File > Exit**

Congratulations! You have now finished learning how Minitab can be used to compare count and numerical summaries for different groups. In the next tutorial you will learn how to use Minitab to explore the relationships between two quantitative variables.

Minitab Command Summary

This section describes the commands introduced in, or related to, this tutorial. Use Minitab's online Help for a complete explanation of all commands.

Minitab Menu Commands

Menu	Command	Description
Stat ➤	Basic Statistics	
	➤ Display Descriptive Statistics	Provides a variety of numerical measures, including ones of central tendency and variability for one or more columns. Various built-in graph options are available. Optionally, provides the same numerical measures for each level of a specified variable.
	Tables	
	➤ Cross Tabulation and Chi-Square	Constructs contingency tables using raw data consisting of numbers or text. These tables may contain counts, row percentages, column percentages, and total percentages.
	➤ Descriptive Statistics	Generates tables containing count statistics for categorical variables and summary statistics for associated numerical variables.
Graph ➤	Dotplot	
	One Y, With Groups	Produces multiple dotplots on the same graph.
	Boxplot	
	One Y, With Groups	Produces multiple boxplots on the same graph.

Menu	Command	Description
Graph ➤	Individual Value Plot	
	One Y, With Groups	Produces multiple individual value plots on the same graph.
	Bar Chart	
	Counts of unique values, Cluster	Produces a cluster bar chart. Side-by-side bars can represent counts or percentages for categories of other variables.
	Counts of unique values, Stack	Produces a stack bar chart. Each bar can represent counts or percentages for categories of other variables.
	Bar Chart	
	A function of a variable, One Y, Simple	Side-by-side bars can represent a function of categories of another variable (such as the mean or standard deviation).

Minitab Right-Hand Mouse Button Commands

Command	Description
Edit Y Scale	Changes the scale on the Y axis of a graph.
Edit Y Axis Label	Changes the label on the Y axis of a graph.
Add ➤ Footnote	Adds a footnote to the graph.
Copy Graph	Creates a duplicate of the graph in the clipboard.

Review and Practice

Matching Problems

Match the following terms to their definitions by placing the correct letter next to the term it describes.

_____ By

_____ Dotplot

_____ Individual Value Plot

_____ Contingency table

_____ Cluster bar chart

_____ Stack bar chart

_____ Footnote

_____ ReportPad

_____ Categorical Variables

_____ Associated Variables

a. Project Manager folder that can contain graphs as well as text

b. The option under the Display Descriptive Statistics command that provides for descriptions of subgroups

c. A display, useful for comparing groups, which by default, represents values by solid circles arranged horizontally

d. A display, useful for comparing groups, which by default, represents values by solid circles arranged vertically

e. Cross Tabulation

f. Bar chart in which each bar of one qualitative variable is separated into bars based upon the distribution of another variable

g. Bar chart in which each bar for one qualitative variable is segmented based upon the distribution of another variable

h. Graph component frequently used to provide a source

i. Quantitative variables that can be summarized in a tabular display

j. Qualitative variables that can be summarized in a tabular display

True/False Problems

Mark the following statements with a *T* or an *F*.

a. ____ Charts can represent means and medians as well as counts.

b. ____ Boxplots are especially useful for comparing subgroups.

c. ____ Charts can be used to explore the relationship between two qualitative variables.

d. ____ The Cross Tabulation and Chi-Square command can be used only with text data.

e. ____ Row percents, Column percents, and Total percents are Minitab options available on the Cross Tabulation and Chi-Square dialog box that produce percentage summaries.

f. ____ Minitab graphs cannot be pasted directly into Microsoft Word.

g. ____ Cluster bar charts are usually preferred over stack bar charts because, with the former, it is easier to compare all the counts/percents for one qualitative variable.

h. ____ The Bar Chart pictorial gallery changes depending upon what the bars represent.

i. ____ The command File > Open Project is the only way to open a project in Minitab.

j. ____ The Cross Tabulation and Chi-Square command by default shows column totals.

Practice Problems

The practice problems instruct you to save your worksheets with a filename prefixed by P, followed by the tutorial number. In this tutorial, for example, use *P4* as the prefix. If necessary, use Help or Appendix A Data Sets, to review the contents of the data sets referred to in the following problems. Interpretations should use the language of the subject matter of the question. If you are using the Student version, be sure to close worksheets when you have completed a problem.

1. Open *EmployeeInfo.mtw*.

   a. Use the Stat > Tables > Tally Individual Variables command to obtain the counts and percentages for the two qualitative variables. How many members of the administration are there? How many Hispanic employees are there? What percentage of the employees is Hispanic? What is the modal category for each variable?

   b. Use the Stat > Tables > Cross Tabulation and Chi-Square command to obtain a contingency table with Job Status as the row variable and Ethnicity as the column variable. What is the modal category for each variable? (Because there are no missing data in this data set, your answers should agree with those in part a.) What do the numbers 3, 303, 36, and 788 represent?

   c. Use the Stat > Tables > Cross Tabulation and Chi-Square command to obtain the percentage of each ethnic group for each job category.

   d. Construct a percentage bar chart of ethnic group clustered by job category.

   e. Write a brief account comparing the distribution of whites across the job categories.

   f. Use the Stat > Tables > Cross Tabulation and Chi-Square command to obtain the count and corresponding percentage for each combination of the categories between the two variables. Explain how you obtained these percentages. What is the modal combination? What is its corresponding percentage?

2. Open *Survey.mtw*.

   a. Change the data type of the variables Exercise and Smoke from numeric to text.

b. Obtain a contingency table with percentages that enables you to compare the percentage of smokers that exercise with the percentage of non-smokers that exercise. What are these respective percentages? In what sense are these values surprising?

c. Construct a percentage stack bar chart showing the result in part b.

3. Open *Compliance.mtw*. You want to construct a display to represents the percentage distribution of educational level for each employment status.

a. Construct a contingency table to represent these percentages.

b. Construct a cluster bar chart to represent these percentages.

c. Construct a stack bar chart to represent these percentages.

d. Which of these three displays do you prefer? Explain briefly.

e. For your preferred method of display provide a brief written explanation of what the display illustrates.

4. Open *MBASurvey.mtw*. You are interested in visualizing the percentage of men and women within each highest degree category.

a. Construct a percentage bar chart of the highest degree with gender as the cluster variable.

b. Construct a percentage bar chart of gender with the highest degree as the cluster variable.

c. Construct a percentage bar chart of the highest degree with gender as the stack variable.

d. Construct a percentage bar chart of gender with the highest degree as the stack variable.

e. Which of these four charts do you prefer? Explain briefly.

f. For your preferred method of display provide a brief written explanation of what the chart displays.

5. Open *PulseA.mtw*

a. Construct a contingency table with Smoke as the row variable and Gender as the column variable. Interpret the counts 8 and 20 in the table. Criticize the following comment: "Because there are two-and-a-half times as many male smokers as female smokers, males are two-and-one-half times more likely to be smokers than are females."

b. Add percentages to the table you obtained in part a, so that you can compare the percentage of males that smoke with the corresponding percentage of females that smoke. What is the difference in percentages?

c. Construct (i) a percentage clustered bar chart of Smoke with Gender as the cluster variable, and (ii) a percentage stack bar chart of Smoke with separate bars for each gender. Which display do you prefer? Explain briefly.

d. Use the Data > Split Worksheet command to split the *PulseA* worksheet into two separate worksheets, one for those that ran in place and one for those that did not.

The following questions apply only to those who ran in place.

e. Use the Calc > Calculator command to obtain, in C9 Increase, the difference between the second and the initial pulse rates (Pulse2 – Pulse1).

f. Obtain boxplots of Increase for each Gender.

g. Obtain the default descriptive statistics of Increase by Gender.

h. Write a brief account of how the increase in pulse rate, after running in place, varies by gender. Your account should discuss measures of central tendency, measures of variability, the shapes of the two distributions, and any outliers.

i. Use the Stat > Tables > Descriptive Statistics command to obtain the mean, median, and standard deviations of Increase for each gender. If you are interested in just these three statistics, what is the advantage of using this approach over the technique you used in part g?

6. Open *Rivers.mtw*.

a. Determine if the median temperature of the river is different at site 2, which is directly up river from the power plant, than at site 3, which is directly down river from the power plant's discharge from its cooling towers. (You may unstack the Temp column to create new columns that contain site 2 and site 3.)

b. Construct two dotplots to examine whether or not the two populations have the same shape. Explain your display.

7. Open *Baby.mtw*.

a. Use the Stat > Basic Statistics > Display Descriptive Statistics command to compute the means and standard deviations for the Time variable for each design. Describe the differences among the means. Describe the difference among the standard deviations.

b. Use the Stat > Basic Statistics > Display Descriptive Statistics command to construct individual value plots of the variable Time for each design. Do these plots support your answers in part b? Explain briefly.

8. Open *BallparkData.mtw*.

a. Construct a new variable equal to the ratio of attendance to capacity for each team expressed as a percentage. Round this percentage to the nearest whole number. Which team has the highest rounded percentage? Which team has the lowest rounded percentage? Which league has the highest average rounded percentage? Does the same league have the highest median rounded percentage?

b. Construct a bar chart with a bar for each league, where the height of the bar is the mean of the variable ParkBlt. Do you find the chart helpful? Explain any problems. Change the scale on the Y axis so that it runs from 1970 to 1980. Is this new display an improvement? Explain briefly.

c. Use the Calc > Calculator command to construct a new variable that is the age of the ballpark in 2002, i.e., 2002 − ParkBlt. Use the Stat > Tables > Descriptive Statistics command to compare the mean age and the standard deviation of age for the two leagues. Summarize your findings in a couple of sentences.

d. Obtain dotplots, individual value plots, and boxplots to compare the distribution of the ages of the parks for the two leagues. Which plot do you prefer? Explain briefly.

e. Copy the plot you preferred in part d into ReportPad. Also copy into ReportPad, your text output from part c. In ReportPad, write a couple of sentences that summarize how the age of the ballpark varies with each league.

f. Repeat part e except copy the graph and the text into Microsoft Word. Write your summary in Word.

g. Did you prefer working with ReportPad or Microsoft Word? Explain briefly.

9. Open *Note02.mtw*.

a. What are the mean and median prices of the laptops for each possible speed? Explain how you obtained these statistics.

b. Use another command to obtain the same two statistics. Explain how you obtained these statistics. Why did you decide to use the command in part a?

c. Construct a bar chart to display the median prices of the laptops for each possible speed.

d. Explain why a reader cannot easily distinguish between the two median prices in your display.

e. Alter your display so that it emphasizes the difference in the median prices. Explain how you changed your display.

f. What are the standard deviations of the prices for each possible speed? Explain why you obtained a missing value for one of the speeds.

g. Construct a bar chart to display these standard deviations. Are you concerned with this display? Explain briefly.

10. Open *TwoTowns.mtw*.

a. Construct dotplots of the price of a home for each town. Based upon these displays which town has the highest median home price? Explain briefly.

b. Construct individual value plots of the price of a home for each town. Based upon these displays which town has the highest median home price? Explain briefly.

c. Construct boxplots of the price of a home for each town. Based upon these displays which town has the highest median home price? Explain briefly.

d. What are the median home prices for each town? Explain which display was most helpful in visualizing those statistics?

e. Edit that display so that it is more appropriate for a report by changing the title, including the units for which price is measured, and including a source. Explain how you performed your edits.

f. Export your edited display to the ReportPad. Explain how you exported the display.

g. Export your edited display to a word-processing package. Explain how you exported the display.

On Your Own

In this tutorial you used the Stat > Tables > Cross Tabulation and Chi-Square command to construct a contingency table from the data in file *Infants.mtw*. Another name for a contingency table is a *two-way table*. The Stat > Tables > Cross Tabulation and Chi-Square command can also be used to produce a one-way table for one qualitative variable (similar to output from the Stat > Tables > Tally Individual Values command) and multiway tables for three to ten qualitative variables. In addition, it can be used to produce separate tables for each category of another qualitative variable, defined as a *layering* variable. For further information about these command options, check the Help in the command's dialog box.

Open *Infants.mtw*. Consider the following three quantitative variables: Ethnic, Smoke, and BreastFed. Construct one-way tables for each of these variables. Construct two-way tables for each pair of these variables. Construct multiway tables with Ethnic and Smoke as the row variables and BreastFed as the column variable, with Ethnic and BreastFed as the row variables and Smoke as the column variable, and with Smoke and BreastFed as the row variables and Ethnic as the column variable. Construct separate two-way tables of Ethnic and Smoke for each category of BreastFed. Which of these displays best explains the relationship between ethnicity and smoking status for each category of breast feeding? Explain briefly. Use your "best" display to compare the relationship between ethnicity and smoking status for each category of breast-feeding.

Answers to Matching Problems

(b) By
(c) Dotplot
(d) Individual Value Plot
(e) Contingency table
(f) Cluster bar chart
(g) Stack bar chart
(h) Footnote
(a) ReportPad
(j) Categorical variables
(i) Associated variables

Answers to True/False Problems

(a) T, (b) T, (c) T, (d) F, (e) T, (f) F, (g) T, (h) T, (i) F, (j) T

Examining Relationships Between Two Quantitative Variables

In this tutorial you will construct graphs and compute statistics designed to examine the relationship between two quantitative variables. As a special case, one of the two variables will be time. Also in this tutorial, you will learn how to export files to and import files from Excel.

OBJECTIVES

In this tutorial, you learn how to:

- Plot one quantitative variable against another using scatterplots
- Use the brushing palette to obtain information about selected points
- Introduce a qualitative variable into a scatterplot
- Create separate scatterplots on the basis of a third variable
- View the History folder
- Obtain marginal plots
- Obtain the covariance and the correlation for two quantitative variables
- Obtain the equation of the regression line and associated results
- Create a fitted line plot
- Create a scatterplot involving a time series
- Construct multiple plots on the same graph
- Export a Minitab worksheet as a text file and as an Excel spreadsheet
- Import a text file and an Excel spreadsheet into a Minitab worksheet

5.1 Creating Scatterplots

> **CASE STUDY**
>
> ## SPORTS—BASEBALL STADIUMS (CONTINUED)
>
> You've been asked to write an article about Major League Baseball ballparks and their effects on the success of the teams that play in them. You've gathered some basic data about the capacity of each of the 30 parks and when each was built, as well as the average attendance at each park for each event in 2001 and the team performance in 2001. You intend to continue your article by analyzing several quantitative variables.

In Tutorial 4, you compared the American and National Leagues on a variety of quantitative variables. Now you want to explore the relationship between pairs of these quantitative variables. Specifically, you want to begin by looking at the relationship, if any, between the age of the ballpark and its capacity.

Begin by opening Minitab:

- Open **Minitab** and maximize the Session window
- If necessary, **Enable Commands** and clear Output Editable from the Editor menu

Recall that Minitab automatically opens a new, untitled project each time you start.

- Click the **Save Project** toolbar button ■ and, in the location where you are saving your work, save this project as *T5.mpj*
- Open the worksheet *BallparkData.mtw*

First, create a new variable that is the ages of the ballparks in 2002:

- Choose *Calc > Calculator*
- Type **AgePark** as the "Store result in variable" column
- Enter, by clicking or typing, **2002 - ParkBlt** as the "Expression" and click **OK**
- Click the **Current Data Window** toolbar button ▦

The new variable, AgePark, is stored in C7.

Generally, the best way to begin an exploration of the relationship between two quantitative variables is to construct a *scatterplot*. This is a two-dimensional graph in which each pair of values is represented by a symbol. To obtain a scatterplot of Capacity against AgePark:

- Choose **Graph > Scatterplot** and double-click *Simple*
- Double-click **C4 Capacity** as the "Y variables" column for Graph 1
- Double-click **C7 AgePark** as the "X variables" column for Graph 1
- Click **Labels**, type **Scatterplot of Capacity Against the Age of the Ballpark** as the "Title", and click **OK**

The completed Scatterplot - Simple dialog box is shown in Figure 5-1.

FIGURE 5-1

*The completed
Scatterplot - Simple
dialog box*

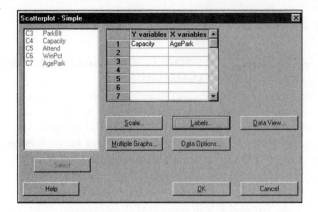

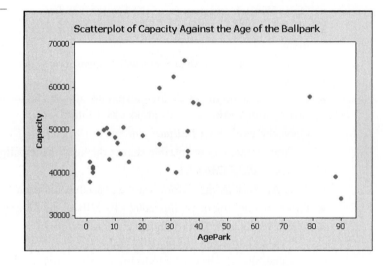

- Click **OK**

   The resulting scatterplot should appear as in Figure 5-2.

FIGURE 5-2

*Scatterplot of Capacity
against AgePark*

   At first glance it is hard to discern a pattern in the plot. However, if you leave aside the two teams in the lower right-hand corner of the plot, you can see that the remaining points form a cloud of points from lower left to upper right—suggesting a moderate positive linear relationship. Newer parks tend to have a smaller capacity than older parks. Also, you can see that no parks were built between about 45 years and 75 years ago.

- Hold your cursor over the lowest right-hand point to identify the row

   The data tip reveals that this point corresponds to row 3. The age of the park is 90 years and the capacity, 33,871. Can you guess which team this is?

   In displays that show individual points, data tips identify the row number associated with the point and the corresponding value (or values, in this case) of the variable (or variables) in the display. If you want more information about the point (in this case, the team), you need to return to the worksheet and search the appropriate row. Minitab has a more sophisticated tool called brushing that allows you to (i) select more than a single point to review and (ii) choose which variables will be used to identify the point(s). As an example, you will identify the teams associated with the three oldest ballparks:

■ Hold your cursor over the scatterplot, click the **right-hand mouse button,** and choose **Brush** (or choose Editor > Brush)

Moving your cursor over the scatterplot note that it has changed to a pointing finger. In addition, a small window appears in the upper left-hand corner of the plot. This is the *brushing palette*. It presents information about the points you select. The word *Row* that appears is the default piece of information that always appears.

Have the brushing palette identify the name of the team in addition to ParkBlt and AgePark:

■ Hold your cursor over the scatterplot, click the **right-hand mouse button**, and choose **Set ID Variables** (or choose Editor > Set ID Variables)

■ Double-click **C1 Team**, **C3 ParkBlt**, and **C7 AgePark** as the "Variables"

The completed Set ID Variables dialog box is shown in Figure 5-3.

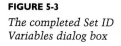

FIGURE 5-3

The completed Set ID Variables dialog box

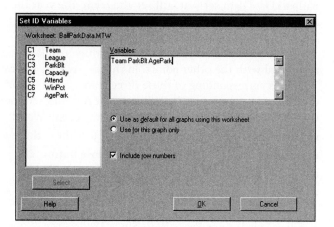

■ Click **OK**

The brushing palette is now much wider in order to accommodate the three variables.

■ Drag the pointing finger to form a box around the three points representing the oldest ballparks

The brushing palette indicates that the three teams are the Boston Red Sox, the New York Yankees, and the Chicago Cubs. The scatterplot and the brushing palette are shown in Figure 5-4.

FIGURE 5-4

Brushed scatterplot of
Capacity against AgePark
showing the Brushing
Palette

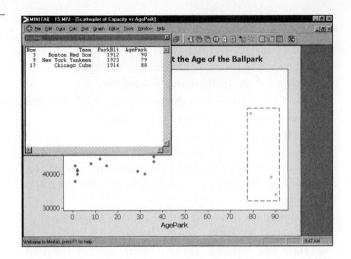

▶ **Note** Notice from the Set ID Variables dialog box in Figure 5-3 that your choice of variables is also used as the default for all graphs based upon the current worksheet. That is, if you brushed any other graphs based upon the data in the BallparkData data set you will see the team, the year the park was built, and the age of the park in 2002. ■

The points corresponding to these three teams do not fit in with the pattern associated with the other points. In this sense, they are called *bivariate outliers*. The rows corresponding to these three teams are 3, 9, and 17. You would like to examine a scatterplot of Capacity against AgePark without these three outliers, that is, for parks built since the end of the Second World War.

- Click anywhere in the scatterplot to clear these three points
- Click the brushing palette window **Close** button ☒
- Choose **Graph > Scatterplot** and double-click *Simple*
- Verify that Capacity is the "Y variables" column and AgePark is the "X variables" column
- Click **Labels**, type **Parks Built Since 1945** as the "Subtitle 1", and click **OK**
    Now exclude the three prewar parks.
- Click **Data Options** and click the **Specify which rows to exclude** option button
- Click the "Specify Which Rows To Exclude" **Row numbers** option button
- In the "Row numbers" text box, type **3 9 17**

The completed Scatterplot - Data Options dialog box should appear as shown in Figure 5-5.

FIGURE 5-5

The completed
Scatterplot - Data Options
dialog box

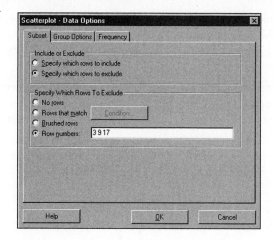

- Click **OK** twice

The new scatterplot, without the older parks and with the subtitle, is shown in Figure 5-6.

FIGURE 5-6

Scatterplot of Capacity
against AgePark without
the three oldest parks and
with a subtitle

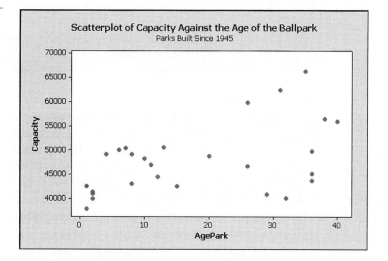

The plot shows a generally positive but weak relationship between the age of the park and capacity. You will further explore the relationship between these two variables later in this tutorial.

One other relationship that interests you is that between attendance (C5 Attend) and the team's winning percentage (C6 WinPct). To obtain a scatterplot of Attend against WinPct:

- Click the **Edit Last Dialog** toolbar button
- Press F3 to clear the previous settings

- Double-click **C5 Attend** as the "Y variables" column and **C6 WinPct** as the "X variables" column
- Click **Labels**, type **Scatterplot of Attendance Against Winning Percentage** as the "Title", and click **OK** twice

The resulting plot is shown in Figure 5-7. The scatterplot suggests a moderate positive linear relationship between the two variables. Not surprisingly, attendance tends to increase as a team's performance improves.

FIGURE 5-7

Scatterplot of attendance against winning percentage

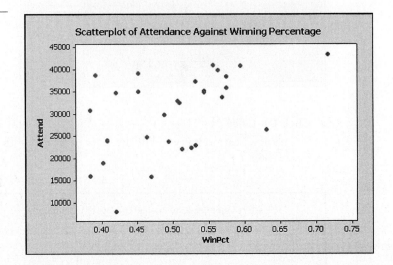

Using Crosshairs

The values marked on the X- and Y-axes of the scatterplot of attendance against winning percentage enable you to get only an approximate sense of the X and Y values for any particular point on the plot. Minitab offers a procedure, *crosshairs*, which allows the user to find the exact coordinates (X and Y values) for any point on the plot. For example, you can easily locate the point on the plot corresponding to $\overline{X}$, $\overline{Y}$. To obtain the crosshairs:

- Hold your cursor over the scatterplot, click the **right-hand mouse button**, and choose **Crosshairs** (or choose Editor > Crosshairs)

You can see the crosshairs—moving horizontal and vertical lines—as you move your cursor over the scatterplot. As you move the cursor the coordinates of the point where the lines cross are recorded in the data tip in the upper left-hand corner of the plot. Recall from Tutorial 4 that the average winning percentage was .5001 and the average attendance was 30,062. To locate the point (.5001, 30,062) on the scatterplot:

- Move the cursor until the coordinates in the data tip are as close as you can get to .5001, 30,062. The authors' best effort is shown in Figure 5-8.

FIGURE 5-8

Scatterplot of attendance against winning percentage, showing the crosshairs

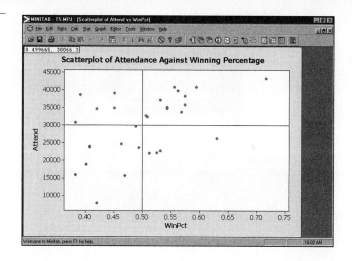

The significance of this point in the scatterplot is that for a positive relationship you would expect the majority of points to fall in the upper right-hand quadrant or the lower left-hand quadrant formed by the crosshairs. You can easily confirm that in this case 21 of the 30 points lie in these two quadrants. (When you count the number of points in the lower left-hand quadrant, notice that two points overlap substantially; both have winning percentages a little over 0.400 and an average attendance of approximately 24,000.)

- Hold your cursor over the scatterplot, click the **right-hand mouse button**, and click **Select** to clear the crosshairs.

5.2 *Adding a Grouping Variable to a Scatterplot*

Is the positive relationship between attendance and winning percentage the same for teams that are in the American League and in the National League? You can examine this question graphically by labeling the points on the scatterplot according to the league to which the team belongs.

- Choose **Graph > Scatterplot** and double-click *With Groups*
- Double-click **C5 Attend** as the "Y variables" column for Graph 1 and **C6 WinPct** as the "X variables" column for Graph 1
- Double-click **C2 League** as the "Categorical variables for grouping (0-3)" value
- Click **Labels**, type **Scatterplot of Attendance Against Winning Percentage By League** as the "Title", and click **OK**

At this point, the completed Scatterplot - With Groups dialog box should look as shown in Figure 5-9.

FIGURE 5-9

*The completed
Scatterplot - With Groups
dialog box*

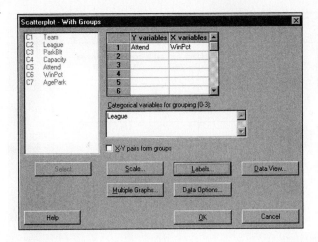

■ Click **OK**

The resulting plot is shown in Figure 5-10.

FIGURE 5-10

*Scatterplot of attendance
against winning
percentage by league*

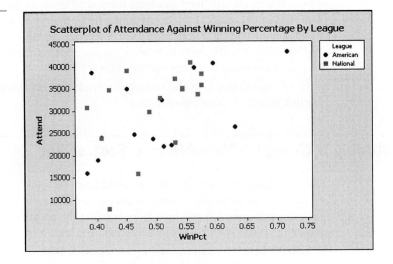

As the legend on the upper right of the graph indicates, Minitab plots American League teams with a black circle and National League teams with a red square. The plot suggests that for both leagues, there is a similar positive linear relationship between attendance and winning percentage.

In this context, the variable League is called a *grouping variable*. A grouping variable is used to define different groups, in this case the American and National Leagues. Again, in this context, a grouping variable may be either text or numeric. As you can see in Figure 5-9, Minitab will accept up to three grouping variables. (Using a grouping variable is comparable to your use of cluster and stack variables in the last tutorial.)

Paneling a Scatterplot

The procedure above involves selecting the grouping variable before the scatterplot is constructed. An alternative approach is to create the scatterplot first and then create separate plots on the basis of a grouping variable. Minitab calls this process *paneling*. To explore this approach, begin with the original scatterplot of Attend against WinPct:

- Choose **Graph > Scatterplot** and double-click *Simple*
- Verify that Attend is the "Y variables" column and WinPct is the "X variables" column and click **OK**

The resulting scatterplot should appear as shown in Figure 5-7. Now, create two separate plots based upon league:

- Hold your cursor over the scatterplot, click the **right-hand mouse button**, and choose **Panel** (or choose Editor > Panel)
- Double-click **C2 League** as the "By variables with groups in separate panels" column

The completed Edit Panels dialog box should look as shown in Figure 5-11.

FIGURE 5-11

The completed Edit Panels dialog box

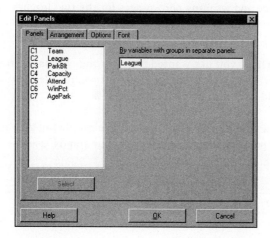

- Click **OK**
- At the warning "This change cannot be undone. Do you want to continue?", click **OK**

The warning here refers to the fact that you cannot return to the original graph. You can, however, always recreate it. You could also save the graph before issuing this command.

The paneled graphs are shown in Figure 5-12.

FIGURE 5-12

*Panels showing the
scatterplot of attendance
against winning
percentage for each league*

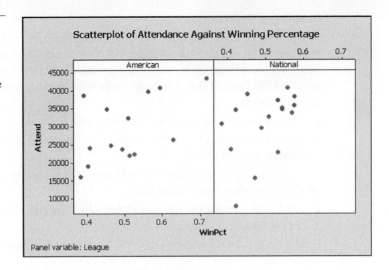

These separate plots show the similar positive linear relationship between attendance and winning percentages for the two leagues more clearly than the single plot with different symbols did.

▶ **Note** Had you not wanted to retain the original scatterplot of attendance against winning percentage, you could have recalled that graph from the Graphs folder and split that graph into panels. ■

You have generated a considerable number of graphs thus far in this tutorial. If at any point you want to close all your graphs, use the Close All Graphs toolbar button ▨.

5.3 Viewing the History Folder

As you progress through a complex series of statistical analyses in Minitab, it is easy to lose track of what you have done. For this reason Minitab includes a History folder as a component of the Project Manager. Before continuing with your examination of relationships among the ballpark variables, view this folder.

■ Click the **Show History** toolbar button ▷

The Session window contains Minitab commands and the resulting output. The History folder records only the commands; it does not show the output. For example, the History folder displays the Session command corresponding to the last command you issued:

```
Plot 'Attend'*'WinPct';
  Title "Scatterplot of Attendance Against Winning Percentage";
  Symbol.
```

The History folder can help you in a number of ways. It allows you to retrace your analysis of a data set. Also, you can copy commands from either the

History folder or the Session window and paste them into the Command Line Editor. There, you can modify these commands and submit the modified commands for execution.

For example, suppose you want a scatterplot of attendance against ballpark capacity (C4 Capacity) instead of against winning percentage. You could use the menu commands as before, but using the Command Line Editor is often quicker.

- In the History folder, drag you cursor over the **last three lines** in order to highlight them (you may have to drag from the right-hand edge of the last commands)

Figure 5-13 shows the current History folder.

FIGURE 5-13

The current History folder

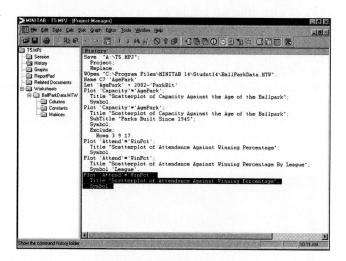

◆ **Note** Your History folder will match that in Figure 5-13 only if you have exactly followed all the steps to this point in the tutorial. ■

- Choose **Edit > Command Line Editor**
- In the first line change **WinPct** to **Capacity**; in the second line change **Winning Percentage** to **Ballpark Capacity** (be sure to leave the quotation marks in place)
- Click **Submit Commands**

The new scatterplot appears in a new graph window.

◆ **Note** The History folder records nearly all the commands you issue, but does not record other changes to the project such as changes to the data set or changes to the value order for a variable. ■

◆ **Note** The Session folder within the Project Manager provides both the commands and the corresponding text output found in the Session window. Use the Show Session Folder toolbar button to view this folder. ■

After this brief application of the History folder, continue with your exploration of the relationship between attendance and winning percentage.

A *marginal plot* combines the features of a scatterplot with some of the one-variable graphs you generated in Tutorial 3. You can examine the relationship between two variables while also viewing the distribution of each variable, all on the same graph.

To create a marginal plot of Attend against WinPct with boxplots on each axis:

■ Choose **Graph > Marginal Plot**

Notice, from the pictorial gallery, that you can produce Marginal Plots with histograms, boxplots, or dotplots on the margins. (You cannot obtain Marginal Plots with different graphs on the two margins.)

■ Double-click ***With Boxplots***

■ Double-click **C5 Attend** as the "Y variable" column

■ Double-click **C6 WinPct** as the "X variable" column

■ Click **Labels**, type **Marginal Plot of Attendance Against Winning Percentage** as the "Title", and click **OK**

At this point, the completed Marginal Plot - With Boxplots dialog box should look as shown in Figure 5-14.

FIGURE 5-14

The completed Marginal Plot - With Boxplots dialog box

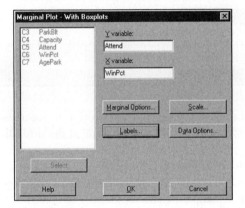

■ Click **OK**

The resulting plot is shown in Figure 5-15.

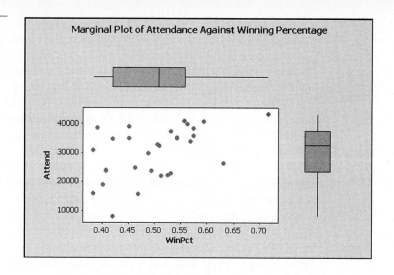

On the top and right margins of the scatterplot, Minitab places a boxplot of the X (WinPct) values and the Y (Attend) values, respectively. The boxplot for Attend is skewed to the right. The shape of the boxplot of WinPct is not as clear. The middle 50% of values (within the box) are skewed to the left, but the right whisker is longer than the left whisker.

5.5 Computing Covariance

The various scatterplots of attendance against winning percentage that you have generated suggest a moderate positive linear relationship between these two variables. The *covariance* is a numerical measure of the strength of the linear relationship between two quantitative variables. To compute the covariance between C5 Attend and C6 WinPct:

- Choose **Stat > Basic Statistics > Covariance**
- Double-click **C5 Attend** and **C6 WinPct** as the "Variables" and click **OK**

The resulting output appears in the Session window and should appear as shown in Figure 5-16.

FIGURE 5-16

Covariance between Attend and WinPct

```
MTB > Covariance 'Attend' 'WinPct'.

Covariances: Attend, WinPct

           Attend    WinPct
Attend   78860148
WinPct        356         0
```

The value 78,860,148 is the variance of Attend and 0, the (rounded) variance of WinPct. (Recall that variance is the square of the standard deviation.) The covariance between the two variables is 356. A positive covariance suggests that high values for one variable tend to be associated with high values for the other. However, because the value for the covariance depends on the units associated with the two variables, it is difficult to determine from the value 356 the exact strength of the relationship. You need a way of measuring the strength of the relationship between two quantitative variables that does not depend on the units associated with the variables.

Before examining such a measure in the next section, notice that Minitab computes the two variances and the covariance with equal precision. As a consequence, because Attend takes values in the range 8,000 to 45,000 and WinPct takes values in the range .3 to .8 the variance of WinPct is reported as 0. It is not really 0, and you can find the variance more accurately by using the Stat > Basic Statistics > Display Descriptive Statistics command.

- Choose **Stat > Basic Statistics > Display Descriptive Statistics**
- Double-click **C6 WinPct** as the "Variables"
- Click **Statistics**, click the **Variance** check box, and click **OK** twice

The output will include the value 0.00648, small but not 0, for the variance of WinPct.

5.6 Computing Correlation

Pearson's correlation coefficient or, simply, the *correlation coefficient*, measures the strength of the linear relationship between two quantitative variables. It does not depend on the units of the two variables and is usually designated by *r*. The correlation always lies between –1 and 1. (In fact, r is the covariance divided by the product of the standard deviations of the two variables.) Suppose you want to obtain, in addition to the correlation between WinPct and Attend, the correlation between WinPct and Capacity and between WinPct and AgePark.

- Choose **Stat > Basic Statistics > Correlation**
- Double-click **C6 WinPct, C5 Attend, C4 Capacity,** and **C7 AgePark** (in that order) as the "Variables"
- Clear the Display p-values check box

The completed Correlation dialog box should look as shown in Figure 5-17.

FIGURE 5-17

The completed Correlation dialog box

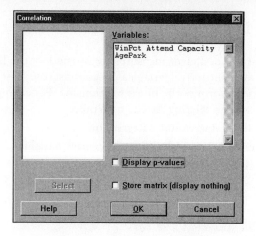

- Click OK

  The correlations appear as shown in Figure 5-18.

FIGURE 5-18

Correlation matrix for four quantitative variables

```
MTB > Correlation 'WinPct' 'Attend' 'Capacity' 'AgePark';
SUBC>    NoPValues.
```

Correlations: WinPct, Attend, Capacity, AgePark

```
            WinPct    Attend   Capacity
Attend       0.499
Capacity     0.359     0.189
AgePark      0.270     0.082      0.059
```

Cell Contents: Pearson correlation

The output consists of a triangle (half-matrix) consisting of six correlations. The ones you are most interested in are in the first column and consist of the correlation between WinPct and, in turn, Attend, Capacity, and AgePark. All three correlations are positive. Attend (attendance) is the variable most highly correlated (r = 0.499) with WinPct. The middle column of correlations shows the small correlation (r = 0.189) between Attend and Capacity, and the even smaller correlation (r = 0.082) between Attend and AgePark. Finally, in the last column, the correlation between Capacity and AgePark (r = 0.059) suggests that these two variables are almost linearly unrelated.

▶ **Note** A complete correlation matrix in this case would consist of an array with four rows and four columns of correlations—each row and column corresponding to one of the four variables. The four diagonal entries would all be one (the correlation of a variable with itself). Minitab provides only the lower left-hand section of the matrix below the diagonal entries. The upper right-hand section of the matrix would match the lower left-hand section because the correlation between X and Y is the correlation between Y and X. ■

5.7 Computing the Least Squares/Regression Line

The correlation coefficient measures the strength of the linear relationship between two quantitative variables. By contrast, the least squares/regression line summarizes the form of the linear relationship. For example, Minitab can compute the regression line relating Attend to WinPct.

- Choose **Stat > Regression > Regression**
- Double-click **C5 Attend** as the "Response" variable
- Double-click **C6 WinPct** as the "Predictors" variable

The completed Regression dialog box should look as shown in Figure 5-19.

FIGURE 5-19

The completed Regression dialog box

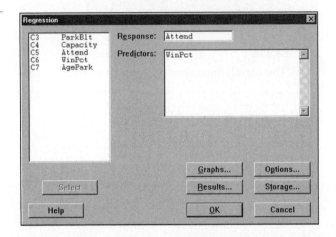

- Click **OK**

By default, Minitab provides a great deal of output for the Stat > Regression > Regression command. The first half of the output is shown in Figure 5-20.

FIGURE 5-20

Regression of attendance against winning percentage

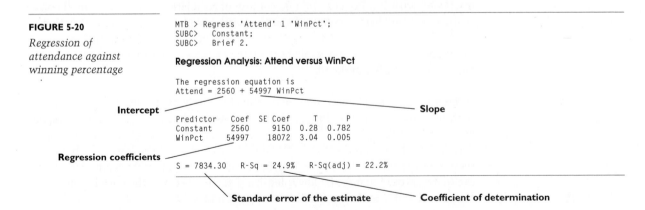

```
MTB > Regress 'Attend' 1 'WinPct';
SUBC>   Constant;
SUBC>   Brief 2.
```

Regression Analysis: Attend versus WinPct

```
The regression equation is
Attend = 2560 + 54997 WinPct
```

Intercept ——   —— Slope

```
Predictor    Coef   SE Coef     T      P
Constant     2560      9150   0.28  0.782
WinPct      54997     18072   3.04  0.005
```

Regression coefficients ——

```
S = 7834.30   R-Sq = 24.9%   R-Sq(adj) = 22.2%
```

Standard error of the estimate   Coefficient of determination

The first piece of output is the equation of the regression line, Attend = 2560 + 54997 WinPct. Predicted attendance is obtained by adding 2560 to the product of 54997 and the team's winning percentage (expressed as a proportion). For example, the predicted attendance for a team with a winning percentage of .55 is 2560 + 54997*.55 = 32808. The slope of the line, 54997, indicates that for each additional 1.00 in winning percentage predicted attendance increases by 54997. This appears absurd until you recall that winning "percentage" is recorded as a number between 0 and 1 (.463, .391, and so on). It would be more appropriate to say that an increase of .1 in winning percentage is associated with a predicted increase in attendance of 5499.7 or approximately 5500. The intercept of the line, 2560 is not helpful. It can be thought of as the predicted attendance for a team with a winning percentage of 0. Because this is not a realistic end-of-season winning percentage, it is not helpful. The intercept and the slope appear again under the heading "Coef" (which is short for *regression coefficients*) in the middle section of Figure 5-20.

Two other numbers in the output have important interpretations. On the last line in Figure 5-20, S = 7834.30 is the estimated standard deviation around the regression line. This statistic is called the *standard error of the estimate*. It can be thought of as an estimate of the standard deviation of attendance among teams with the same winning percentage. *R-Sq*, the *coefficient of determination*, is simply 100 times the square of the correlation coefficient (r). The value 24.9 indicates that approximately one quarter (24.9%) of the variation in attendance among Major League Baseball teams can be associated with its linear relationship to winning percentage. In general, the higher the R-Sq value, the better the estimated regression line fits the data. In Tutorials 10 and 11 you will return to the subject of correlation and regression. In those tutorials you will consider the interpretation of both R-Sq and R-Sq(Adj).

5.8 | Displaying the Least Squares/Regression Line

Minitab offers two simple ways of displaying the least squares/regression line. One method is based upon the scatterplot you are familiar with.

- Choose **Graph > Scatterplot** and double-click *With Regression*
- Double-click **C5 Attend** as the "Y variables" column and **C6 WinPct** as the "X variables" column
- Click **Labels**, type **Scatterplot of Attendance Against Winning Percentage** as the "Title", type **With Regression Line** as "Subtitle 1", and click **OK** twice

The resulting plot is shown in Figure 5-21.

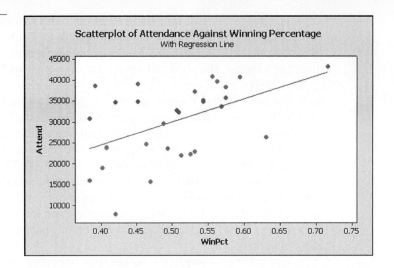

The plot in Figure 5-21 shows that there is quite a bit of variation around the regression line. Attendance at Major League Baseball games is not determined solely by winning percentage.

A second way to display the regression line is to use a *fitted line plot*:

- Choose **Stat > Regression > Fitted Line Plot**
- Double-click **C5 Attend** as the "Response (Y)" variable and **C6 WinPct** as the "Predictor (X)" variable
- Click **Options** and type **Fitted Line Plot of Attendance Against Winning Percentage** as the "Title"
- Click **OK**

The completed Fitted Line Plot dialog box should look like Figure 5-22.

FIGURE 5-22

The completed Fitted Line Plot dialog box

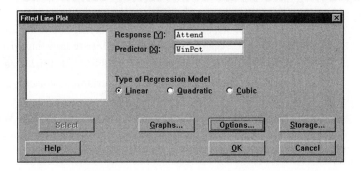

The Fitted Line Plot dialog box differs from that for other graphs. The fitted line plot is an example of a graph, designed for use in conjunction with a specific statistical technique (in this case, regression analysis). Though the default option is to fit a linear model to the data, you also have the option of fitting a quadratic or a cubic model. You will explore one of these options in later tutorials.

- Click **OK**

  The fitted line plot is shown in Figure 5-23.

FIGURE 5-23

Fitted line plot of attendance against winning percentage

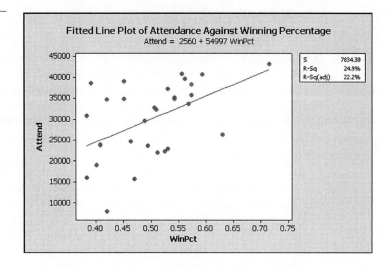

The plot is similar to that in Figure 5-21. However, in the fitted line plot the equation of the regression line, Attend = 2560 + 54997 WinPct, appears as a subtitle and the box to the right of the plot contains the value for the standard error of estimate, S = 7834.30, the value R-Sq = 24.9%, and, finally, the value R-Sq(adj) = 22.2%.

The Stat > Regression > Fitted Line Plot command produces not only the plot shown in Figure 5-23, but also most of the regression output in the Session window. To see this:

- Click the **Session Window** toolbar button 🔲

  The output is similar to the output you obtained with the Stat > Regression > Regression command shown in Figure 5-20. You may have to scroll to see it all.

▶ **Note** In both the scatterplot with the regression line and the fitted line plot, Minitab does not extend the line beyond the smallest and the largest values for X and Y. This is good statistical practice. ■

Before beginning the next case study save the project *T5* again.

- Click the **Save Project** toolbar button 🔲

  In the next case study you revisit scatterplots but in the context in which the X-variable represents the passage of time in equally spaced intervals.

5.9 Creating Plots on Which X Represents Time

CASE STUDY | **BUSINESS—COMPETITION**

In your Modern Business Competition course you are working on a paper comparing the competitive strategies of two companies, Microsoft and IBM. As background material you would like to trace their respective, recent stock market performances. Fortunately, your instructor has created a Minitab worksheet, *DJC20012002.mtw*, containing the value of the Dow Jones Composite (DJC) price index for each of the 518 working days in the years 2001 and 2002. Also included in the file is the value for each of the 30 stocks that made up the DJC, including Microsoft and IBM at that time.

A special case of the scatterplot occurs when the horizontal axis represents the passage of time in equally spaced intervals. In this case the graph is called a *time series plot*. The first variable in the data set is C1-D Day, the day. You will use this variable on the horizontal axis of your plots. The second variable, C2 DJC, is the daily value for the DJC price index. The remaining 30 variables are the daily values for the 30 stocks, which make up the DJC. The value of IBM stock is in C19 IBM and the value for Microsoft is in C25 MSFT.

- Open the worksheet *DJC20012002.mtw*

If you are using the Student version of Minitab you will be immediately informed by the Erase Columns dialog box that the worksheet contains more data, 16,576 elements (cells), than can be processed by the Student version of Minitab, which has a limit of 10,000 elements. (If you are using the Professional version of Minitab you will not see this dialog box and you can resume this tutorial two paragraphs below—where the variable C1-D Day is introduced.)

The dialog box is shown in Figure 5-24.

FIGURE 5-24

The Erase Columns dialog box

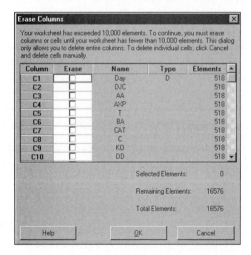

Erase Columns

Your worksheet has exceeded 10,000 elements. To continue, you must erase columns or cells until your worksheet has fewer than 10,000 elements. This dialog only allows you to delete entire columns. To delete individual cells, click Cancel and delete cells manually.

Column	Erase	Name	Type	Elements
C1	☐	Day	D	518
C2	☐	DJC		518
C3	☐	AA		518
C4	☐	AXP		518
C5	☐	T		518
C6	☐	BA		518
C7	☐	CAT		518
C8	☐	C		518
C9	☐	KO		518
C10	☐	DD		518

Selected Elements: 0

Remaining Elements: 16576

Total Elements: 16576

Help | OK | Cancel

Because each variable in the worksheet contains 518 elements, you need to erase at least 13 columns (16,576 – 13*518 = 9,842). In this case, erase all the variables except for the ones you will need for this case study, C1-D Day, C19 IBM, and C25 MSFT.

■ In the "Erase" column click the check boxes for each of the columns *except* C1 Day, C19 IBM, and C25 MSFT
■ Click **OK**

Notice that the variable C1-D Day is a date/time variable—hence the D after the C1. (You have now been introduced to the three types of data in Minitab—Numeric, Text, and Date/Time.) Now, try plotting IBM's stock value against the date. At this point you could change the data type of C1-D Day from a date/time variable to a numeric variable, but there is no need. Minitab can accept a date/time variable for a scatterplot.

■ Choose **Graph > Scatterplot** and double-click *Simple*
■ Press F3 to clear the previous settings
■ Double-click **C19 IBM** as the "Y variables" column and **C1 Day** as the "X variables" column for Graph 1
■ Click **Labels**, type **Scatterplot of the Value of IBM Stock Against Day** as the "Title", and type **2001–2002** as the "Subtitle 1"
■ Click **OK** twice

The resulting plot is shown in Figure 5-25.

FIGURE 5-25

Scatterplot of the value of IBM stock against Day

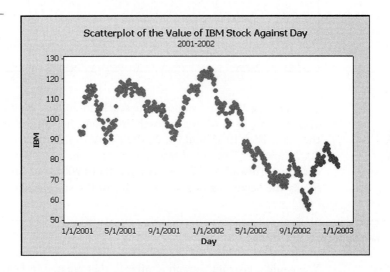

Making sense of exactly how IBM's stock value varies over this time period is not easy because it is difficult to distinguish consecutive points. This can be remedied by connecting the points. To do this:

■ Click the **Edit Last Dialog** toolbar button 🔳
■ Click **Data View**, clear the Symbols check box, and click the **Connect line** check box

Note In the Scatterplot - Data View dialog box, "Symbol" and "Connect line" are only two of the four options. "Project lines" will represent each point by a vertical line (projection) and "Area" will shade in the area under the lines that connect the points. ∎

■ Click **OK** twice

You could have connected the points and retained the symbols. But when, as in this case, there are many points, the symbols obscure rather than clarify the appearance of the plot. The new version of the plot, shown in Figure 5-26, de-emphasizes the individual points and clarifies the pattern in the changing values of the IBM stock.

FIGURE 5-26

Scatterplot of the value of IBM stock against Day with the points connected

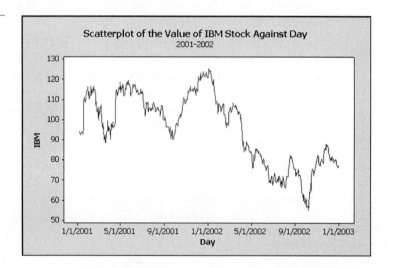

During 2001 the value of the stock (in dollars) moves through three cycles, averaging about four months each, in which the value climbs to around 120 and then falls back to around 90. At the end of the third cycle at the beginning of 2002, however, instead of recovering again, the value for the stock continues to decline until well into the fourth quarter of 2002, reaching a low of approximately 55. By the beginning of 2003 the value of the stock had recovered to almost 80.

5.10 Overlaying Plots

You could construct a graph similar to that created in Figure 5-26 in order to examine the behavior of the value of Microsoft's stock. However, if your goal is to compare the two graphs and you can do this most easily when both plots are on the same graph. Minitab allows you to do this.

■ Click the **Edit Last Dialog** toolbar button ▣
■ Double-click **C25 MSFT** as the "Y variables" column for Graph 2
■ Double-click **C1 Day** as the "X variables" column for Graph 2

- Click **Labels** and change the "Title" to **Scatterplot of the Values of IBM and Microsoft Stock Against Day**, and click **OK**
- Click **Multiple Graphs**, click the "Show Pairs of Graph Variables" **Overlaid on the same graph** check box, and click **OK** twice

The resulting graph is shown in Figure 5-27.

FIGURE 5-27

Scatterplot of the values of IBM and MSFT stock against Day

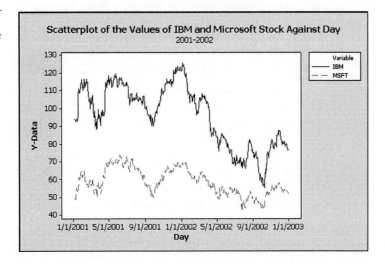

The legend on the right of the graph indicates that the upper plot is that for IBM and the lower plot is that for Microsoft. Though the value for Microsoft stock is always less than that for IBM the pattern of changes over the two-year period are remarkably similar. The increases and the declines in Microsoft's stock prices are not as steep as are those for IBM, but they happen at roughly the same times.

One problem with this *overlaid scatterplot* is that Minitab has provided a default label, Y-Data, for the vertical axis. Something more helpful, such as "Stock Value", would be preferable.

- Hold your cursor over the Y-Data label, click the **right-hand mouse button**, and choose **Edit Y Axis Label**
- In the "Text" text box change the default label, Y-Data, to **Stock Value** and click **OK**

    The new label will appear on the vertical axis.

    Before beginning the next case study save the project *T5* again.
- Click the **Save Project** toolbar button ▣

The next two sections deal with exporting data files from Minitab and importing such files into Minitab. If you don't wish to cover these topics, then you have completed this tutorial and may exit Minitab. Congratulations! In the next tutorial you will work with randomly generated data and probability distributions.

5.11 Exporting Data

SPORTS—BASEBALL STADIUMS (CONTINUED)

You have now written an article about Major League Baseball parks and their effects on the success of the teams that play in them. You've gathered some basic data about the capacity of each of the 30 parks and when each was built, as well as the average attendance at each park for each event in 2001 and the team performance in 2001. You intend to send these data upon request.

Earlier in this tutorial you analyzed the relationship between the attendance at Major League Baseball games and the teams' winning percentages. A friend who has read the article you wrote based upon your analysis asks for a copy of these two variables. You agree to provide him with the relevant two columns in a file so that he can perform his own analyses.

First, return to the BallparkData worksheet in your project. To make this the current worksheet:

- Click the **Current Data Window** toolbar button ▦ (twice if necessary)

  For convenience:

1. Erase all but the name of the team and the two variables of interest, Attend and WinPct

2. Move these latter two variables to columns C2 and C3.

- Choose **Data > Erase Variables**
- Double-click **C2 League**, **C3 ParkBlt**, **C4 Capacity**, and **C7 AgePark** as the "Columns, constants, and matrices to erase"
- Click **OK**

  To move Attend and WinPct to C2 and C3:

- Highlight **C5 Attend** and **C6 WinPct** by dragging your cursor over the two columns
- Choose **Editor > Move Columns**
- In the "Before column" list box, highlight **C2**, and click **OK**

  Your worksheet now contains the three variables you want to provide to your friend. It should look as shown in Figure 5-28.

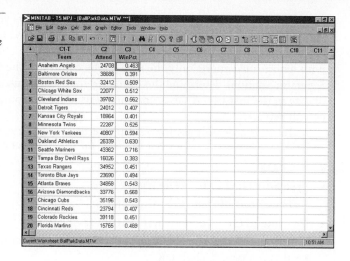

You can save these data in a text file. A simple way to do this is to use the copy-and-paste approach that you used in Tutorial 1. A second approach, useful with large data sets, is to use the File > Save Current Worksheet As command to save the worksheet as a text file.

- Choose **File > Save Current Worksheet As**
- Click the "Save as type" **drop-down list arrow**

    As you can see from the list of file types, several options are available for saving data. You can save a file in a format compatible with earlier versions of Minitab (for example, Minitab 13) or the Minitab Portable format, a file with the *mtp* extension that can be opened by several versions of Minitab. In addition, you can save the file as a Web page, or in popular spreadsheet and database formats, as you will see later in this tutorial.

    To save the worksheet as a text file:

- Click **Text (ANSI)**
- Click **Options**

    The Save Worksheet As - Options dialog box opens, as shown in Figure 5-29.

FIGURE 5-29

The Save Worksheet As - Options dialog box

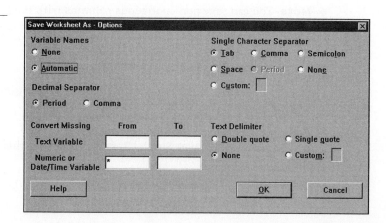

The Save Worksheet As - Options dialog box allows you to control the way Minitab saves the worksheet data in the text file. The "Automatic" option places the variable names in the first row of the text file, while "None" omits the column names and exports only the data in the columns. The "Single Character Separator" options determine the character that will be used to separate columns of data in the text file. A tab is commonly used for this purpose and is Minitab's default setting.

To use the default settings, simply return to the Save Current Worksheet As dialog box, designate a name and file type for the file, and save the text file.

- Click **OK**
- Type *5BPDText* as the "File name"

  The letters BPD are the initials for BallparkData.

- Navigate to the location where you are saving your work and click **Save**
- In response to the warning that "The Worksheet Description cannot be saved for this file type!", click **OK**

  Notice that Minitab has changed the name of the current worksheet to *5BPDText.txt*.

Exporting Formatted Data

Another method for exporting these three variables to a text file is to use the File > Other Files > Export Special Text command. This method allows you to save your data with a user-specified (FORTRAN-type) format, but without the corresponding variable names.

You specify an appropriate format by observing that the first column, Team, is a text (alphanumeric) variable with a maximum width of 21 characters—designated by A21 (Philadelphia Phillies—the blank counts as one character). Attend is an integer with a maximum width of 5 digits (I5 where "I" stands for *integer*) and WinPct consists of 5 symbols (including a decimal place) with three significant digits after the decimal place (F5.3 where "F" stands for *floating point*). The specification of the form of each of the three variables should be separated by one blank space, designated by 1X. Thus, for these three columns of data you would use the format A21,1X,I5,1X,F5.3.

▶ **Note** You can save non-contiguous columns with the File > Other Files > Export Special Text command, but not with the File > Save Current Worksheet As command. ■

- Choose **File > Other Files > Export Special Text**
- Double-click each of **C1 Team**, **C2 Attend**, and **C3 WinPct** as the "Columns to export"
- Type **A21,1X,I5,1X,F5.3** as the "User-specified format"

  The completed Export Special Text dialog box for this command is shown in Figure 5-30.

FIGURE 5-30

*The completed Export
Special Text dialog box*

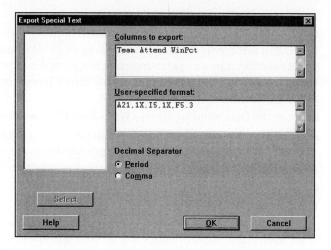

- Click **OK**

    In the Export Data To File dialog box, you need to specify the name, type, and location in which you wish to export these data:

- Click the "Save as type" **drop-down list arrow** and click **ANSI Text Files (*.TXT)**

- Type **5BPDForm** as the "File Name"

- Click **Save**

    Now examine the file's contents using Notepad:

- Choose **Tools > Notepad**

- From the location where you are saving your work, open the file **5BPDForm.txt**

    The opened file is shown in Figure 5-31.

FIGURE 5-31

*The contents of
5BPDForm.txt in Notepad*

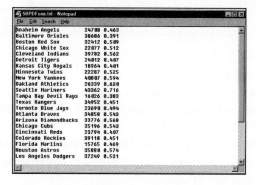

    Notice that the formatting in *5BPDForm.txt* is exactly as you specified. In Notepad, you could insert the three variable names above the columns, but now leave this file and explore another approach to exporting data.

- Choose **File > Exit**

Exporting Data to an Excel Spreadsheet

One application that your friend is considering using with your data is Microsoft Excel. You can export the three variables to an Excel spreadsheet, too. In Tutorial 1 you used the copy-and-paste approach to export data from Minitab to Excel. Here, you will again use the File > Save Current Worksheet As command to save the worksheet as a Excel spreadsheet:

■ From the Data window, choose **File > Save Current Worksheet As**

■ Click the "Save as type" **drop-down list arrow** and click **Excel 97 – 2000** (or **Excel 5.0 – 7.0**, or **Excel 2.0 – 4.0**, depending upon the version of Excel that you have access to)

■ Type **5BPDExcel** as the "File name"

■ Click the **Options** button

The Save Current Worksheet As - Options dialog box, shown in Figure 5-32, is set so that the variable names will be stored in the first row of the Excel spreadsheet (Automatic) and that the standard defaults are employed. You don't need a character separator between the columns for the export because each column of data in Minitab will become a column on the Excel spreadsheet.

FIGURE 5-32

The default Save Worksheet As - Options dialog box

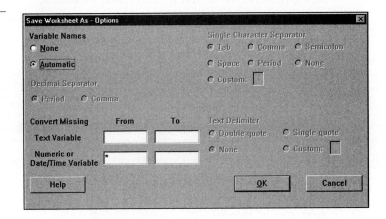

To create the Excel file:

■ Click **OK** and then click **Save**

■ In response to the warning that "The Worksheet Description cannot be saved for this file type!", click **OK**

■ If necessary, click the **Current Data Window** toolbar button 🖼

Notice that, once again, Minitab has changed the name of the current worksheet, this time to *5BPDExcel.xls*.

You have transferred your data from Minitab into both a text file and an Excel file. Your friend should now be able to analyze the data for the ballparks.

At this point you can close the current worksheet, *5BPDExcel.xls*:

■ Choose **File > Close Worksheet**

◆ **Note** In this section you have saved a data file in a variety of formats. One useful format that you might want to explore on your own is HTML. In this format, your data can be saved as a Web page. ∎

5.12 Importing Data

The friend who wanted the ballpark data promised to send you a couple of data sets that he thought might interest you. Anticipating this, practice importing both text and Excel files into Minitab. Begin by importing the text file *5BPDText.txt* into Minitab using the File > Open Worksheet command.

- Choose **File > Open Worksheet**
- From the location where you are saving your work, click the "Files of Type" **drop-down list arrow**, and click **Text (*.txt)**
- Click *5BPDText.txt* from the list of text files that appear, but don't open the file yet
- Click **Options**

The default Open Worksheet - Options dialog box opens, as shown in Figure 5-33. The defaults match those you used to export your text file earlier. You can change these settings to match the way a text file has been saved so that it will open correctly in Minitab.

FIGURE 5-33

The default Open Worksheet - Options dialog box

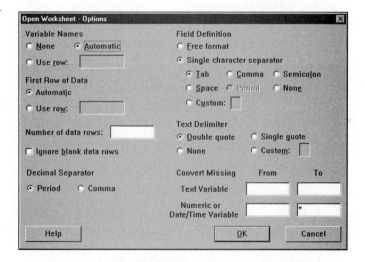

Now, return to the Open Worksheet dialog box and preview the contents of this text file:

- Click **OK**
- Click **Preview**

The Open Worksheet - Preview dialog box opens, as shown in Figure 5-34.

FIGURE 5-34

The Open Worksheet -
Preview dialog box

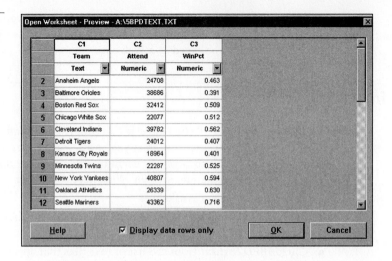

The Open Worksheet - Preview dialog box allows you to examine your data file before you open it in Minitab. It displays the column number, name, and data type for each variable. In addition, you can scroll through the first 100 rows of the data set. If the data are not being read correctly, you can return to the Options dialog box and adjust the settings so that the file is read correctly. This preview can save you a lot of time when you have to import a large data set and/or a data set not in Minitab format.

You now are ready to import the data:

- Click **OK** and click **Open**

  Your data set is placed in a new worksheet *5BPDText.txt.*

Renaming a Worksheet in Minitab

You prefer the name *5BallparkDataText* to the simpler *5BPDText* for this new file. It is simple to change the name of a worksheet in Minitab. To make this change:

- Locate the worksheet icon that is to the left of the File menu if the data window is maximized and to the left of the name *5BPDText.txt* if the screen is split
- Hold your cursor over this worksheet icon, click the **right-hand mouse button**, and choose **Rename Worksheet**
- In the "Rename Worksheet" text box type *5BallparkDataText.txt* and click **OK**

  Minitab immediately changes the name of the worksheet.

Importing Formatted Text Data Files

A second approach to importing data from a text file is to use the File > Other File > Import Special Text command. This approach is analogous to the File > Other File > Export Special Text command and is most helpful if you need to import data contained in a text file that is formatted in an unusual way. You can reference Minitab's Help for more details about such formatting, but at this time, you will use this Import command on the *5BPDForm.txt* file that you formatted earlier in this tutorial.

First, you need to create a new Minitab worksheet to receive your imported data:

- Choose **File > New** and click **OK** to open a new worksheet

    Next, you alert Minitab to expect formatted data:

- Choose **File > Other Files > Import Special Text**

- Type **c1-c3** as the "Store data in column(s)" value

    The Import Special Text dialog box now looks as shown in Figure 5-35.

FIGURE 5-35

The completed Import Special Text dialog box

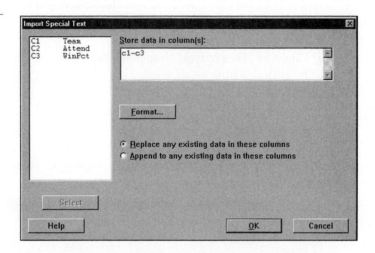

> ◆ **Note** By default Minitab will replace any existing data in these columns with the imported data; that is why you opened a new Minitab worksheet. There is, however, an option available in this dialog box to append the data to any existing data in these columns. ■

To import the text, open the Format dialog box:

- Click **Format**

    The default settings in the Import Special Text - Format dialog box are not correct for the file you are about to import. The data format of the *5BPDForm.txt* file is not blank delimited; nor is it is tab delimited. It has the user-specified format that you specified when you created the file.

- Click the **User-specified format** option button

- Type **A21,1X,I5,1X,F5.3** in the "User-specified format" text box

- Click **OK** twice

    The Import Text From File dialog box appears. You are now ready to import the data.

- From the location where you are storing your work, click **Text Files (*.TXT)** as the "Files of type" and *5BPDForm.txt* as the "File name"

- Click **Open**

    The three columns of data will appear, without names, in C1, C2, and C3.

Because you are simply practicing importing data at this point, you can close this worksheet and answer "No" to any messages about saving it in a separate file.

■ Choose **File > Close Worksheet**

Importing an Excel File

Now suppose that your friend has decided to provide you with data in an Excel spreadsheet, such as *5BPDExcel.xls*. In Tutorial 1 you used the copy-and-paste approach to import data from an Excel spreadsheet into Minitab. An alternative way to bring data from an Excel spreadsheet into Minitab is to use the File > Open Worksheet command:

■ Choose **File > Open Worksheet**

■ Click the "Files of type" **drop-down list arrow** and click *Excel (*.xls)*

■ From the location where you are saving your work, click *5BPDExcel.xls* as the "File name"

Before you open this Excel spreadsheet, explore the Options and Preview dialog boxes:

■ Click **Options**

The Open Worksheet - Options dialog box opens, as shown in Figure 5-36.

FIGURE 5-36

The default Open Worksheet - Options dialog box

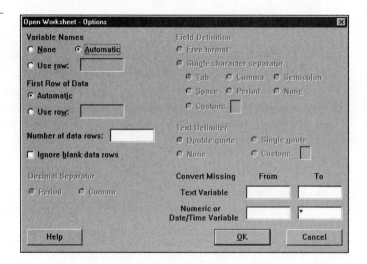

The defaults of Automatic Variable Names and First Row of Data are appropriate because your data have variable names in the first row and the first row of data is in the second row. Note that the upper right-hand corner is shaded because "Field Definition" and "Text Delimiter" are not needed with data from an Excel spreadsheet.

■ Click **OK** to return to Open Worksheet dialog box

■ Click **Preview**

Note, as before, that the Preview dialog box allows you to confirm the column number, name, and data type of each variable. It also allows you to preview a portion of the data before you enter the data into Minitab.

- Click **OK** and then click **Open**

The results of your efforts are shown in Figure 5-37.

FIGURE 5-37

The data imported into Minitab from an Excel spreadsheet

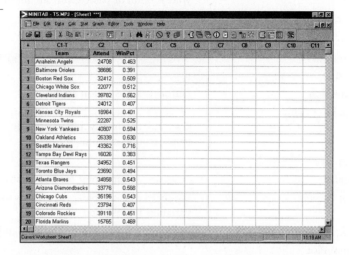

> ➧ **Note** If you import a Microsoft Excel workbook with more than one worksheet into Minitab, each worksheet is placed in a different data window. ■

- Click the **Save Project** toolbar button 🖫
- Choose **File > Exit**

You have covered a lot of ground in this tutorial and have now completed all the data analysis topics in the book. Congratulations! In the next tutorial you will work with randomly generated data and probability distributions.

Minitab Command Summary

This section describes the commands introduced in, or related to, this tutorial. Use Minitab's online Help for a complete explanation of all commands.

Minitab Menu Commands

Menu	Command	Description
File ➢	Open Worksheet	Opens a previously saved worksheet, including an Excel spreadsheet or a text file.
	Save Current Worksheet As	Saves a copy of the current worksheet to a file, including an Excel spreadsheet and a text file, with a new filename.

Menu	Command	Description
File ➤	Other Files	
	➤ Export Special Text	Exports data in the specified columns to a text file.
Edit ➤	Command Line Editor	Provides a quick way to execute commands used earlier in your session or pasted in from another source.
Stat ➤	Basic Statistics	
	➤ Correlation	Computes the correlation between pairs of variables.
	➤ Covariance	Computes the covariance between pairs of variables.
	Regression	
	➤ Regression	Performs a regression analysis.
	➤ Fitted Line Plot	Graphs a regression line and a scatterplot of the regression data and, optionally, places confidence bands and prediction bands around the line.
Graph ➤	Scatterplot	
	Simple	Produces a scatterplot for two quantitative variables.
	With Groups	Produces a scatterplot with values for up to three other variables represented.
	With Regression	Produces a scatterplot with the regression line plotted.
	Marginal Plot	
	With Histograms	Produces a scatterplot in which one or both of the margins forms the base for a histogram.
	With Boxplots	Produces a scatterplot in which one or both of the margins forms the base for a boxplot.
	With Dotplots	Produces a scatterplot in which one or both of the margins forms the base for a dotplot.
Editor ➤	Brush	When a graph window is active (showing individual points), opens the Brushing palette.
	Crosshairs	When a graph window is active, causes crosshairs to appear on the graph.
	Panel	When a graph window is active, allows the creation of separate graphs based upon a By variable.
	Set ID Variables	In Brush mode only, adds more information to the brushing palette for each brushed point.
	Move Columns	Allows selected columns to be moved.
Window ➤	Project Manager	Opens the Project Manager, providing access to the History folder.

Minitab Toolbar Button Commands

Toolbar Button	Icon	Description
Show History	▷	Opens the Project Manager, providing access to the History folder.
Show Session Folder	⊕	Opens the Project Manager, providing access to the Session folder.
Close All Graphs	▣	Closes all the graphs that are present in the current session.

Minitab Right-Hand Mouse Button Commands

Command	Description
Brush	When a graph window is active (showing individual points), opens the brushing palette.
Crosshairs	When a graph window is active, causes crosshairs to appear on the graph.
Panel	When a graph window is active, allows the creation of separate graphs based upon a By variable.
Set ID Variables	Adds the selected variables to the brushing palette for each brushed point.
Rename Worksheet	When the Worksheet icon is clicked, renames the current worksheet.

Review and Practice

Matching Problems

Match the following terms to their definitions by placing the correct letter next to the term it describes.

_____ Brushing Palette

_____ Grouping

_____ Scatterplot

_____ Covariance

_____ Correlation

_____ Paneling

_____ R-Sq

_____ Response

_____ Predictor

_____ History

a. The construction of separate graphs in the same graph window, based upon the values for another variable

b. Minitab's notation for the coefficient of determination

c. The window that displays the values for specified variables for brushed points

d. Minitab's name for the dependent variable

e. A folder that records the commands issued

f. A measure of the strength of the linear relationship between two quantitative variables whose value depends on the units for the two variables

g. A type of variable used to assign different symbols in a scatterplot

h. A graph designed to show the relationship between two quantitative variables

i. A measure of the strength of the linear relationship between two quantitative variables whose value lies between –1 and 1

j. Minitab's name for an independent or explanatory variable

True/False Problems

Mark the following statements with a *T* or an *F*.

a. ____ In a scatterplot, where the X-axis represents time measured in equally spaced intervals, it is helpful to connect the points.

b. ____ There are several ways to produce a scatterplot with the least-squares/regression line.

c. ____ The History folder records the commands, the subcommands, and the output associated with these commands.

d. ____ Minitab does not permit multiple plots on the same graph.

e. ____ Text files can only be imported into Minitab if they are FORTRAN-formatted.

f. ____ Minitab allows up to four grouping variables in a scatterplot.

g. ____ The default output for the Regression command includes the standard error of the estimate and the coefficient of determination.

h. ____ When used on a scatterplot, crosshairs allow you to identify the row number associated with a particular point.

i. ____ You cannot use a stemplot on the margins of a Marginal Plot.

j. ____ Text files can be exported from Minitab only if they are FORTRAN-formatted.

Practice Problems

If necessary, use Help or Appendix A Data Sets, to review the contents of the data sets referred to in the following problems. Interpretations should use the language of the subject matter of the question. Be sure to provide all graphs with a suitable title. If you are using the Student version, be sure to close worksheets when you have completed a problem.

1. Open *PulseA.mtw*. This problem explores the relationship between weight and height for males and females.

   a. Produce a scatterplot of weight against height. What does the plot suggest about the relationship between these two variables in this case?

   b. Panel the scatterplot by gender. Explain how the relationship between weight and height differs for males and females?

   c. Obtain a marginal plot of weight against height with boxplots on the margins. What feature of the weights is highlighted in the boxplot? Can you obtain a marginal plot with a graph on only one of the two margins/axes?

   d. Use the Data > Unstack command to unstack the heights and weights by gender. For the male students only, obtain the regression line relating weight to height. Interpret the slope of the line. What fraction of the variation in weight can be associated with differences in height?

   e. Obtain (i) a Fitted Line Plot of the male weights against the male heights, and (ii) a scatterplot with the regression line. What do you see as the advantages of each plot?

   f. Reviewing your answer in part b, why should you be reluctant to compute the regression line relating weight to height for female students? Explain briefly.

2. Open *Textbooks.mtw*.

   a. Is there any relationship between a text's reading ease score and the corresponding grade level? Produce a scatterplot of ReadEase against GradeLev. What does the plot suggest about the relationship? The Microsoft Company uses quite a complex procedure to compute appropriate grade level for a piece of prose. Why do you think that so many of the texts scored at the 12th grade level?

b. What is the correlation between the price of the textbooks at Amazon.com and the publisher's list price? Does the result surprise you? Why?

c. Obtain a scatterplot of the price at Amazon.com against the list price using the type of textbook as the grouping variable. There are some interesting features in the plot. Write a brief account of these features. Do these features explain the correlation in part b?

d. Open the History folder and highlight the command that produced the correlation in part b. Copy this command into the Command Line Editor. Modify the command to ask for the correlation between the list price and the number of pages in the text. Submit this new command. What does this value suggest about the strength of the linear relationship between these two variables?

3. Open *CollMass.mtw*.

a. Produce a scatterplot of Tuition against average SAT verbal scores. How would you describe the relationship between these two variables?

b. A clump of three colleges falls outside the pattern exhibited by the other colleges. Use the brushing palette to identify the names, the SAT verbal scores, and the tuition for these three colleges. Explain, in terms of SAT scores and tuition amounts, the manner in which these three schools differ from the other colleges.

c. Produce a marginal plot of tuition against average SAT verbal scores with dotplots on the margins. Are these three aberrant points obvious from the dotplots? Do they become obvious if you use either histograms or boxplots on the margins?

d. What is the covariance between these two variables? Why is it difficult to tell the strength of the relationship between these two variables from your answer?

e. What is the correlation between these two variables? Is this value consistent with your description in part a? Explain briefly.

f. What is the equation of the regression line relating tuition to average SAT verbal scores? Suppose you were told that the average SAT verbal score at college A exceeded that at college B by ten points. Use the slope of the regression line to predict how much higher or lower the tuition at A would be relative to B.

g. Obtain a fitted line plot showing the regression line you obtained in part f.

h. Interpret the coefficient of determination given above the fitted line plot. What is the relationship between this coefficient and the correlation coefficient you obtained in part e?

4. Open *TwoTowns.mtw*. This problem explores the impact of an outlier on a data analysis.

a. Obtain a scatterplot of the list price against the number of rooms in the home. Use Town as a grouping variable. How would you characterize the relationship between price and the number of rooms? Which town seems to have the higher prices?

b. One point on the scatterplot is an outlier. In which town is this home located? Explain, in terms of price and number of rooms, how this home differs from the other homes. By brushing this point, determine the corresponding row number, the number of rooms in the home, and the list price.

c. Compute the correlation between list price and the number of rooms. Also, obtain the equation of the regression line relating list price to the number of rooms. Interpret the slope of the line in this case. How would you interpret the coefficient of determination in this case?

d. For the outlier, replace both the list price and the number of rooms with an asterisk (*)—(remember that an asterisk is Minitab's notation for missing values in numeric columns). Recompute the correlation and the regression line without this outlier.

e. Explain the impact of removing this outlier on (i) the correlation coefficient, (ii) the slope of the regression line, and (iii) the coefficient of determination.

5. Open *Salary02.mtw*.

a. As a member of the college faculty, you are interested in the relationship between salary and the length of time that an instructor has been at a particular rank. (i) Obtain a scatterplot of the 2002 salaries of full professors (Salary02Prof) against their corresponding years in rank (YrsinRankProf). (ii) How would you characterize the strength of the relationship? (iii) Compute the corresponding correlation coefficient. (iv) Is this value for r consistent with your answer in part a(ii)?

b. Repeat part a for both associate and assistant professors. For which of the three professorial ranks is the relationship between these two variables the strongest? The weakest?

c. As you did in part a, obtain a scatterplot of the 2002 salaries of full professors against years in rank, but use different symbols for male and female professors. Does the relationship between salary and years in rank seem stronger for males or for females?

d. As you did in part a, obtain a scatterplot of the 2002 salaries of full professors against years in rank. Panel the plot by gender. Does the relationship between salary and years in rank seem stronger for males or for females?

e. When comparing the relationship between salary and years in rank for males and female, do you prefer the approach in part c or in part d? Explain.

f. Use the File > Save Current Worksheet As command to save these salary data in a text file called *P5Salary2001/02.txt*. Open the file in Notepad or in Microsoft Word to check that it was saved correctly. Also, import this file back into Minitab. When you have imported it, change the name of the worksheet to *P5Salaries.mtw*.

6. Open *NHL2003.mtw*. This problem analyzes National Hockey League statistics for the 2002–2003 season.

a. You are interested in the relationship between the number of points a team has at the end of the season, and both goals for (which measures offensive power), and goals against (which measures defensive power). Obtain a correlation matrix for the three variables Points, GoalsFor, and GoalsAgainst. Which of the latter two variables is more highly correlated with Points?

b. (i) Based upon the matrix you obtained in part a, what can you say about the covariances between these three variables? (ii) Obtain the covariance matrix for these three variables. (iii) Are the values for the three covariances consistent with your answers in part b(i)?

c. Obtain a scatterplot of Points against GoalsAgainst? Is the plot consistent with the corresponding correlation from part a?

d. Is the relationship between Points and GoalsAgainst the same for the two divisions? Answer this question (i) by paneling the scatterplot by Division, and (ii) using a scatterplot of Points against GoalsAgainst with Division as the grouping variable. Which plot did you find the most helpful in comparing the two divisions? Explain briefly.

e. Use the Calc > Column Statistics command to compute the mean number of points per team and the mean GoalsAgainst. Again, obtain a scatterplot of Points against GoalsAgainst? Use crosshairs to divide the plot into four quadrants based upon the mean points and mean GoalsAgainst. What fraction of points on your plot lie in the upper-left or the lower-right quadrants? Explain the significance of your answer.

f. What kind of relationship would you expect between the number of points a team has and the team's player payroll? Obtain a scatterplot of Points against Payroll. Is the plot consistent with your expectations?

g. One of the NHL teams falls well outside the pattern exhibited by the other teams? Use the brushing palette to identify the team. Explain, in terms of points and payroll amounts, how this team differs from the other teams.

h. A friend has asked you for the data on team name, points, goals for, goals against, and division in a formatted text file. (i) Use the Data > Erase Variables and Editor > Move Columns commands to arrange the worksheet so that these variables occupy columns C1–C5. (ii) Use the File > Other Files > Export Special Text command to save these five variables in a suitably formatted text file. (iii) Make sure you can import the same file back into Minitab.

i. Use the File > Save Current Worksheet As command to save this entire data set as a Web page.

7. Open *MnWage2.mtw.*

   a. Obtain a scatterplot of the minimum wage against year. How would you characterize the overall trend in the plot?

   b. Since 1950, what is the longest period of time without an increase in the minimum wage? What was the approximate minimum wage during that period? What was the approximate minimum wage in 2001?

8. Open *Election2.mtw.*

   a. Obtain a scatterplot of the Democratic percentage of the vote against year. Connect the points. Write a brief interpretation of the behavior of this percentage over time.

   b. Overlay the plot of the Republican percentage of the vote against year on the plot of the Democratic percentage of the vote against year. What can you say about the relationship between these two plots?

   c. Create a new variable 100 – Dem% – Rep%. Name this variable Other%. What does this new variable represent? Obtain a scatterplot of Other% against Year. Do you see any striking pattern in the plot?

   d. A colleague would like a copy of these data in the form of an Excel spreadsheet. Prepare a file for her. Also, make sure that you can import the Excel spreadsheet back into Minitab.

9. Open *DJC20012002.mtw*. If necessary, eliminate all the columns except Day, IBM, and MSFT.

   a. Create a new variable that is the amount by which the value for IBM stock exceeds the value for Microsoft stock? Call this new variable IBM–MSFT.

   b. Construct a scatterplot of IBM–MSFT against Day. Describe the pattern in the plot.

   c. Open the History folder and copy into the Command Line Editor the commands that created the differences in part a and the plot in part b. Modify the command you used in part a so that you compute the amount by which the value for Microsoft stock exceeds the value for IBM stock. Call this new variable MSFT–IBM. Submit these modified commands.

   d. Do you prefer your graph in part b or in part c? Explain briefly.

10. Open *Marathon2.mtw*.

   a. On the same graph, obtain a plot of the winning times for males against year and a plot of the winning times for females against year. In each case, connect the points, but do not show symbols.

   b. Describe the two time series plots. Do they have features in common? Explain briefly.

On Your Own

Minitab comes with a series of tutorials—not to be confused with the tutorials in this book—that provide excellent overviews of different aspects of the software. If you have progressed to this point in the book, you will have a sound understanding of how to make effective use of Minitab to do data analysis. However, to reinforce your understanding of the package, work through the first two Minitab Tutorial Sessions, which are called Graphing Data and Entering and Exploring Data.

From within Minitab, you can access the tutorials with the command Help > Tutorials. If Minitab 14 is installed in your computer's Programs folder, you can easily locate the tutorials from the Desktop with the command Start > Programs > Minitab 14 > Minitab Tutorials.

Were there any commands in the Tutorial Sessions that you had not seen before? If so, what are they? Conversely, do you think that there are important commands that you have used, that were not included in these tutorials? If so, what are they?

Answers to Matching Problems

(c) Brushing Palette
(g) Grouping
(h) Scatterplot
(f) Covariance
(i) Correlation
(a) Paneling
(b) R-Sq
(d) Response
(j) Predictor
(e) History

Answers to True/False Problems

(a) T, (b) T, (c) F, (d) F, (e) F, (f) F, (g) T, (h) F, (i) T, (j) F

Distributions and Random Data

Probability distributions are the foundation of inferential statistics. In this tutorial, you will use Minitab to compute probabilities and percentiles for various distributions, including the normal distribution. You will also generate random samples from these distributions and learn how to use Minitab to check whether a data set comes from a normal population.

OBJECTIVES

In this tutorial, you learn how to:

- Compute individual and cumulative binomial probabilities
- Generate random samples from a discrete population with a specified probability distribution
- Generate random samples from a normal population
- Produce normal probability plots
- Obtain cumulative probabilities for a normal distribution
- Use the inverse cumulative probability procedure to obtain percentiles for a normal distribution
- Sample from a column, with and without replacement

6.1 | *Calculating Binomial Probabilities*

CASE STUDY	**BIOLOGY— BLOOD TYPES**

You have been a Red Cross volunteer for the last few years. The volunteer coordinator, Paul Van Vleck, hears that you are learning how to use Minitab. Eager to put your skills to work, he tells you that the Type O blood supply is running dangerously low and that he anticipates needing from ten to 12 additional pints for surgeries scheduled in the upcoming week. Paul has already recruited 25 unrelated potential donors. He wants to know if it is likely that next week's need for Type O blood can be met using these donors or if he will need to recruit more.

Individual Binomial Probabilities

You know that, on average, 45 out of 100 people have Type O blood, so the probability of a randomly selected individual having Type O blood is .45. To help Paul, you need to compute $P(X = K)$—the probability of K people having Type O blood— for the values of K = 0, 1, 2, ..., 25. In this case, $P(X = K)$ is a *binomial probability function* with n = 25 trials and p = .45. You could compute these probabilities using a calculator, but it would be very tedious. You also could obtain these probabilities from a table, but sets of probability tables cannot cover all values of n and p. Finding these probabilities using Minitab is simple.

To get started:

- Open **Minitab** and maximize the Session window
- If necessary, click **Enable Commands** and clear Output Editable from the Editor menu
- Click the **Save Project** toolbar button ▣ and in the location where you are saving your work, save this project as *T6.mpj*
- Click the **Current Data Window** toolbar button ▣

Minitab lets you compute at once all of the probabilities $P(X = 0)$, $P(X = 1)$, ..., $P(X = 25)$, but first you must create a column containing the values 0, 1, 2, ..., 25 for K. Although you could type all these numbers directly into a column, there is a much easier way to enter them using Minitab's *Autofill* feature:

- Type **0** in as the first value in column C1
- While that cell is highlighted, move your cursor to the lower right-hand corner of the cell until you see a heavy, black cross (+), Minitab's Autofill symbol
- Press Ctrl, click, and drag down the column

As you drag down the column a small window just to the right of the column informs you of the value that Minitab will insert into that cell. The window will show the successive values 1, 2, 3, and so on.

- Stop dragging when the small window shows 25

The column C1 should now contain the values 0, 1, 2, ..., 25.

Note If you had not pressed the Ctrl key, Minitab would have copied the value 0 into each of the 26 cells. (Use Help to learn more about this and other Autofill options.) ∎

Note Another way to enter these 26 values into C1 is to use the Calc > Make Patterned Data > Simple Set of Numbers menu command. (Use Help to learn more about this alternative approach.) ∎

- Name column C1 **K**

    Now, obtain the binomial probabilities for each value of K in C1.

- Choose **Calc > Probability Distributions > Binomial**
- Click the **Probability** option button
- Type **25** as the "Number of trials" and **.45** as the "Probability of success"
- Double-click **C1 K** as the "Input column"

    The completed Binomial Distribution dialog box looks as shown in Figure 6-1.

FIGURE 6-1

The completed Binomial Distribution dialog box

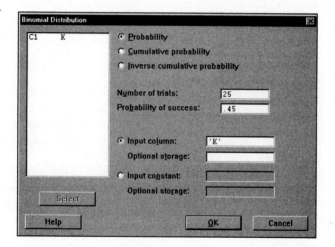

- Click **OK**

    Minitab lists the complete binomial probability distribution for n = 25 and p = 0.45 in the Session window, as shown in Figure 6-2. Scroll the Session window to see the complete listing.

FIGURE 6-2

*Binomial probability
distribution in the
Session window*

```
MTB > PDF 'K';
SUBC>   Binomial 25 .45.

Probability Density Function

Binomial with n = 25 and p = 0.45

   x    P( X = x )
   0     0.000000
   1     0.000007
   2     0.000065
   3     0.000407
   4     0.001830
   5     0.006290
   6     0.017155
   7     0.038097
   8     0.070133
   9     0.108387
  10     0.141889
  11     0.158306
  12     0.151110
  13     0.123636
  14     0.086705
  15     0.052023
  16     0.026603
  17     0.011523
  18     0.004190
  19     0.001263
  20     0.000310
  21     0.000060
  22     0.000009
  23     0.000001
  24     0.000000
  25     0.000000
```

Minitab displays the probability distribution in two columns. The first column contains the values for the possible number of successes (people with Type O blood). The second lists the corresponding probability of obtaining exactly that number of successes. For example, the probability of having exactly 11 people in this sample with Type O blood is 0.158306.

These probability values are rounded to the sixth decimal place, so probabilities less than 0.0000005 are listed as 0.000000. For example, the probability of obtaining exactly one Type O individual in your sample isn't really 0.000000—it's less than 0.0000005.

▶ **Note** The menu under Probability Distributions lists three groups of probability distributions, many of which you may not have heard of. The first group consists of the most widely used continuous distributions. The second group consists of the most popular discrete distributions, including the binomial. The final group consists of less frequently used continuous distributions. The Minitab at Work vignette at the end of this tutorial provides an application of a discrete distribution—*the Poisson distribution*—which has almost as many applications as the binomial distribution. ■

For discrete random variables, the Calc > Probability Distributions command can produce:

1. Individual probabilities of the form $P(X = K)$

2. Cumulative probabilities of the form $P(X \leq K)$

3. Percentiles or inverse cumulative probabilities (the approximate value, K, such that $P(X \leq K)$ is a specified probability)

For continuous random variables (such as the normal), the Calc > Probability Distributions command produces:

1. Density function values
2. Cumulative probabilities of the form $P(X \leq K)$
3. Percentiles or inverse cumulative probabilities (the value, K, such that $P(X \leq K)$ is a specified probability)

You can tell Minitab to store the probabilities in a column in the worksheet:

- Click the **Edit Last Dialog** toolbar button
- Click the first "Optional storage" text box
- Type **'P(X = K)'** (be sure to type the single quotation marks) as the "Optional storage" for the output column and click **OK**
- Click the **Current Data Window** toolbar button

Column C2, with the name $P(X = K)$, lists the binomial probabilities, as shown in Figure 6-3. You may have to scroll down to see the entire distribution in the Data window.

FIGURE 6-3

Binomial probability distribution in the Data window

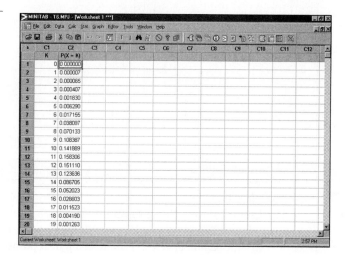

Note that these probabilities are identical to those probabilities that were previously printed in the Session window. According to these probabilities, you are most likely to obtain 10, 11, or 12 people with Type O blood in your sample of 25 (these are the values with the highest probabilities). However, these values will occur only 14.2%, 15.8%, and 15.1% of the time, respectively. Collectively, though, you have almost a 50% chance that the 25 donors will yield 10, 11, or 12 people with Type O blood (14.2 + 15.8 + 15.1 = 45.1%).

If you don't need all values of the probability distribution, you can compute the probability for a single value. For example, if you want only $P(X = 10)$:

- Click the **Edit Last Dialog** toolbar button
- Click the **Input constant** option button and type **10** as the "Input constant"
- Click **OK**

The Session window shows that the probability that X = 10 is 0.141889. Verify that you obtained identical results in Figures 6-2 and 6-3.

To confirm that the most likely values (highest probabilities) are 10, 11, and 12, produce a plot of the binomial distribution:

- Choose **Graph > Scatterplot >** *Simple*
- Double-click **C2 P(X = K)** as the "Y variables" column and **C1 K** as the "X variables" column
- Click **Labels**, type **Binomial (25, .45) Probabilities** as the "Title", and click **OK**
- Click **Data View** and verify that Project lines is checked
- Click **OK** twice

The plot appears as shown in Figure 6-4.

FIGURE 6-4

Plot of binomial probabilities against the number of successes, K

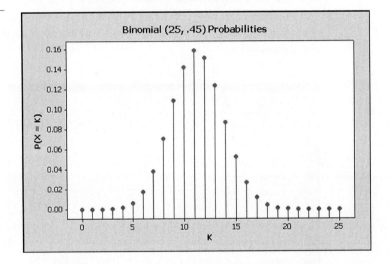

The heights of each line (projection) over each value for K correspond to the probability that K Type O blood donors will be found in the sample of 25. The peak of your plot occurs at K = 11; the next highest points are at K = 10 and K = 12. Indeed, the most likely numbers of Type O donors are 10, 11, and 12. Also, note that the shape of your plot is approximately bell-shaped.

Cumulative Binomial Probabilities

In addition to calculating individual probabilities, Minitab can compute cumulative probabilities for either an entire column or an individual value. You can do this by using the cumulative probability option in the Binomial Distribution dialog box. After talking with the surgeons who will perform next week's surgery, Paul informs you that at least 13 pints of Type O blood will be needed. He needs to know the probability of getting 13 or more Type O donors in his sample of 25.

▶ **Note** For discrete random variables, individual probabilities are of the form P(X = K), whereas cumulative probabilities are of the form P(X ≤ K). For binomial random variables, this is the probability of K or fewer successes. ■

You can find the probability of 13 or more Type O donors by first determining the probability of the complementary event, obtaining 12 or fewer individuals with Type O blood from the sample of 25. To do this:

- Choose **Calc > Probability Distributions > Binomial**
- Click the **Cumulative probability** option button
- Verify that "Number of trials" is 25 and "Probability of success" is .45
- Click the **Input constant** option button
- Type **12** as the "Input constant" and click **OK**

The output appears in the Session window as shown in Figure 6-5.

FIGURE 6-5

Cumulative binomial probability

```
MTB > CDF 12;
SUBC>   Binomial 25 .45.
```

Cumulative Distribution Function

```
Binomial with n = 25 and p = 0.45

  x  P( X <= x )
 12    0.693676
```

You tell Paul that there is a probability of .693676 (a 69% chance) that 12 or fewer people in this group of 25 will have Type O blood. Therefore, the likelihood of getting the 13 or more pints he needs is less than one-third (1 − .69376 = .306334 = a 31% chance).

Paul is concerned. The chances are quite good that he won't be able to provide enough Type O blood with only the current donors. He asks you to find the fewest number of unrelated donors he must have to be reasonably certain (probability ≥ .75) there will be at least 13 with Type O blood. In other words, you need to find the smallest value of n so that $P(X \leq 12)$ is less than .25.

You must repeat the previous actions, substituting different values for n. To save time, start with the value n = 33.

- Click the **Edit Last Dialog** toolbar button 📭
- Type **33** as the "Number of trials" and click **OK**

In the Session window, Minitab reports that a pool of 33 donors gives Paul a probability of .206228 of getting 12 or fewer Type O donors (and hence a probability of .793772 of getting 13 or more). This is fine. But to determine if you can get by with fewer donors, try n = 32:

- Click the **Edit Last Dialog** toolbar button 📭
- Type **32** as the "Number of trials" and click **OK**

When 32 donors are considered, the probability of 12 or fewer is .251178 (which means a probability of only .748822 of getting 13 or more). This doesn't quite meet Paul's requirements. You tell Paul that he will need at least 33 unrelated donors to obtain 13 or more pints of Type O blood with a probability of .75 or more. Minitab saved you quite a bit of calculation time.

Returning to the case in which Paul has only 25 volunteer donors, you can easily construct and display the complete set of cumulative probabilities for the values K = 0, 1, 2, ... , 25.

- Click the **Edit Last Dialog** toolbar button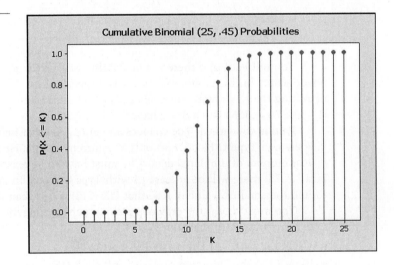
- Type **25** as the "Number of trials"
- Click the **Input column** option button and verify that the "Input Column" is K
- Type **'P(X <= K)'** as the "Optional storage" value and click **OK**
- Click the **Current Data Window** toolbar button

The complete set of cumulative probabilities should appear in C3 P(X <= K) in the Data window. To plot these probabilities:

- Choose **Graph > Scatterplot > *Simple***
- Double-click **C3 P(X <= K)** as the "Y variables" column and verify that **C1 K** is the "X variables" column
- Click **Labels**, type **Cumulative Binomial (25, .45) Probabilities** as the "Title", and click **OK** twice

The graph of cumulative probabilities should look as shown in Figure 6-6.

FIGURE 6-6

Plot of cumulative binomial probabilities against the number of successes, K

As the number of successes, K, increases, the cumulative probability of getting that number or fewer, $P(X \le K)$, increases. Notice that the scale on the vertical axis (y-axis) of the cumulative probabilities runs from 0 to 1. By contrast, a review of the plot of individual binomial probabilities in Figure 6-4 shows that the vertical axis runs from 0 to .15.

You have completed your volunteer research on blood type samples. Before beginning the next case study save the project.

- Click the **Save Project** toolbar button

6.2 | Generating Random Data from a Discrete Distribution

CASE STUDY	MANAGEMENT— ENTREPRENEURIAL STUDIES

Your management professor, Dr. Michaels, has assigned a term paper in which you are to report on the pattern of sales for small-scale retail outlets for an industry of your choice. You select the music industry. As part of your research, you discover the fact that your neighborhood music outlet sells approximately 1,000 items of recorded music per week.

One of the references used in your Sales Management course, *The New York Times 2003 Almanac*, presents the percentages of recorded music sales for 2001 associated with different music genres. (The percentages are adjusted slightly from the original so that they add to 100%.)

Genre	Percentage of Sales
1. Rock	27.1
2. Pop	13.4
3. Rap/Hip-Hop	12.6
4. R&B/Urban	11.8
5. Country	11.6
6. Gospel	7.4
7. Jazz	3.8
8. Classical	3.5
9. Other	8.8
	100.0

According to these data, the probability that a random music sale in 2001 is rock music is .271. The corresponding probability for classical music is .035. You decide to use Minitab to generate a random sample of sales for one week. Specifically, you will *simulate* a week of music sales at the outlet and examine the differences between the population percentages and the sample percentages.

Minitab allows you to use your computer to obtain a random sample of data and simulate an experiment. This saves both time and money.

In a new worksheet, you will enter three columns:

1. C1 contains the number (1 to 9) associated with each genre.
2. C2 contains the name of the genre.
3. C3 contains the theoretical probabilities for each genre.

The numbers in C1 for each genre are needed because the Minitab command you use to simulate the sales requires you to describe each genre with a numeric value.

Before entering this information, you need to create a new worksheet:

- Choose **File > New** and click **OK**
- Maximize the Data window, if necessary

Now, enter the data into C1–C3 using the case data as a guide:

- Assign the names **GenreN, Genre,** and **Probability** to columns C1, C2, and C3, respectively
- Use Autofill to enter **1, 2, ... , 9** as numeric values to represent the nine genres in C1 GenreN
- Type **Rock, Pop, ... , Other** in C2 Genre as the names of the genres
- Type **.271, .134, .126, .118, .116, .074, .038, .035, .088** in C3 Probability as the probabilities corresponding to each genre

As part of your simulation, you will tell Minitab to produce a batch of random numbers. If you don't specify a *base*, or starting point, for the simulation, when generating random numbers Minitab chooses its own. When you set a base, you can generate the same random data again simply by entering the same "seed" number. The choice of number for the base is arbitrary. However, it will determine the final sample, although not the randomness of the sample. Because the data were recorded in 2001, specify 2001 as the base Minitab should use when generating these random numbers.

- Choose **Calc > Set Base**
- Type **2001** as the "Set base of random data generator to" value to enter it as the base and click **OK**

You can now generate a random sample of 1,000 music sales using the theoretical (population) probabilities provided by the almanac, which can be duplicated for future use:

- Choose **Calc > Random Data > Discrete**
- Type **1000** as the "Generate [ ] rows of data" value
- Type **Sample** as "Store in column(s)"
- Double-click **C1 GenreN** as the "Values in" column
- Double-click **C3 Probability** as the "Probabilities in" column

The completed Discrete Distribution dialog box should look as shown in Figure 6-7.

FIGURE 6-7

The completed Discrete Distribution dialog box

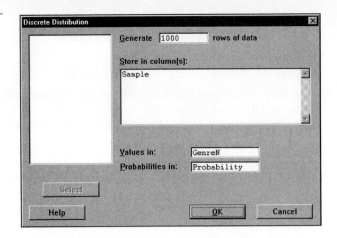

- Click **OK**

    The column C4 Sample now lists 1,000 genre numbers, 1, 2, ..., or 9, in random order, generated using the probabilities for each genre.

    Presume these 1,000 numbers are the results of recording the genres for 1,000 sales of recorded music at your neighborhood music store. Before asking for the count for each genre, you can convert the 1,000 sales to text and determine the order in which the genres will be listed. To convert the genre numbers in C4 Sample to their corresponding names:

- Choose **Data > Code > Use Conversion Table**
- Double-click **C4 Sample** as the "Input column"
- Type **TxSample** as the "Output column"
- Double-click **C1 GenreN** as the "Column of Original Values"
- Double-click **C2 Genre** as the "Column of New Values" and click **OK**

    The original genre names will appear in C5-T TxSample next to the corresponding numbers in C4 Sample. The Data > Code > Use Conversion Table command was used rather than the Data > Code > Numeric to Text command because the latter is limited to eight values and you had nine.

    To guarantee that the nine genres will be listed in the same order in later analysis as they were in the almanac:

- Click anywhere in the C5-T TxSample column
- Choose **Editor > Column > Value Order**
- Click the **User-specified order** option button
- Arrange the nine genre names in the user-specified order text box in the following order: **Rock**, **Pop**, **Rap/Hip-Hop**, **R&B/Urban**, **Country**, **Gospel**, **Jazz**, **Classical**, **Other**
- Click **OK**

    When you obtain output involving this column, the values will be ordered in the sequence you listed. (Recall that the default ordering is alphabetical.)

    You can now use Tally to obtain the count with which each genre occurs in the sample.

- Choose **Stat > Tables > Tally Individual Variables**
- Double-click **C5 TxSample** as the "Variables"
- Verify that Counts is checked
- Click the **Percents** check box and click **OK**

The count distribution for your sample appears in the Session window as shown in Figure 6-8. (If you didn't set the base to 2001, your random sample Count and Percent values will differ from those in Figure 6-8.)

FIGURE 6-8

Count distribution of genres in the sample

```
MTB > Tally 'TxSample';
SUBC>    Counts;
SUBC>    Percents.
```

Tally for Discrete Variables: TxSample

TxSample	Count	Percent
Rock	295	29.50
Pop	137	13.70
Rap/Hip-Hop	110	11.00
R&B/Urban	111	11.10
Country	125	12.50
Gospel	80	8.00
Jazz	41	4.10
Classical	23	2.30
Other	78	7.80
N=	1000	

Together, the music genres in C1 and C2-T and the probabilities in C3 constitute the population of genres. Minitab selected 1,000 genres at random using the probabilities in C3. The percentages of music genres in this simulation are quite close to the percentages presented in the almanac. For example, in your simulated week of sales, 29.5% of sales were for rock music compared to 27.1% in the almanac. Similarly, in the sample 2.3% of sales were for classical music; the corresponding figure in the almanac was 3.5%. You can see that in a random sample of size n = 1,000, the percentages in the sample will not differ a great deal from those for the population. If you had selected a sample of only 100, the percentages in the sample would likely have deviated much more from the population percentages than in this sample.

Before moving on to the next case study, save the project.

- Click the **Save Project** toolbar button

6.3 | *Generating Random Data from a Normal Distribution*

The family of *normal distributions* (bell-shaped curves) plays a central role in statistics. In this and the next two sections you will generate a sample from a normal distribution and compute probabilities and percentiles for a normal distribution.

Dr. Wei is a professor in the Physiology Department of the medical school on campus. She hopes that you can help her with a lecture she is preparing for her physiology class. She would like to explore with her class how to check whether certain physiological variables have a normal distribution. As one example, she would like to use the heights of college-age women, which, she believes, have close to a normal distribution with a mean of 64 inches and a standard deviation of 3.1 inches. You suggest that she should illustrate methods for checking for normality by using both a random sample of heights from a population known to be normal and the actual heights for the 60 women students in her class. Dr. Wei will get the heights of her students from a questionnaire, while you agree to simulate the selection of a random sample of 60 heights from a normal population.

Your immediate task is to simulate the selection of a random sample of heights from the college-age women using the values μ = 64 inches and σ = 3.1 inches.

To obtain a new worksheet:

- Choose **File > New** and click **OK**

So that you can replicate the simulation, use the base value 333 (Dr. Wei's office number). Now generate a sample of 60 observations from a normal population with a mean of 64 and a standard deviation of 3.1.

- Choose **Calc > Set Base**
- Type **333** as the "Set base of random data generator to" value and click **OK**
- Choose **Calc > Random Data > Normal**
- Type **60** as the "Generate [ ] rows of data" value
- Type **'RandomHts'** (don't forget the single quotes) as the "Store in column(s)"
- Type **64** as the "Mean" and **3.1** as the "Standard deviation" to replace the default values of 0 and 1

The completed Normal Distribution dialog box should look as shown in Figure 6-9.

FIGURE 6-9

The completed Normal Distribution dialog box

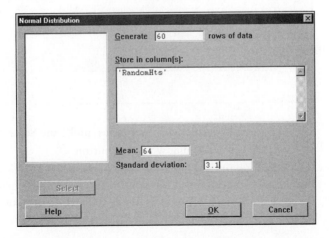

- Click **OK**

Column C1 RandomHts in the worksheet contains 60 random values from a normal (μ = 64 and σ = 3.1) distribution.

Dr. Wei will present these data to her class as a random sample from the normal distribution of the heights of college-age women. She hopes to demonstrate that even though the population is normal, a randomly selected sample from such a population may not be exactly bell-shaped.

6.4 | Checking Data for Normality

You agree to help Dr. Wei compare the sample results to the normal population from which they were selected. You can use Minitab to obtain the sample mean and standard deviation and a histogram of the sample heights with a superimposed normal curve.

- Choose **Stat > Basic Statistics > Display Descriptive Statistics**
- Double-click **C1 RandomHts** as the "Variables"
- Click **Graphs** and then click the **Histogram of data, with normal curve** check box
- Click **OK** twice

The histogram of the 60 heights, as shown in Figure 6-10, is only approximately bell-shaped.

FIGURE 6-10

Built-in histogram for RandomHts with normal curve

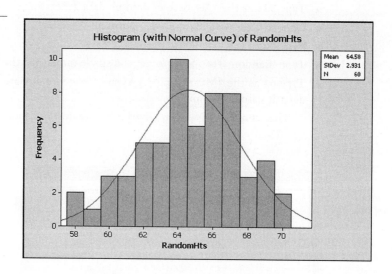

To view the numerical summaries, make the Session window active:

- Click the **Session Window** toolbar button

Scroll up to the results shown in Figure 6-11. The sample mean is 64.584 inches; the standard deviation is 2.931 inches. The normal curve that was superimposed on the histogram in Figure 6-10 has the mean of 64.584 inches and the standard deviation of 2.931 inches.

FIGURE 6-11

Descriptive statistics for RandomHts

```
MTB > Describe 'RandomHts';
SUBC>   Mean;
SUBC>   SEMean;
SUBC>   StDeviation;
SUBC>   QOne;
SUBC>   Median;
SUBC>   QThree;
SUBC>   Minimum;
SUBC>   Maximum;
SUBC>   N;
SUBC>   NMissing;
SUBC>   GNHist.
```

Descriptive Statistics: RandomHts

Variable	N	N*	Mean	SE Mean	StDev	Minimum	Q1	Median	Q3
RandomHts	60	0	64.584	0.378	2.931	58.051	62.734	64.523	66.640

Variable	Maximum
RandomHts	70.316

While the graph might be sufficient to convince Dr. Wei's class of the data's normality, you can use another Minitab display, the *normal probability plot (NPP)*, to determine whether normality is a reasonable assumption.

The Normal Probability Plot

A common graphical technique for checking whether a sample comes from a normal population is a normal probability plot. You can generate an NPP in Minitab with the following sequence:

- Choose **Stat > Basic Statistics > Normality Test**
- Double-click **C1 RandomHts** as the "Variable" and click **OK**

The resulting NPP should look as shown in Figure 6-12.

FIGURE 6-12

Normal probability plot of the random heights

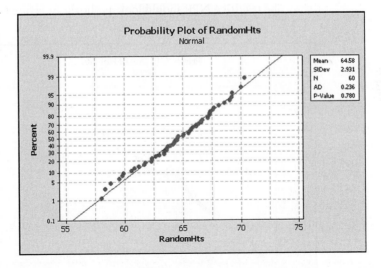

The horizontal axis represents the random heights. The construction of the values on the vertical axis is quite complex and is outlined in the note below. You should know that the closer the heights are to being normal the closer the 60 points will lie to a straight line. Minitab shows the least-squares line fitted to the data. In this case you can see that the points lie very close to a straight line, which suggests that the random data are very close to being normal. This graph will make a good illustration for Dr. Wei's lecture on how to show that a variable is approximately normally distributed.

◆ **Note** For each height, Minitab computes the corresponding cumulative probability of occurrence, assuming a random sample of size 60 from a normal distribution. For instance, the smallest cumulative probability corresponding to the smallest height is about .01 (close to 1% on the chart). The logarithms of these cumulative probabilities are plotted on the log-scaled vertical axis, paired with the corresponding random height. Further information about this NPP may be found by examining the StatGuide associated with this plot. ■

Dr. Wei can produce a similar NPP for the self-reported heights she collected from her class. She stored the female students' heights and an ID variable in a Minitab file called *Height.mtw*, which can be found in the Studnt14 folder. To preview the results of plotting the data before Dr. Wei does so in class:

■ Open the worksheet *Height.mtw*
■ Click the **Current Data Window** toolbar button, if necessary

The worksheet contains the ID number and the self-reported height for each of the 60 female students in the class.

To produce the NPP for these 60 heights:

■ Choose **Stat > Basic Statistics > Normality Test**
■ Double-click **C2 Heights** as the "Variables" and click **OK**

The resulting NPP should look as shown in Figure 6-13.

FIGURE 6-13

Normal probability plot of self-reported female heights

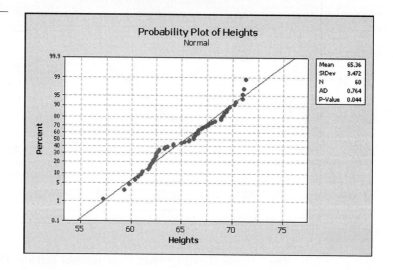

This plot isn't as straight as your NPP with simulated data. There is a bump in the middle and the points at either end of the plot deviate from linear. You report to Dr. Wei that the NPP suggests that self-reported heights of college-age women may not be exactly normally distributed.

You can investigate the distribution of the self-reported heights further by constructing a stem-and-leaf display of C2 Heights.

- Choose **Graph > Stem-and-Leaf**
- Double-click **C2 Heights** as the "Graph variables" and click **OK**

The resulting stem-and-leaf display is shown in Figure 6-14.

FIGURE 6-14

Stem-and-leaf display for the self-reported heights

```
MTB > Stem-and-Leaf 'Heights'.

Stem-and-Leaf Display: Heights

Stem-and-leaf of Heights   N  = 60
Leaf Unit = 0.10

  1    57   2
  1    58
  3    59   38
  5    60   47
 10    61   01789
 21    62   01234555678
 24    63   346
 26    64   22
 30    65   0378
 30    66   2223466679
 20    67   1467
 16    68   0399
 12    69   012457
  6    70   23
  4    71   0013
```

The bump in the NPP in Figure 6-13 corresponds to the two peaks (bimodality) in the center of the display. Why do you think that these peaks occurred? The right-hand portion of the NPP corresponds to the "heavy" right-hand tail of the display. The isolated point on the left of the NPP corresponds to the outlying value on the left tail of the stem-and-leaf display.

◆ **Note** NPPs are frequently used in analyses that involve the assumption of normal data. In future tutorials you will use NPPs. ■

6.5 Determining Cumulative Probabilities and Inverse Cumulative Probabilities for the Normal Distribution

Impressed with your work, Dr. Wei returns a few weeks later to show you her follow-up work on self-reported heights. Using extensive information from all of her classes, she found that self-reported heights of college-age women appear to be approximately normally distributed with a mean of 64.25 inches and a standard deviation of 3.28 inches.

Dr. Wei has two questions about this distribution for you:

1. For a study of unusually tall female students, she would like to know what percentage of such students are above 74 inches

2. She wants to compare the bone structure distribution in four height groups

These height groups will be based upon the quartiles of the N(64.25, 3.28) distribution. She asks you to find the first and third quartiles of this distribution.

Minitab makes it simple to find the percentage of students who are shorter than or equal to 74 inches, from which the percentage of students who are taller than 74 inches can easily be found.

- Choose **Calc > Probability Distributions > Normal**
- Verify that the Cumulative probability option button is selected
- Type **64.25** as the "Mean" and **3.28** as the "Standard deviation"
- Click the **Input constant** option button and type **74** as the "Input constant"

The completed Normal Distribution dialog box should look as shown in Figure 6-15.

FIGURE 6-15

The completed Normal Distribution dialog box

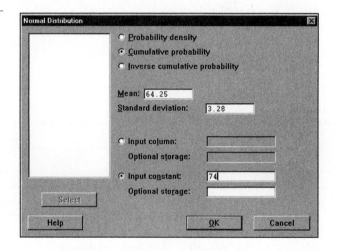

- Click **OK**

The Session window displays the results, as shown in Figure 6-16.

FIGURE 6-16

Cumulative normal probability value

```
MTB > CDF 74;
SUBC>   Normal 64.25 3.28.
```

Cumulative Distribution Function

```
Normal with mean = 64.25 and standard deviation = 3.28

  x   P( X <= x )
 74      0.998523
```

The results indicate that 99.85% of the female heights are less than or equal to 74 inches. So, only 0.15% of female college students are taller than 74 inches.

To find the first quartile, Q1 (the 25th percentile), of the N(64.25, 3.28) distribution:

- Click the **Edit Last Dialog** toolbar button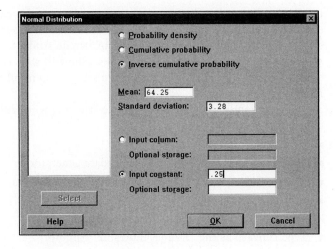
- Click the **Inverse cumulative probability** option button
- Verify that the Input constant option button is selected and type **.25** as the "Input constant"

The completed Normal Distribution dialog box should look as shown in Figure 6-17.

FIGURE 6-17

The completed Normal Distribution dialog box

- Click **OK**

The Session window displays the results, as shown in Figure 6-18.

FIGURE 6-18

Inverse cumulative normal probability value

```
MTB > InvCDF .25;
SUBC>   Normal 64.25 3.28.
```

Inverse Cumulative Distribution Function

```
Normal with mean = 64.25 and standard deviation = 3.28

P( X <= x )        x
     0.25  62.0377
```

The results indicate that 25% of the female heights are less than or equal to 62.04 inches. The first study group, therefore, consists of women who are less than or equal to 62.04 inches tall. The second group consists of women whose heights falls between 62.04 inches and the median of 64.25 inches. (Recall that for a normal distribution, the mean is equal to the median. So, in this case, the median is 64.25 inches.)

To obtain the third quartile, Q3 (the 75th percentile):

- Click the **Edit Last Dialog** toolbar button
- Type **.75** as the "Input constant" and click **OK**

The output indicates that the 75th percentile of female heights is 66.4623 inches. So, the third group consists of women with heights between 64.25 and 66.46 inches. The last group contains those women who are taller than 66.46 inches.

Tutorial 7 introduces other uses for the cumulative and the inverse cumulative probability functions.

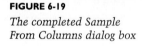

6.6 Sampling from a Column

Dr. Wei will be presenting a paper at a convention in Chicago and wants eight students in the class to accompany her—four men and four women. To be fair, she asks you to use Minitab to select four female students at random from those whose heights are reported in C2 Heights. (She will then select the male students similarly.) Recall that Dr. Wei included a student ID number in C1 ID in the *Height.mtw* worksheet.

To randomly select the four female students who will travel to the Chicago convention:

- Choose **Calc > Set Base**
- Type **444** as the "Set base of random data generator to" value and click **OK**
- Choose **Calc > Random Data > Sample From Columns**
- Type **4** as the "Sample [ ] rows from column(s)" value
- Double-click **C1 ID** as the column to sample
- Type **Chicago** as the "Store samples in" column

The completed Sample From Columns dialog box resembles Figure 6-19.

FIGURE 6-19

The completed Sample From Columns dialog box

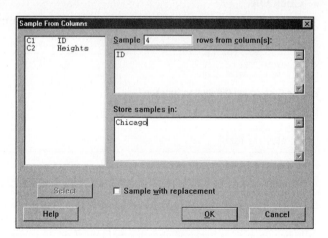

The default sampling option simulates an actual sample by eliminating an observation from the population as it is selected for the sample. This is called sampling *without replacement*. To sample *with replacement* (so that the same population unit can be chosen more than once), you can click the "Sample with replacement" check box in the Sample From Columns dialog box.

- Click **OK**
- Click the **Current Data Window** toolbar button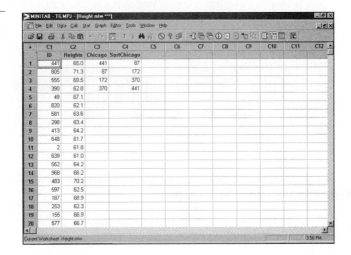

Minitab places the ID numbers of the four randomly selected students in C4 Chicago.

You want to give Dr. Wei the four ID numbers arranged from smallest to largest. To sort the column you just created:

- Choose **Data > Sort**
- Double-click **C3 Chicago** as the "Sort column(s)"
- Double-click **C3 Chicago** as the first "By column" variable
- Verify that the first Descending check box is not checked
- Click the **Column(s) of current worksheet** option button
- Type **SortChicago** as the "Column(s) of current worksheet" and click **OK**

The ID numbers of the four lucky students are displayed in C4 SortChicago, and in Figure 6-20 in ascending order.

FIGURE 6-20

The Data window showing four randomly selected ID numbers

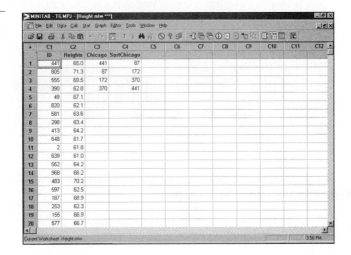

You can now inform Dr. Wei that the students with the ID numbers 87, 172, 370, and 441 were selected. Minitab also allows you to sample from a column containing text data, so if Dr. Wei had included names instead of ID numbers you could have selected a random sample of four names.

- In the location where you are saving your work, save the current project **T6** and exit Minitab

Congratulations on finishing Tutorial 6! In this tutorial, you worked with several of the most commonly used distributions. In later tutorials, you will use the Calc > Probability Distributions command to help solve other statistical problems.

PUBLIC SAFETY

Some years after a rural university in Pennsylvania installed a nuclear reactor for educational purposes, concerned residents of the surrounding town claimed that the presence of the reactor was increasing infant mortality in the area. To support their contention, they presented data that compared infant mortality in their town to that in another town that was similar in size and character.

A statistician examined the data to help determine whether the infant mortality rate over the nine-year period was unusually high. The statistician focused on several aspects of the claim. One was whether the town's 1968 rate of ten infant deaths was unusually high in a community in which the average yearly mortality rate was six deaths.

The statistician decided to see if randomly generated samples of similar data would contain any values as high as the 1968 value. He had Minitab randomly generate 100 Poisson distributions of nine-year spreads. He found that 58% of those randomly generated spreads contained one or more years with an infant mortality rate of at least ten. Because a majority of the spreads included a value as high as the 1968 rate, the statistician decided that the peak of ten in 1968 was not uncommon in a situation with six deaths in an average year.

In the end, no dangerous environmental effects were linked to the presence of the reactor and it was allowed to remain. Although the causes of infant mortality remain a vital concern, it is important to recognize that peaks and valleys normally occur in a nine-year period.

Minitab Command Summary

This section describes the commands introduced in, or related to, this tutorial. Use Minitab's online Help for a complete explanation of all commands.

Minitab Menu Commands

Menu	Command	Description
Data ➤	Code	
	➤ Conversion Table	Codes numeric values, text values, or date/time values, to new values based upon a conversion table set up in the worksheet.

Menu	Command	Description
Calc ➤	Make Patterned Data	
	➤ Simple Set of Numbers	Generates sequences or repetitive patterns of numbers and stores the results in a column. (Use as an alternative to Autofill.)
	Set Base	Allows the user to specify the "seed" number used to generate random data.
	Random Data	Produces random samples from specified columns or from discrete and continuous distributions (given below).
	➤ Sample From Columns	Takes random samples, with or without replacement, from a given column or columns.
	➤ Chi-Square	
	➤ Normal	
	➤ Multivariate Normal	
	➤ F	
	➤ t	
	➤ Uniform	
	➤ Bernoulli	
	➤ Binomial	
	➤ Hypergeometric	Various continuous and discrete distributions from which random samples can be generated.
	➤ Discrete	
	➤ Integer	
	➤ Poisson	
	➤ Beta	
	➤ Cauchy	
	➤ Exponential	
	➤ Gamma	
	➤ Laplace	
	➤ Largest Extreme Value	
	➤ Logistic	
	➤ Loglogistic	
	➤ Lognormal	
	➤ Smallest Extreme Value	
	➤ Triangular	
	➤ Weibull	

Menu	Command	Description
Calc ➤	Probability Distributions	Calculates density function values, probabilities, cumulative probabilities, and percentiles for the continuous and discrete random variables given below.
	➤ Chi-Square	
	➤ Normal	
	➤ F	
	➤ t	
	➤ Uniform	
	➤ Binomial	
	➤ Hypergeometric	Various continuous and discrete distributions upon which probability calculations are based.
	➤ Discrete	
	➤ Integer	
	➤ Poisson	
	➤ Beta	
	➤ Cauchy	
	➤ Exponential	
	➤ Gamma	
	➤ Laplace	
	➤ Largest Extreme Value	
	➤ Logistic	
	➤ Loglogistic	
	➤ Lognormal	
	➤ Smallest Extreme Value	
	➤ Triangular	
	➤ Weibull	
Stat ➤	Basic Statistics	
	➤ Normality Test	Generates a normal probability plot (NPP) and performs a hypothesis test to decide whether observations follow a normal distribution

Review and Practice

Matching Problems

Match the following terms to their definitions by placing the correct letter next to the term it describes.

_____ Autofill

_____ Binomial

_____ 0

_____ Random Data

_____ Sample From Columns

_____ Sample with replacement

_____ Inverse cumulative probability

_____ NPP

_____ Base

_____ 1

a. A widely used discrete probability distribution

b. The Minitab option in the Sample From Columns dialog box that can result in observations being sampled more than once

c. The "seed" number used to generate random data

d. The Minitab option that can provide percentiles, quartiles, or deciles

e. A graph in which points lying close to a straight line indicate normality

f. A Minitab feature that allows you to easily enter sequences and/or repetitive patterns of data into a column

g. The Minitab command that generates data at random from a large number of discrete or continuous distributions

h. The Minitab command you use to obtain a subset of values from a variable, with or without replacement

i. The default value for a normal distribution standard deviation when generating random numbers

j. The default value for a normal distribution mean when generating random numbers

True/False Problems

Mark the following statements with a *T* or an *F*.

a. ____ For both discrete and continuous random variables, Minitab can compute cumulative probabilities for individual values and for columns of values.

b. ____ Minitab can compute probability density function values for individual values and for entire columns of values for continuous random variables.

c. ____ Minitab can sample both with and without replacement from any column.

d. ____ Minitab can generate random data from over 20 continuous and discrete distributions.

e. ____ Under the Probability Distribution menu, the default option for the binomial distribution is to compute individual probabilities.

f. ____ In probability distributions you use the Inverse cumulative probability option to determine quartiles for random variables.

g. ____ The results of the Probability Distributions command is always displayed in the Session window.

h. ____ For discrete random variables, Minitab can generate cumulative probabilities but not individual probabilities.

i. ____ The Random Data command allows you to simulate the outcomes of experiments without actually performing those experiments.

j. ____ Minitab calculates probabilities for the Bernoulli random variable.

Practice Problems

The practice problems instruct you to save your worksheets with a filename prefixed by P, followed by the tutorial number. In this tutorial, for example, use *P6* as the prefix. If necessary, use Help or Appendix A Data Sets, to review the contents of the data sets referred to in the following problems. Interpretations should use the language of the subject matter of the question. If you are using the Student version, be sure to close worksheets when you have completed a problem.

1. A commuter airline flies 38-seat planes. The airline knows that, on average, 5% of passengers with reservations fail to show up for the flight. To compensate, the airline sells 40 tickets for each flight. Assume that whether or not one ticket-holder shows up is independent of whether or not any other ticket-holder shows up.

   a. Generate individual binomial probabilities for n = 40 and p = .05. (Begin by using the Autofill feature (or Calc > Set Patterned Data > Simple Set of Numbers command) to generate the integers between 0 and 40 in C1.) Based upon these probabilities find the following:

      i. The most likely number of passengers that will show up for a flight,

      ii. The probability that all 40 ticket-holders will show up for a flight, and

      iii. The probability that all those who show up for a flight can be accommodated. Can you suggest a reason why airlines overbook flights?

   b. Plot the individual binomial probabilities for the binomial distribution with n = 40 and p = .05. Comment on the shape of the distribution.

2. Mr. Sims, the executive director of a professional organization, plans to contact 18 firms in hopes of recruiting five as members. He estimates that the probability of recruiting any firm is 0.6. Assume recruiting one firm is independent of recruiting another firm.

   a. What is the probability that he will recruit five or more firms? What is the most likely number of firms that he will recruit?

   b. Based upon your analysis, has the executive director overestimated or underestimated his ability to recruit new members? Briefly explain your answer.

3. Open *Murders.mtw.*

   a. With a base of 3, use the Calc > Random Data > Sample From Columns command to obtain a random sample of five states and the corresponding murder rates and regions. Which states were selected? From which region(s) did your states come?

   b. Could all of the randomly selected states have come from one region? Briefly explain your answer.

4. Open *MusicData.mtw.* This worksheet contains a replica of the worksheet you created earlier in this tutorial. There, you used the Calc > Random Data > Discrete command to simulate the selection of 1,000 music genres. Your sample is in TxSample. Here you will do something superficially similar.

   a. With a base of 4, use the Calc > Random Data > Sample From Columns command to take a sample of 1,000, with replacement, from the nine genres listed in Genre. Put the sample in Sample2.

b. Obtain the percentage distribution of genres in TxSample and Sample2. How do they differ?

c. Based upon your results in part b, was the population from which you took the sample in part a the same population as the one you used in the case study of the tutorial? Explain briefly.

5. Among humans, there are four blood types, O, A, B, and AB. The approximate proportions of the United States population having each type are:

1	O	.45
2	A	.43
3	B	.09
4	AB	.03

a. Enter these three columns of data into C1–C3. With a base of 5, simulate the selection of a random sample of 200 people (actually, their blood groups).

b. Obtain and record the percentage distribution of blood types in your sample.

c. Repeat the previous two parts but for a sample size of 2,000 people.

d. For the two sample sizes, compare how close the sample percentages are to the population percentages.

6. In this problem you will review the distribution of the sample mean.

a. With a base of 6, use the Calc > Random Data > Normal command to generate 500 values from the N(100, 15) distribution in each of the ten columns C1–C10. Do this by typing C1–C10 as the "Store in column(s)".

b. Regard each of the 500 rows as a random sample of size n = 10 from the N(100, 15) distribution. Use the Calc > Row Statistics > Normal command to store the mean of each sample in C12. Name this column SampleMean.

c. Obtain a description of the 500 sample means. What is the mean of the sample means? The standard deviation of the sample means? Explain why these values make sense.

7. Open *Infants.mtw*.

a. Describe BthWeight, the birth weight of the infant. What is the mean and standard deviation of infant birth weight?

b. With a base of 7, generate 68 observations from a normal distribution with the same mean and standard deviation. Place these data in C11. Name the column RandomWt.

c. Obtain normal probability plots for BthWeight and C11. Which variable is more normal? Briefly explain your answer.

8. Open *Grades.mtw*. This data set is located in the Data folder not the Studnt14 folder.

a. Obtain normal probability plots for each of these three variables.

b. Based upon these displays do you consider each of these variables to be normally distributed? Explain briefly.

9. This problem explores the effect of the mean and standard deviation on the shape of the normal distribution.

   a. Begin by entering the two values 0 and .2 into C1. Use the Autofill feature (or Calc > Set Patterned Data > Simple Set of Numbers command) to generate the values between 0 and 20 in increments of 0.2 in C1.

   b. For the values in C1 obtain the cumulative probabilities for a normal distribution with a mean of 10 and a standard deviation of 2. Store your results in C2. (For each value for X, the probability density value is the height of a normal curve.)

   c. For the values in C1 use the Calc > Probability Distributions > Normal command with the Probability density option, to obtain the probability density values for a normal distribution with a mean of 10 and a standard deviation of 5. Store your results in C3.

   d. For the values in C1 obtain the probability density values for a normal distribution with a mean of 5 and a standard deviation of 2. Store your results in C4.

   e. Name C1 X, C2 N(10, 2), C3 N(10, 8), and C4 N(5, 2).

   f. Construct three scatterplots. Plot each set of probability density values in C2, in C3, and in C4, in turn, against C1. Use the "Overlaid on the same graph" option under Multiple Graphs to produce all three plots on the same graph. What happens to the shape of the normal distribution if the standard deviation increases? What happens to the shape of the normal distribution function if the mean changes?

10. This problem explores the effect of degrees of freedom on the shape of the chi-square distribution.

    a. Begin by entering the two values 0 and .2 into C1. Use the Autofill feature (or Calc > Set Patterned Data > Simple Set of Numbers) to generate the values between 0 and 20 in increments of 0.2 in C1.

    b. Use the Calc > Probability Distributions > Chi-Square command with the Probability density option, to store in C5 the chi-square density values for a chi-square distribution with three degrees of freedom.

    c. Plot the density values against C1. Does your plot resemble a plot of a chi-square distribution? In which direction is the distribution skewed?

    d. Repeat parts a and b using a chi-square distribution with ten degrees of freedom. Is your curve more symmetrical? What appears to happen to the chi-square distribution as the degrees of freedom increase?

11. A normal probability distribution with a mean of 0 and a standard deviation of 1 is called the Z distribution. In statistical inference three inverse cumulative Z probabilities are often used. These are the inverse cumulative Z probabilities for cumulative probabilities of .95, .975, and .995. Obtain these inverse cumulative probabilities. Why do these inverse cumulative probabilities increase as the cumulative probabilities increase?

12. Open *TwinsYankees.mtw*. Suppose you have been asked to randomly select two Twins and two Yankees so that you write a report contrasting what has happened to members of each team.

    a. Explain why you should not use the Calc > Random Data > Sample From Columns command with replacement to select your four players.

b. Explain why you should not use the Calc > Random Data > Sample From Columns command without replacement on this worksheet to select your four players.

c. Use Data > Split Worksheet into to create two worksheets, one with Twins and one with Yankees. With a base of 12, use the Calc > Random Data > Sample From Columns command without replacement to select two Twins. Which players did you select? With a base of 122, use the Calc > Random Data > Sample From Columns command without replacement to select two Yankees. Which players did you select?

On Your Own

You serve in the Office of Institutional Research at a state college. The president of the college has asked your office to conduct a telephone survey of approximately 200 residents of the area in order to investigate attitudes toward the college. Your task is to generate 1,000 "random" telephone numbers. The hope is to complete approximately 200 completed interviews.

The area of interest falls within a single area code so you only need to generate 1,000 seven-digit numbers. The first three digits will indicate the local exchange. You find out from the telephone company that the area is served by six exchanges: 220, 341, 387, 461, 559, and 585, and that the approximate percentage of subscribers in the area served by each exchange is as follows:

Exchange	Percentage
220	27%
341	4%
387	12%
461	14%
559	36%
585	7%

Use this table to generate 1,000 three-digit "exchanges" in C3. Also generate 1,000 numbers (with replacement) from the integers 0000 to 9999 in C4. The combination of C3 and C4 will generate your 1,000 telephone numbers.

Answers to Matching Problems

(f) Autofill
(a) Binomial
(j) 0
(g) Random Data
(h) Sample From Columns
(b) Sample with replacement
(d) Inverse cumulative probability
(e) NPP
(c) Base
(i) 1

Answers to True/False Problems

(a) T, (b) T, (c) T, (d) T, (e) F, (f) T, (g) F, (h) F, (i) T, (j) F

Inferences from One Sample

Normally, it is either impossible or impractical to conduct a study of an entire population to obtain the value of a parameter, such as a population mean or a population proportion. Statisticians have developed techniques that allow them to make inferences about parameters based upon sample statistics. There are two forms of statistical inference. *Hypothesis testing* is used to assess the plausibility of theories about parameters. *Confidence intervals* are used to estimate the value of a parameter.

In this tutorial, you will use Minitab to make inferences about the mean of a population and about the proportion of the population that has a particular characteristic. In the former case, you will examine situations in which the population standard deviation is known and those in which it is not. You also will determine the sample size for estimating the population mean with a specified margin of error. Finally, you will compute the power of a test.

OBJECTIVES

In this tutorial, you learn how to:

- Test a hypothesis about a *population mean*, μ, when the *population standard deviation*, σ, is known
- Compute a *confidence interval* for μ when σ is known
- Determine the sample size for estimating μ when σ is known
- Test a hypothesis about μ when σ is unknown
- Compute a confidence interval for μ when σ is unknown
- Test a hypothesis about a population proportion, p
- Compute a confidence interval for p
- Compute the power of a test for a proportion

Testing a Hypothesis About μ When σ Is Known

CASE STUDY	SOCIOLOGY— AGE AT DEATH

A sociology professor, Dr. Ford, is interested in examining the extent to which women living in affluent areas of the United States live longer than women in the country as a whole. Dr. Ford has obtained the age at death for all of the residents of an affluent suburb of Boston, who died in the year 2001. These data are stored in a file called *AgeDeath.mtw*.

Dr. Ford indicates that the mean age at death was 81.9 years for all females and 75.5 years for all males. The standard deviation was approximately 15 years for both females and males. Dr. Ford has asked your class to test the hypothesis that the mean age at death for women living in affluent areas is significantly greater than 81.9 years. You are to assume that the standard deviation of the age at death for such women is also 15 years. You are also to assume that the female deaths in this Boston suburb can be regarded as a random sample of female deaths over all affluent areas of the country. Dr. Ford has instructed you to use a level of significance (usually indicated by the symbol α) of .05.

Begin by opening Minitab:

- Open **Minitab** and maximize the Session window
- If necessary, **Enable Commands** and clear Output Editable from the Editor menu
- Click the **Save Project** toolbar button 🖫 and, in the location where you are saving your work, save this project as *T7.mpj*
- Open the worksheet *AgeDeath.mtw*
- If necessary, click the **Current Data Window** toolbar button 🖽

The worksheet shows Gender in C3-T and Age (at death) in C5.

Because you are interested in only the female ages, begin by using the Data > Unstack Columns command to isolate the female ages and the male ages in separate columns.

- Choose **Data > Unstack Columns**
- Double-click **C5 Age** as the "Unstack the data in" column
- Double-click **C3 Gender** as the "Using subscripts in" column
- Click the **After last column in use** option button and click **OK**

Minitab places the ages of the females in C6 and the ages of the males in C7. As you saw earlier, Minitab provides default names for the columns created with the Data > Unstack Columns command—in this case Age_Female and Age_Male. Before doing the test, examine the female ages.

- Choose **Stat > Basic Statistics > Display Descriptive Statistics**
- Double-click **C6 Age_Female** as the "Variables"
- Click **Graphs**, click the **Histogram of data** check box, and click **OK** twice

The resulting histogram of Age_Female is shown in Figure 7-1.

FIGURE 7-1

Built-in histogram for Age_Female

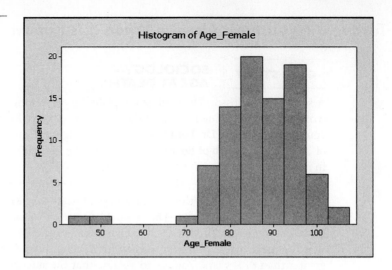

FIGURE 7-1

Built-in histogram for Age_Female

The distribution of the ages is skewed substantially to the left due to the deaths of a few relatively young women.

To see the text output in the Session window:

■ Click the **Session Window** toolbar button ▣

The numerical output from the Stat > Basic Statistics > Display Descriptive Statistics command, shown in Figure 7-2, indicates that the sample mean age at death for the 86 females is 87.09 years. This is certainly considerably greater than 81.9 years, the mean age at death for all females in the United States. But perhaps this large sample mean can be accounted for by chance.

FIGURE 7-2

Descriptive statistics for Age_Female

```
MTB > Describe 'Age_Female';
SUBC>    Mean;
SUBC>    SEMean;
SUBC>    StDeviation;
SUBC>    QOne;
SUBC>    Median;
SUBC>    QThree;
SUBC>    Minimum;
SUBC>    Maximum;
SUBC>    N;
SUBC>    NMissing;
SUBC>    GHist.
```

Descriptive Statistics: Age_Female

Variable	N	N*	Mean	SE Mean	StDev	Minimum	Q1	Median	Q3
Age_Female	86	0	87.09	1.03	9.57	47.00	81.75	87.00	94.00

Variable	Maximum
Age_Female	105.00

You will use the *one-sample Z-test* to examine this question. This test is appropriate when testing a hypothesis about a population mean, μ, when the value for the population standard deviation, σ, is known. (This is one of those rare examples where you don't know the value for μ but the value for σ is given.)

Start by formulating the appropriate hypotheses.

Null Hypothesis, H_0: $\mu = 81.9$

Alternative Hypothesis, H_1: $\mu > 81.9$

where μ is the (unknown) mean age at death of all females living in affluent areas of the country.

Now perform the test:

- Choose **Stat > Basic Statistics > 1-Sample Z**
- Double-click **C6 Age_Female** as the "Samples in columns"
- Type **15** as the "Standard deviation"
- Type **81.9** as the "Test mean" and click **Options**
- Click the "Alternative" **drop-down list arrow**, click **greater than**, and click **OK**

The completed 1-Sample Z (Test and Confidence Interval) dialog box should look as shown in Figure 7-3.

FIGURE 7-3

The completed 1-Sample Z (Test and Confidence Interval) dialog box

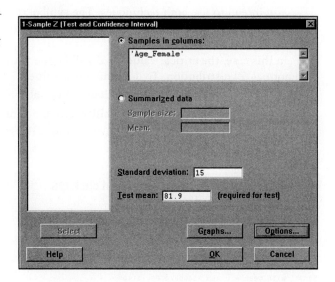

- Click **OK**

The test output is shown in Figure 7-4.

FIGURE 7-4

One-sample Z-test for Age_Female

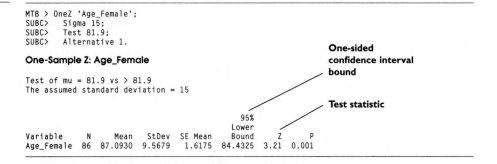

```
MTB > OneZ 'Age_Female';
SUBC>   Sigma 15;
SUBC>   Test 81.9;
SUBC>   Alternative 1.
```

One-Sample Z: Age_Female

```
Test of mu = 81.9 vs > 81.9
The assumed standard deviation = 15
```

One-sided confidence interval bound

Test statistic

Variable	N	Mean	StDev	SE Mean	95% Lower Bound	Z	P
Age_Female	86	87.0930	9.5679	1.6175	84.4325	3.21	0.001

The output for the Z-test includes the sample size (notice again, that Minitab uses N to indicate the sample size rather than the more usual n), the sample mean (87.0930 years), sample standard deviation (9.5679 years), and $\sigma/\sqrt{n} = 1.6175$. Minitab refers to this last value as the *standard error of the mean* (SE Mean). The test statistic corresponding to the sample mean is

$$Z = \frac{\overline{X} - 81.9}{\sigma/\sqrt{n}} = \frac{87.0930 - 81.9}{1.6175} = 3.2$$

The output also includes the p-value (0.001) associated with the Z value 3.21.

You can test the null hypothesis above in two (exactly equivalent) ways:

1. By comparing the value of the test statistic Z, with the critical value for Z corresponding to $\alpha = .05$.

2. By comparing the p-value associated with the value of the test statistic to $\alpha = .05$.

You intend to use both techniques. The symbol α is called the *level of significance* for the test. It is the probability of rejecting H_0 when H_0 is true. Falsely rejecting H_0 is called a *Type I error*.

In this case, the critical value for Z will be the 95th percentile of the standard normal (Z) distribution. To determine this value:

- Choose **Calc > Probability Distributions > Normal**
- Click the **Inverse cumulative probability** option button

Notice that the default normal distribution is the *standard normal* with a mean of 0.0 and a standard deviation of 1.0, which is exactly what you want.

- Click the **Input constant** option button
- Type **.95** as the "Input constant" and click **OK**

FIGURE 7-5

95th percentile of the Z distribution

```
MTB > InvCDF .95;
SUBC>    Normal 0.0 1.0.
```

Inverse Cumulative Distribution Function

Normal with mean = 0 and standard deviation = 1

```
P( X <= x )      x
       0.95  1.64485 ──────────── Critical Z-value
```

The output in Figure 7-5 indicates that the 95th percentile of Z is 1.64485. (You may be familiar with the more common value, 1.645.) The value 1.64485 is shown in Figure 7-6. It is the value for the standard normal (Z) distribution that has an area of .95 to the left, and an area .05 to the right. (This figure was created in MINITAB Release 14.)

FIGURE 7-6

1.64485 is the 95th percentile of the Z distribution

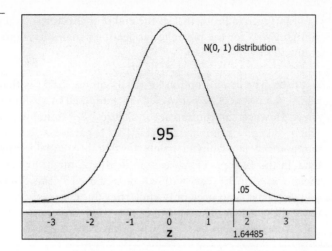

You will reject H_0 only if the value for the test statistic exceeds this *critical value*. Because Z = 3.21 greatly exceeds 1.64485, you can reject the null hypothesis at the .05 level of significance. The data suggest that the mean age at death of women living in affluent areas of the country is significantly greater than 81.9.

You can also base your decision on the *p-value* in the Z-test output in Figure 7-4. The p-value is the probability of getting a result at least as extreme as that obtained in the sample, assuming the null hypothesis is true. You reject H_0 when the p-value is less than your chosen level of significance. In this case because the p-value is 0.001, you can again, reject H_0 at the .05 level of significance. Many statisticians believe that reporting test results in terms of the p-value avoids confusion by letting readers use their own judgment as to whether the p-value is small enough to reject the null hypothesis.

◆ **Note** Minitab does not ask you to specify a level of significance and does not include one in the output. The user is free to compare the p-value to a level of significance of their choice. ■

You may worry about the validity of the Z-test in view of the pronounced skewness evident in the histogram in Figure 7-1. Dr. Ford assures you that in samples as large as n = 86, the Central Limit Theorem suggests that the test statistic

$$Z = \frac{\overline{X} - 81.9}{\sigma/\sqrt{n}}$$

will have close to an N(0, 1) distribution (assuming H_0 is true) even when the population (of ages at death) is quite skewed. The other assumption needed for the Z-test (and every other test) to be valid is that the sample be drawn randomly from the population. The question as to whether this group of 86 females can be viewed as a random sample of affluent women is a matter of judgment.

In Practice Problem 1 at the end of this tutorial, you are asked to perform a similar Z-test for the mean age at death for males living in affluent areas of the country.

▶ **Note** The default option for the 1-Sample Z-test is that the (raw) data on which the test will be performed are contained in a column in the current worksheet. However, as you can see in Figure 7-3, Minitab will also perform the test when you enter summarized data. This is particularly useful when you obtain the summarized data from an article, for instance, and don't have access to the raw data. In the example of the ages at death you might have read that the sample mean was 87.0930 years with n = 86 and σ = 15 years. To perform the same test as you did above, you need only click the "Summarized data" option button and then enter these values. ■

7.2 Computing a Confidence Interval for μ When σ Is Known

Notice in Figure 7-4 that, as part of the output for the Z-test, Minitab includes a 95% lower bound for μ (84.4325 years). This is an example of a *one-sided confidence interval* for μ. In this case the confidence interval is 84.4325 years to ∞. Minitab produces an upper one-sided confidence interval for μ when the alternative hypothesis is of the form $H_1: \mu > \mu_0$. When the alternative is of the form $H_1: \mu < \mu_0$, Minitab will produce a lower one-sided confidence interval for μ.

A $(1 - \alpha)100\%$ upper one-sided confidence interval for μ has the form $\bar{X} - Z(\sigma/\sqrt{n})$ to ∞ where Z is the $(1 - \alpha)100$th percentile of the standard normal distribution.

A $(1 - \alpha)100\%$ lower one-sided confidence interval for μ has the form $-\infty$ to $\bar{X} + Z(\sigma/\sqrt{n})$ where Z is the $(1 - \alpha)100$th percentile of the standard normal distribution.

This upper one-sided confidence interval can be used to test $H_0: \mu = 81.9$ at the .05 level of significance by checking whether the hypothesized value (81.9) is in the interval or not. Because this 95% confidence interval does not include the value 81.9 years, you can reject the null hypothesis at the .05 level of significance.

As part of your sociology project, Dr. Ford asks you to estimate the mean age at death for all women living in affluent areas by forming a (more conventional) 90% two-sided confidence interval.

- Choose **Stat > Basic Statistics > 1-Sample Z**
- Verify that the "Samples in columns" is Age_Female and that the "Standard deviation" is 15
- Clear 81.9 as the "Test mean" value (this is not necessary, but produces cleaner output)
- Click **Options** and type **90** (instead of the default value 95.0) as the "Confidence level"
- Click the "Alternative" **drop-down list arrow**, click **not equal**, and click **OK** twice

The output, shown in Figure 7-7, includes the 90% two-sided confidence interval 84.4325 years to 89.7536 years.

FIGURE 7-7

90% confidence interval for the mean age at death for females living in affluent areas of the country

```
MTB > OneZ 'Age_Female';
SUBC>   Sigma 15;
SUBC>   Confidence 90.
```

One-Sample Z: Age_Female

```
The assumed standard deviation = 15


Variable     N    Mean   StDev  SE Mean       90% CI
Age_Female  86  87.0930  9.5679  1.6175  (84.4325, 89.7536)
```

A $(1 - \alpha)100\%$ two-sided confidence interval for μ has the form $\bar{X} \pm Z(\sigma/\sqrt{n})$ where Z is the $(1 - \alpha/2)100$th percentile of the standard normal distribution.

Based upon this result, you can be 90% confident that the population mean age at death lies between 84.4 years and 89.8 years.

▶ **Note** As you have seen, Minitab ties the form of the confidence interval to the form of the alternative hypothesis for the same parameter. If you want to obtain a two-sided confidence interval, you must specify a two-sided alternative. ■

▶ **Note** When you enter a confidence level Minitab does not mind if you enter it as a percentage (90, for instance), or as a decimal (.9, for instance). The confidence level for confidence intervals can be set to any value from 0 to 100, not including these end points. ■

7.3   *Sample Size for Estimating* μ *When* σ *Is Known*

The half-width of a two-sided confidence interval is frequently called the *margin of error*. For instance, the margin of error when estimating the mean age at death for all women living in affluent areas of the country (89.7536 – 84.4325) / 2 is 2.66055 years. You can be 90% confident that the population mean lies within 2.66 years of the sample mean, 87.093 years. One way to improve (reduce) the margin of error is by increasing the sample size, n.

As a footnote to your work for Dr. Ford, use Minitab to compute the sample sizes needed to be 90% confident of achieving the following margins of error: .5 year, .75 year, 1.0 year, 1.25 years, …, 2.5 years. The formula for the sample size needed to achieve a margin of error of M with 90% confidence is, in this case,

$$n = \frac{1.645^2 \sigma^2}{M^2} = \frac{(1.645)^2 (15)^2}{M^2} = \frac{608.856}{M^2}$$

The value 1.645 is used here because approximately 90% of the Z distribution lies between –1.645 and 1.645.

- Click the **Current Data Window** toolbar button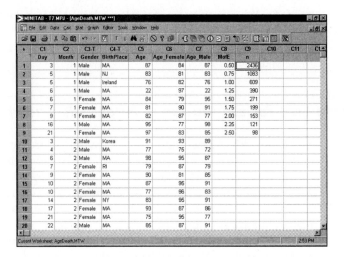

  First, enter the values .5, .75, ..., 2.5 into C8 using Autofill.
- Type **.50** and **.75** into the first two cells of C8 and highlight these cells
- Hold the cursor over the lower right-hand corner of the second cell until you see the heavy, black cross (+), Minitab's Autofill symbol
- Click and drag the cursor down the column until the small moving window reads **2.50**
- Name C8 **MofE**

  You will compute the sample size needed to obtain each of these margins of error by having Minitab calculate the value for $608.856/M^2$ and, conservatively, rounding up to the next highest integer. You can obtain the next highest integer by using Minitab's *ceiling* function.
- Choose **Calc > Calculator**
- Type **n** as the "Store result in variable"
- Enter, by clicking or typing, **Ceiling(608.856/MofE**2)** as the "Expression"
- Click **OK**
- Click the **Current Data Windo**w toolbar button , if necessary

  The values for C8 MofE and C9 n are shown in the Data window in Figure 7-8.

FIGURE 7-8

Values for the margin of error and the corresponding sample size

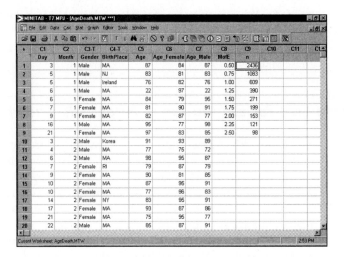

The more accurately you want to estimate the population mean—that is, the smaller you want the margin of error to be—the larger the sample size must be. To obtain a margin of error of 1.0 year, a sample size of 609 women is necessary. For a margin of error of 2.5 years, however, a sample size of n = 98 is sufficient.

You can plot the required sample sizes against the desired margin of error.
- Choose **Graph > Scatterplot** and double-click **With Connect Line**
- Double-click **C9 n** as the "Y variables"column and **C8 MofE** as the "X variables" column

- Click **Labels**, type **Plot of Sample Size Against the Margin of Error** as the "Title", and click **OK** twice

The resulting plot, shown in Figure 7-9, clearly shows that as the desired margin of error increases the required sample size decreases.

FIGURE 7-9

Plot of required sample size against desired margin of error

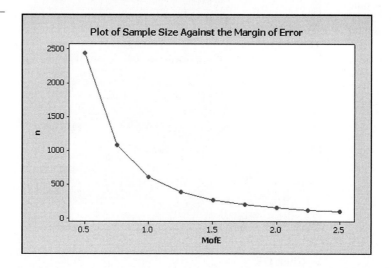

7.4 Inferences About μ When σ Is Unknown

Dr. Ford has discovered that the presumption that the value for σ is known to be 15 years is incorrect. In fact, the value for σ is not known. In this case you should do a one-sample t-test in place of the Z-test. The hypotheses are the same as before.

Null Hypothesis, H_0: μ = 81.9

Alternative Hypothesis, H_1: μ > 81.9

The t-test uses the fact that if the population is normally distributed and the null hypothesis is true, the test statistic

$$t = \frac{\overline{X} - 81.9}{\sigma / \sqrt{n}}$$

has a t distribution with n – 1 = 86 – 1 = 85 degrees of freedom. To perform the test:
- Choose **Stat > Basic Statistics > 1-Sample t**
- Double-click **C6 Age_Female** as the "Samples in column"
- Type **81.9** as the "Test mean" and click **Options**
- Click the "Alternative" **drop-down list arrow**, click **greater than**, and click **OK**
  In this example, examine Minitab's built-in Histogram.
- Click **Graphs**, click the **Histogram of data** check box, and click **OK**

The completed 1-Sample t (Test and Confidence Interval) dialog box should look as shown in Figure 7-10.

FIGURE 7-10

The completed 1-Sample t (Test and Confidence Interval) dialog box

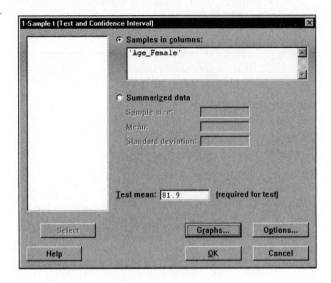

- Click **OK**

The histogram is shown in Figure 7-11.

FIGURE 7-11

Built-in histogram for Age_Female

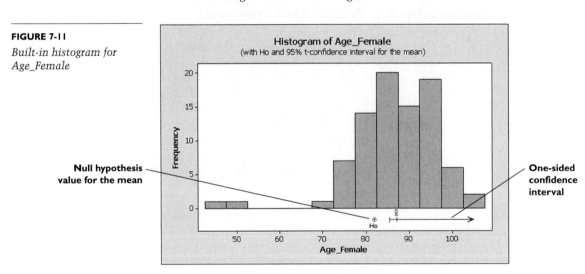

The plot shows the ages concentrated in the range slightly less than 70 to slightly over 100 but with the several outliers. The plot also shows the location of the hypothesized value by H_0 (81.9 years), the sample mean by $\overline{X}$ (87.0930 years), and the 95% upper one-sided confidence bound for μ by the line with the arrowhead.

- Click the **Session Window** toolbar button [⊞]

The numerical output for the t-test is shown in Figure 7-12.

FIGURE 7-12

Results of the t-test

```
MTB > Onet 'Age_Female';
SUBC>    Test 81.9;
SUBC>    Alternative 1;
SUBC>    GHistogram.
```

One-Sample T: Age_Female

Test of mu = 81.9 vs > 81.9

Variable	N	Mean	StDev	SE Mean	95% Lower Bound	T	P
Age_Female	86	87.0930	9.5679	1.0317	85.3773	5.03	0.000

The output is similar to that given for the Z-test. Minitab provides the 95% lower bound (85.3773 years) for μ, the value for the test statistic (t = 5.03), and the corresponding p-value, 0.000. The p-value is tiny, but it is not really 0. Minitab uses the notation 0.000 when the p-value is less than 0.0005. Because the p-value is certainly less than any reasonable level of significance, you can, as with the Z-test, reject the null hypothesis. The data suggest that the mean age at death for all women living in affluent areas of the country exceeds 81.9 years.

▶ **Note** Minitab uses the more common lowercase t in the menu command (Stat > Basic Statistics > 1-Sample t) but uses the uppercase T in the output. ■

▶ **Note** In the output in Figure 7-12 the term SE Mean refers to $s/\sqrt{n} = 9.5679/\sqrt{86} = 1.0317$. Minitab used the same term in the Z-test output (Figure 7-4) to refer to $\sigma/\sqrt{n} = 1.6175 = 15/\sqrt{86} = 1.6175$. There is no agreement among statisticians about whether SE Mean should refer to $s/\sqrt{n}$ or $\sigma/\sqrt{n}$, but you should keep in mind that Minitab uses the term to indicate these two different quantities. ■

The two assumptions underlying the use of one-sample t inferences are:

1. A random sample
2. The population from which the sample is selected is at least approximately normal

The skewed nature of the female ages at death, as shown in Figure 7-11 cast some doubt on the validity of the latter assumption. However, these t procedures are fairly *robust* (insensitive to departures from the normality assumption) particularly when n is large. Our tests and confidence intervals will be valid even if the distribution of age-at-death is skewed to the left.

The t-Test and Confidence Interval with Summarized Data

A friend in the class asks for your help with her project. She has also collected ages at death, but from a random sample of "famous" women whose obituaries appeared in *The Boston Globe*. She needs help performing a two-sided t-test of whether all such women have a mean age at death (μ) significantly different from

81.9 years. She would also like to get an 80% confidence interval for μ. Your friend has only the summarized data: n = 19, $\overline{X} = 78.94$ years, and s = 13.15 years.

Recall that the formula for a $(1 - \alpha)100\%$ two-sided confidence interval for μ based upon the t distribution has the form $\overline{X} \pm t\left(s/\sqrt{n}\right)$ where t is the $(1 - \alpha/2)100$th percentile of the t_{n-1} distribution.

- Click the **Edit Last Dialog** toolbar button 🖰
- Press F3 to eliminate the previous settings
- Click the **Summarized data** option button and type **19** as the "Sample size"
- Type **78.94** as the "Mean" and **13.15** as the "Standard deviation"
- Type **81.9** as the "Test mean"
- Click **Options**, type **80** as the "Confidence level", verify that the "Alternative" is not equal, and click **OK**

The completed 1-Sample t dialog box should look as shown in Figure 7-13.

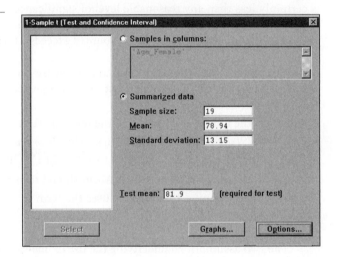

- Click **OK**

The output is shown in Figure 7-14.

FIGURE 7-14

One-sample t-test with summarized data

```
MTB > Onet 19 78.94 13.15;
SUBC>    Test 81.9;
SUBC>    Confidence 80.
```

One-Sample T

```
Test of mu = 81.9 vs not = 81.9


  N    Mean    StDev   SE Mean      80% CI           T      P
 19   78.9400  13.1500  3.0168  (74.9265, 82.9535)  -0.98  0.340
```

Two-sided confidence interval

Notice that the form of the output is similar to that of raw data. The value for the test statistic is –0.98, and the corresponding p-value is 0.340. With such a large p-value your friend should not reject the null hypothesis. If the null hypothesis is true, a sample mean of 78.9400 years may well be due solely to chance. The mean age at death for so-called "famous" women is not significantly different from 81.9 years. Your friend can be 80% confident that the mean age at death of "famous" women lies between 74.9265 years and 82.9535 years.

◆ **Note** The p-value (0.340) in the previous two-sided t-test can be written as the sum of two equal probabilities: $P(t < -0.98)$ and $P(t > 0.98)$, where t has the t distribution with 18 degrees of freedom. You can confirm the fact that $P(t < -0.98)$ is 0.170 (= 0.340/2) by choosing the command Calc > Probability Distributions > t, using 18 as the number of degrees of freedom and –0.98 as the input constant. ■

◆ **Note** You cannot obtain a built-in graph when you perform a test and/or obtain a confidence interval for μ from summarized data because Minitab does not have access to raw data. As a consequence, you cannot check the normality assumption with summarized data. ■

Before going on to the next case study, save your project:

■ Click the **Save Project** toolbar button 🖬

7.5 Inferences About a Population Proportion

CASE STUDY	HEALTH CARE— WORK DAYS LOST TO PAIN

As a researcher at a large Health Maintenance Organization (HMO), you have been asked to collaborate with a nurse who is completing her master's degree to select a random sample of 300 patients who have reported low-back pain in the past six months. Federal guidelines suggest that with the appropriate treatment, less than 60% of patients with low-back pain should lose work days. You are to use the data to test whether this HMO has met this guideline. Because of incomplete records, you have an effective sample size of 279.

The relevant data are in a file called *Backpain.mtw*.

■ Open the worksheet **Backpain.mtw**
■ Use Help to review the contents of the data

For each of the 279 patients, the number of days lost as a result of low-back pain is in C3 LostDays. The Minitab commands to obtain confidence intervals and hypothesis tests for a single proportion require that the sample contains only numeric or text values that stand for "successes" and "failures". If the column contains numeric values, the larger of the two is considered a success. (If the column contains text values, the text value beginning with the letter later in the alphabet is considered a success. For instance, with the text values tea and coffee, tea would be considered a success.) Create a new variable that takes the value 0 if

C3 LostDays is 0 and the value 1 if C3 LostDays takes any non-zero value. Such a variable is often called an *indicator variable*. You can use the Calc > Calculator command to create such a variable, LostInd (short for LostDays Indicator).

- Choose **Calc > Calculator**
- Press F3 to clear the dialog box
- Type **LostInd** as the "Store result in variable"
- Enter, by clicking or typing, **LostDays > 0** as the "Expression"

    The completed Calculator dialog box should look as shown in Figure 7-15.

FIGURE 7-15

The completed Calculator dialog box

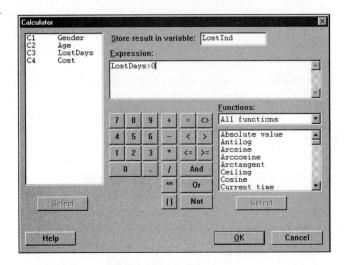

- Click **OK**

    The condition you entered in the "Expression" text box automatically creates an indicator variable in C5 LostInd, which takes the value 1 when the condition is met and 0 otherwise. The 1s in C5 LostInd represent those patients in the sample who lost at least one day of work as a result of low-back pain; the 0s represent those patients who did not. You are ready to test hypotheses about the proportion (p) of all patients at the HMO who missed at least one day's work with low-back pain.

    Recall that federal guidelines suggest that, with appropriate treatment, less than 60% (or a proportion of .6) of patients with low-back pain should lose work days. Do these data suggest that this HMO has met this guideline? Set up the following hypotheses:

    Null hypothesis, H_0: p = .6

    Alternative hypothesis, H_1: p < .6

    To test H_0:

- Choose **Stat > Basic Statistics > 1 Proportion**
- Double-click **C5 LostInd** as the "Samples in columns"
- Click **Options** and type **0.6** as the "Test proportion"

- Click the "Alternative" **drop-down list arrow** and click **less than**

The completed 1 Proportion - Options dialog box should look as shown in Figure 7-16.

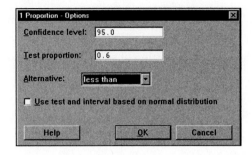

- Click **OK** twice

The resulting output is shown in Figure 7-17.

```
MTB > POne 'LostInd';
SUBC>    Test 0.6;
SUBC>    Alternative -1.
```

Test and CI for One Proportion: LostInd

Test of p = 0.6 vs p < 0.6

Event = 1

				95% Upper	Exact
Variable	X	N	Sample p	Bound	P-Value
LostInd	154	279	0.551971	0.602144	0.058

According to the output, 154 patients, or approximately 55% of the 279 patients in the sample, lost at least one day's work. The p-value associated with the sample proportion, 0.551971, is 0.058. This is small, suggesting that the data provide little support for the null hypothesis that p = .6. However, if you insist on a level of significance of α = .05, you cannot quite reject the null hypothesis at the .05 level of significance.

As well as the hypothesis test, Minitab automatically computes a lower one-sided confidence interval for p. You can be 95% confident that the unknown p is less than 0.602144.

By default, Minitab uses a procedure to compute test hypotheses and confidence intervals about a population proportion that differs from the normal approximation method used in most introductory texts. The procedure used by Minitab—called the *"Exact" Method*—does not require the usual assumption that (sample size)*(sample proportion) ≥ 5 and (sample size)*(1 – sample proportion) ≥ 5. (Some texts suggest other guidelines.) As with any other inference procedure, the sample of 300 patients who have reported low-back pain is assumed to have been selected at random from the population.

As you can see in Figure 7-16, using the normal approximation method is an option. Check what impact it makes if you use the normal approximation method. For the null hypothesis H_0: p = .6, the form of the test statistic is:

$$Z = \frac{\hat{p} - p}{\sqrt{(p)(1-p)/n}} = \frac{\hat{p} - .6}{\sqrt{(.6)(1-.6)/n}}$$

where $\hat{p}$ is the sample proportion of successes. The form of the lower one-sided upper confidence interval for p is

$$-\infty \text{ to } \hat{p} + Z\sqrt{(\hat{p})/(1-\hat{p})/n},$$

where Z is the $(1 - \alpha)$100th percentile of the standard normal distribution.

- Click the **Edit Last Dialog** toolbar button 🖼
- Click **Options** and click the **Use test and interval based on normal distribution** check box
- Click **OK** twice

The output is shown in Figure 7-18.

FIGURE 7-18

Testing the hypothesis p = 0.6 versus p < 0.6 using the normal approximation

```
MTB > POne 'LostInd';
SUBC>    Test 0.6;
SUBC>    Alternative -1;
SUBC>    UseZ.
```

Test and CI for One Proportion: LostInd

Test of p = 0.6 vs p < 0.6

Event = 1

				95% Upper		
Variable	X	N	Sample p	Bound	Z-Value	P-Value
LostInd	154	279	0.551971	0.600942	-1.64	0.051

The use of the normal approximation has made little difference. The 95% upper bound changes from 0.602144 to 0.600942 and the p-value declines slightly from 0.058 to 0.051.

Suppose you are interested primarily not in testing a hypothesis about p (the proportion of all patients at the HMO who missed at least one day's work with low-back pain) but in estimating p with a 90% two-sided confidence interval.

A $(1 - \alpha)$100% two-sided confidence interval for p has the form

$$\hat{p} \pm Z\sqrt{(\hat{p})(1-\hat{p})/n}$$

where Z is the $(1 - \alpha/2)$100th percentile of the standard normal distribution.

To obtain this confidence interval:

- Click the **Edit Last Dialog** toolbar button 🖼
- Click **Options** and type **90** for the "Confidence level"
- Click the "Alternative" **drop-down list arrow** and click **not equal**
- Click **OK** twice

The output is shown in Figure 7-19. The 90% confidence interval for p is 0.503001 to 0.600942 (or, more usefully, .50 to .60). You can be 90% confident that the proportion of all members with low-back pain who have lost work days lies in this interval.

```
MTB > POne 'LostInd';
SUBC>    Confidence 90;
SUBC>    Test 0.6;
SUBC>    UseZ.
```

Test and CI for One Proportion: LostInd

```
Test of p = 0.6 vs p not = 0.6

Event = 1


Variable    X    N  Sample p       90% CI         Z-Value  P-Value
LostInd    154  279  0.551971  (0.503001, 0.600942)   -1.64   0.102
```

7.6 *Computing the Power of a Test*

In the previous example, you could not quite reject the null hypothesis at the .05 level of significance. Thus, you may have made what is known as a *Type II error*, that of failing to reject the null hypothesis (p = .60) when, in fact, the alternative hypothesis (p < .60) is true. What is the probability of such an error in this example? There is no single answer to this question because your chance of failing to reject the null hypothesis depends on the value for the unknown p.

Suppose, for instance, p = .55; that is, 55% of all patients in the HMO who are suffering from low-back pain lose work days. What is the probability that in this case, you would (incorrectly) fail to reject the null hypothesis based upon your sample? Minitab does not compute this probability, but it can compute the probability of the complementary event—that is, the probability of rejecting the null hypothesis correctly if p = .55. This probability is called the *power* of the test when p = .55.

The power of a test is the probability of rejecting the null hypothesis when the alternative hypothesis is true. The value for this probability varies with the value of the parameter (p = .55 above).

To compute the probability of rejecting H_0 when p = .55:

- Choose **Stat > Power and Sample Size > 1 Proportion**
- Type **279** as the "Sample sizes"
- Type **.55** as the "Alternative values of p" and **0.6** as the "Hypothesized p"
- Click **Options**, click the "Alternative Hypothesis" **Less than** option button, verify that the "Significance level" is 0.05, and click **OK**

The completed Power and Sample Size for 1 Proportion dialog box should look as shown in Figure 7-20.

FIGURE 7-20

The completed Power and Sample Size for 1 Proportion dialog box

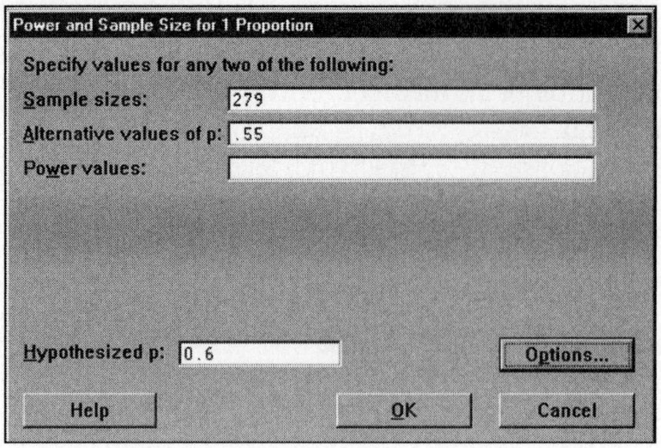

```
Power and Sample Size for 1 Proportion                    [X]
Specify values for any two of the following:
Sample sizes:           279
Alternative values of p: .55
Power values:

Hypothesized p: 0.6                          Options...

    Help                    OK            Cancel
```

■ Click **OK**

The output is shown in Figure 7-21.

FIGURE 7-21

Calculating power for a single proportion

```
MTB > Power;
SUBC>   POne;
SUBC>     Sample 279;
SUBC>     PAlternative .55;
SUBC>     PNull 0.6;
SUBC>     Alternative -1.
```

Power and Sample Size

Test for One Proportion

Testing proportion = 0.6 (versus < 0.6)
Alpha = 0.05

Alternative Sample
 Proportion   Size   Power
    0.55   279   0.523526

With a sample size of 279 patients the test has a power of .523526. This is the probability of correctly rejecting the null hypothesis when p is, in fact, .55. This is not very encouraging. It means that if p is .55, your test is only slightly better than a coin toss at making the correct decision to reject the null hypothesis. The difference between the hypothesized value for p (0.6) and the alternative proportion (0.55) is 0.05. This difference is frequently referred to as the desired *effect size*.

In this context, there is a close relationship between the power of the test, the sample size, and the alternative value for p. If you specify the sample size and the alternative value for p, Minitab will compute the power of the test for that alternative (as you did above). Also, if you specify an alternative value for p and a value for the power of the test, Minitab will compute the sample needed to achieve this power.

You are interested in examining how the power of your test varies with how far the true value for p is from the hypothesized value (p = .60). So you will have Minitab compute the power of the test against the alternatives p = .50, .51, ..., .60.

- Click the **Edit Last Dialog** toolbar button 🖻
- Type **.50:.60/.01** as the "Alternative values of p"(or you could enter all the values .50, .51, .52, ..., .60)
- Click **Options** and type **Proportions** as the "Store alternatives in" column
- Type **Power** as the "Store power values in" column and click **OK** twice

The values for p and the corresponding values for the power of the test are reported in the Session window and are stored in the two new columns—C6 Proportions and C7 Power in the Data window. The output in the Session window is shown in Figure 7-22.

FIGURE 7-22

Power of the test of
H_0: p = .6 for different
values for p

```
MTB > Name c6 "Proportions" c7 "Power"
MTB > Power;
SUBC>    POne;
SUBC>       Sample 279;
SUBC>       PAlternative .50:.60/.01;
SUBC>       PNull 0.6;
SUBC>       Alternative -1;
SUBC>       SPAlternative 'Proportions';
SUBC>       SPower 'Power'.
```

Power and Sample Size

Test for One Proportion

Testing proportion = 0.6 (versus < 0.6)
Alpha = 0.05

Alternative Proportion	Sample Size	Power
0.50	279	0.958099
0.51	279	0.918530
0.52	279	0.855827
0.53	279	0.766740
0.54	279	0.653223
0.55	279	0.523526
0.56	279	0.390751
0.57	279	0.269117
0.58	279	0.169586
0.59	279	0.097012
0.60	279	0.050000

The output contains the corresponding powers of the test for the alternative proportions and the sample size. Notice that the power of the test increases the further the (true) value for p is from the hypothesized value (0.6 in this case). If p is only 0.5 the power of the test is 0.958099. However, if p is actually .59 the power of the test is only 0.097012. If the null hypothesis is true and p = .6, then the power of the test is precisely the level of significance, α = .05. Both are the probability of rejecting the null hypothesis.

Return to the Data window:

- Click the **Current Data Window** toolbar button 🖩

The values for Proportions and Power are stored in C6 and C7, respectively. The relationship between these two variables can be illustrated by a plot of C7 Power against C6 Proportions.

- Choose **Graph > Scatterplot** and double-click *With Connect Line*

- Double-click **C7 Power** as the "Y variables" column and **C6 Proportions** as the "X variables" column
- Click **Labels**, type **Plot of Power Against the Value for p** as the "Title", and click **OK** twice

The resulting plot should look as shown in Figure 7-23.

FIGURE 7-23

Plot of power against p

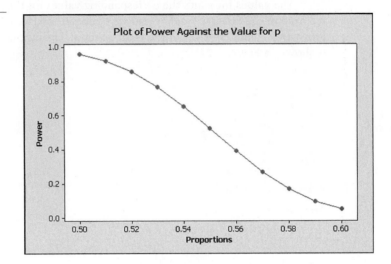

The graph shows quite dramatically that the power of the test decreases the closer the value for the population proportion (p) is to the hypothesized value for p (0.6).

If you choose Stat > Power and Sample Size, you will see that Minitab can perform these kinds of power calculations for the 1-Sample Z, the 1-Sample t, the 2-Sample t, the 2 Proportions, and the One-Way ANOVA procedures, as well as for the 1 Proportion procedure used previously. You will encounter the 2-Sample t and the 2 Proportions procedures in the next tutorial and the One-Way ANOVA in Tutorial 9.

- Click the **Save Project** toolbar button 🖫
- Choose **File > Exit**

▶ **Note** Under certain circumstances you can compute a confidence interval or test a hypothesis for the population standard deviation σ. However, such inferences tend to be much more sensitive to departures from normality than the procedures you have used for making inferences about μ in this tutorial and so are used relatively rarely. Minitab does not include a command to perform such inferences. ■

Well done! You have successfully learned how to use Minitab to perform one-sample inferences for means and proportions. In the next tutorial, you will cover two-sample techniques.

Frozen foods—especially frozen pizzas—are a common staple in many American households. Pizza producers must carefully monitor how bacteria respond to prolonged refrigeration temperatures. Microbiologists used Minitab to help determine how long pizza can be refrigerated without becoming a threat to health.

To examine the effect of freezing on specific bacteria, researchers inoculated the pizzas with measured amounts of *Escherichia coli* (*E. coli*) and other potential contaminants. They stored the pizzas at various temperatures, retrieving samples every other day for microbiological analysis. They recorded the number of colonies of bacteria at each testing and entered that data into Minitab. When they compared each day's frequency with its initial frequency, using a one-sample t-test on the differences in frequencies, they found that the *E. coli* colonies increased significantly (p-value < 0.05).

This result suggests E. coli will grow when acceptable refrigeration temperatures are not maintained. Pizza producers could best safeguard public health by carefully monitoring bacteria levels before beginning the freezing process, while the vendors must ensure adequate refrigeration and must carefully monitor shelf dates.

Minitab Command Summary

This section describes the commands introduced in, or related to, this tutorial. Use Minitab's online Help for a complete explanation of all commands.

Minitab Menu Commands

Menu	Command	Description
Stat ➤	Basic Statistics	
	➤ 1- Sample Z	Performs inferences (tests and confidence intervals) on a population mean when σ is known.
	➤ 1-Sample t	Performs inferences (tests and confidence intervals) on a population mean when σ is unknown.
	➤ 1 Proportion	Performs inferences (tests and confidence intervals) on a population proportion.
	Power and Sample Size	
	➤ 1-Sample Z	Computes the power for different sample sizes or for different values for μ, based upon the Z distribution. Also computes sample sizes for different power values.
	➤ 1-Sample t	Computes the power for different sample sizes or for different values for μ, based upon a t distribution. Also computes sample sizes for different power values.
	➤ 1 Proportion	Computes the power for different sample sizes or for different values for p. Also computes sample sizes for different power values.
Graph ➤	Scatterplot	
	With Connect Line	Produces a scatterplot with successive points connected with a straight line.

Review and Practice

Matching Problems

Match the following terms to their definitions by placing the correct letter next to the term it describes.

_____ Power

_____ 1-Sample t

_____ 1 Proportion

_____ Alternative

_____ α

_____ Confidence level

_____ p-value

_____ 1-Sample Z

_____ σ

_____ p

a. The probability of getting a result at least as extreme as that obtained in the sample, assuming the null hypothesis is true

b. The Minitab command used to perform a one-sample inference about a mean, assuming σ is unknown

c. The level of significance

d. The probability of rejecting the null hypothesis correctly

e. The unknown population proportion

f. The Minitab command you use to perform a one-sample inference about a mean, assuming σ is known

g. The Minitab option that allows you to change how sure you are about an interval estimate

h. The Minitab command you use to perform a one-sample inference about a population proportion

i. The Minitab option that enables you to specify the direction of one of the two hypotheses

j. The population standard deviation

True/False Problems

Mark the following statements with a *T* or an *F*.

a. ____ Confidence levels can be entered in Minitab as a percentage (99) or as a decimal (.99).

b. ____ The population standard deviation must be known to perform a one-sample test of a mean in Minitab.

c. ____ When σ is unknown, one-sample confidence intervals and hypothesis tests for μ are performed using the same Minitab command.

d. ____ Minitab's Power and Sample Size for 1 Proportion command will calculate the sample sizes corresponding to different power values.

e. ____ The confidence level for confidence intervals can be set to any value from 0 to 100 not including these end points.

f. ____ Minitab allows you to specify a significance level for a hypothesis test.

g. ____ Minitab provides histograms of the data as an option when the 1-Sample t command is used with summarized data.

h. ____ In Minitab's one-sample confidence intervals for μ and p, the default confidence level is 90%.

i. ____ Minitab selects the smaller of the two different values in a column to represent a success.

j. ____ Minitab has a command to provide the sample size corresponding to a specified margin of error and a specified level of confidence.

Practice Problems

The practice problems instruct you to save your worksheets with a filename prefixed by P, followed by the tutorial number. In this tutorial, for example, use *P7* as the prefix. If necessary, use Help or Appendix A Data Sets, to review the contents of the data sets referred to in the following problems. Interpretations should use the language of the subject matter of the question. If you are using the Student version, be sure to close worksheets when you have completed a problem.

1. Open *AgeDeath.mtw*

   a. Perform a Z-test of whether the mean age at death of males living in affluent areas of the country is greater than the average for all males, which is 75.5 years. Assume that σ is 15 years. Further assume that these data may be regarded as a random sample of male deaths over all affluent areas of the United States. Your output should include a histogram of the male ages.

   b. State your conclusion.

   c. How would you characterize the shape of the histogram? Does your answer invalidate the test in any way?

2. Open *Infants.mtw*. The infant birth weights in this data set are for infants born to low-income mothers. You are interested in whether, on average, infants born to such mothers weigh less than the national average. The average birth weight over all infants born in the United States is approximately 3350 grams. The standard deviation is approximately 475 grams. Assume that these data are a random sample.

   a. Set up the appropriate null and alternative hypotheses in this case. Be sure to define μ in words.

   b. Perform a Z-test of your null hypothesis. What conclusion seems appropriate in this case? Use $\alpha = .05$.

   c. Interpret the default 95% confidence interval in this case.

3. Open *Homes.mtw*. You are asked to investigate changes in house prices in a nearby county over the last two years for a real estate company operating in the county. Two years ago the mean selling price was $150,350. You have collected data on a random sample of 150 recent home sales. These data are contained in the column Price. Use these data to test whether there has been a significant change in the mean selling price over the last two years.

   a. Define μ in this context and write down the appropriate hypotheses.

   b. What is the p-value in this case? What conclusion seems appropriate given this value?

   c. Obtain a histogram of house prices and describe its shape.

   d. Did you use a one-sample Z-test or a one-sample t-test in this case? Briefly explain your choice of test.

4. A sociologist is interested in estimating the mean time (μ) spent watching television per day for sixth graders in her town. She intends to select a random sample of sixth graders and will ask the selected children to keep (with the help of adults) a log of their TV time over a three-week period. A median time will be computed for each child. But she needs your help in determining the number of children to select. She thinks that the standard deviation of daily TV time is about 90 minutes. You decide to compute the sample size needed to achieve a

margin of error ranging from 10 minutes to 50 minutes (at intervals of five minutes), with 95% confidence.

a. Enter the values 10, 15, …, 50 into C1. Name this column MarginError.

b. Use the Calc > Calculator command, as you did in Section 7.3, to compute (in C2) the sample size needed to estimate m for each margin of error with 95% confidence. Don't forget to round up. Name C2 SampleSize.

c. Create a plot of SampleSize against MarginError.

d. Write a brief explanation of how sample size varies with the margin of error.

e. The sociologist selects a sample of 100 sixth-graders. What does your plot suggest as the corresponding margin of error?

5. Open *Candyb.mtw*. The mean net weight of a package of candy is advertised as 20.89 grams. Use the data in NetWgt to determine if the mean net weight is less than the advertised value at a significance level of .05. Assume that the candy packages in the sample represent a random sample of all such packages.

a. Why is this one-sided alternative appropriate?

b. The p-value in this case is rather large. Provide an explanation for the large p-value and for what it suggests about the mean weight of the candy packages.

6. Open *OldFaithful.mtw*.

a. Compute a 90% two-sided confidence interval for (i) the mean duration of the eruptions, (ii) the mean interval between eruptions, and (iii) the mean height of eruptions.

b. Interpret these intervals for a representative from the National Park Service. Be sure that your interpretations anticipate the question: "90% of what?"

c. What assumptions were implicit in your computation? Obtain what evidence you can to check these assumptions.

7. By law, an industrial plant can discharge no more than 500 gallons of wastewater per hour, on the average, into a neighborhood lake. Based upon infractions they have noticed, an environmental action group believes that this limit is being exceeded. Monitoring the plant is expensive and only a small sample is possible. A random sample of 14 hours is selected and the amount of water discharged is recorded for each hour.

a. Define what μ is in this context and set up the two hypotheses.

b. The group obtained the summarized data below. Use these data to test your null hypothesis.

```
 n  Mean   SD
14 576.2 147.5
```

c. Briefly explain your conclusion for the state Department of the Environment.

d. In reaching your conclusion in part c, what hypothesis testing error might you have made?

e. What does the one-sided confidence interval that comes with the test result tell you about μ?

8. Open *Prof.mtw*. Assume the surveys were distributed to a random sample from all sections at the college. Determine if the proportion of surveyed courses with senior numbers (400s) differs significantly from .25. Use a level of significance of .05.

9. Open *PulseA.mtw*. Assume that these 92 students can be considered a random sample of 18- to 21-year-olds in college.

   a. Some years ago it was believed that approximately one-quarter of such students smoked. There is some concern that the current proportion (p) who smoke is substantially higher. Perform the appropriate test at the .05 level of significance.

   b. How powerful is your test in part a against the alternative that the true proportion of such students who smoke is .3?

   c. Answer part b for the following alternative values: .25, .26, ..., .35. Produce a plot of power against the value for p and comment on what your plot shows.

   d. Answer part b for the following levels of significance: .1, .05, .01, and .005. Briefly explain how the power of the test varies with the level of significance.

10. Use Help and StatGuide to investigate the exact methodology that Minitab uses to compute a confidence interval and test a hypothesis for a population proportion. Compare the explanations given by each of these sources. If possible, consult your instructor for details.

11. During the year 2002 the state Department of Justice introduced reforms designed to reduce the mean waiting time between the date of charging and the date of trial for criminal defendants. A state audit determined that the mean waiting time in 2001 was 142 days. The standard deviation was 41 days. The Department of Justice intends to select a random sample of criminal cases in early 2003 in order to test whether the reforms have significantly reduced the mean waiting time. Your consulting company has been hired to help them do this.

   a. Define μ in this case and state the null and alternative hypotheses.

   b. Compute the sample sizes needed to obtain a t-test with α = .05 and power = .8 against the alternatives μ = 137, 132, 127, 122, 117, 112, 107, and 102. Choose Stat > Power and Sample Size > 1-Sample t. Under "Differences" enter the values –5, –10, –15, –20, –25, –30, –35, and –40 (you can type –5:–40/–5 if you wish). Enter .8 as the power and assume that the standard deviation of waiting time is still 41 days. Store the sample sizes in a column. Plot the needed sample sizes against μ.

   c. If the state Department of Justice decides on a random sample of size n = 50 cases, compute the power of the test against each of the alternative values for μ listed in part b. Plot the power of the test against μ.

   d. Write a paragraph explaining your results.

On Your Own

A recent study suggested that too much sleep or too little sleep could be detrimental to one's health. The study suggested that approximately 7.5 hours of sleep per day was optimal for adults. However, with the pressures of modern life it is frequently difficult to get as much sleep as is needed. Select a group of people who study or work with you and ask each of them for the number of hours of sleep they get on a typical weeknight. Get as large a sample of responses as you can. Enter these times into Minitab and use them to address the following issues.

a. For the people you selected, is there evidence that the mean number of hours of sleep is significantly less than 7.5?

b. Estimate, with a 90% two-sided confidence interval, the mean hours of sleep for such people.

c. Estimate, with a 90% two-sided confidence interval, the proportion of such people who get less than 6 hours of sleep per night.

d. Comment on the normality assumption in your tests.

Answers to Matching Problems

(d) Power
(b) 1-Sample t
(h) 1 Proportion
(i) Alternative
(c) α
(g) Confidence level
(a) p-value
(f) 1-Sample Z
(j) σ
(e) p

Answers to True/False Problems

(a) T, (b) F, (c) T, (d) T, (e) T, (f) F, (g) F, (h) F, (i) F, (j) F

Inferences from Two Samples

The one-sample procedures you used in Tutorial 7 are not used as frequently as two-sample procedures, which involve making inferences about the differences between two population means or between two population proportions. In this tutorial, you will examine Minitab's various procedures for doing two-sample t-tests and t-confidence intervals for comparing means, as well as procedures for doing Z-tests and Z-confidence intervals for comparing two proportions. You will also investigate Minitab's procedure for inference on the difference between paired population means. All of these hypothesis tests and confidence intervals can be computed with raw or with summarized data. Finally, you will explore the relationship between sample size and power when comparing two means and when comparing two proportions.

OBJECTIVES

In this tutorial, you learn how to:

- Test for the significance of the difference between two population means when two *independent* samples are, first, stacked in a single column and, then, recorded in separate columns

- Obtain a confidence interval for the difference between two population means when two independent samples are, first, stacked in a single column and, then, recorded in separate columns

- Test for the significance of and obtain a confidence interval for the difference between two population means when the samples are *paired* (or *dependent*)

- Compute the power of, and the sample size for, the two-sample t-test for the difference between two population means when the two samples are independent

- Test for the significance of, and obtain a confidence interval for, the difference between two population proportions when the two samples are independent

- Compute the power of, and the sample size for, the two-sample Z-test for the difference between two population proportions when the two samples are independent

Comparing Population Means from Two Independent Samples

CASE STUDY | **SOCIOLOGY—AGE AT DEATH (CONTINUED)**

A sociology professor, Dr. Ford, is interested in examining the extent to which women living in affluent areas of the United States live longer than women in the country as a whole. Dr. Ford has obtained the age at death for all of the residents of an affluent suburb of Boston who died in the year 2001. These data are stored in a file called *AgeDeath.mtw*.

When your sociology class was discussing the results of the mean age at death of females and males in affluent areas of the United States, the question arose whether there was a difference between the means of these two distinct populations. Actuarial studies predict that women will live longer than men, so the class was curious whether the same was true for people living in affluent areas—the group from which the samples were presumed to have been drawn. Dr. Ford, your professor, suggests that the class attempt to settle this question by performing a two-independent-sample t-test.

Begin by opening Minitab:

- Open **Minitab** and maximize the Session window
- If necessary, **Enable Commands** and clear Output Editable from the Editor menu
- Click the **Save Project** toolbar button 🖫 and, in the location where you are saving your work, save this project as *T8.mpj*

    You will be working with the *AgeDeath* data, as you did in the last tutorial.

- Open the worksheet *AgeDeath.mtw*

Two-Sample t-tests Using Stacked Data

Before performing the test, examine the data by looking at numerical summaries:

- Choose **Stat > Basic Statistics > Display Descriptive Statistics**
- Double-click **C5 Age** as the "Variables"
- Double-click **C3 Gender** as the "By variables"
- Click **OK**

    The output is shown in the Session window and in Figure 8-1.

FIGURE 8-I

Descriptive statistics for Age by Gender

```
MTB > Describe  'Age';
SUBC>    By  'Gender';
SUBC>    Mean;
SUBC>    SEMean;
SUBC>    StDeviation;
SUBC>    QOne;
SUBC>    Median;
SUBC>    QThree;
SUBC>    Minimum;
SUBC>    Maximum;
SUBC>    N;
SUBC>    NMissing.
```

Descriptive Statistics: Age

Variable	Gender	N	N*	Mean	SE Mean	StDev	Minimum	Q1	Median	Q3
Age		1	0	92.000	*	*	92.000	*	92.000	*
	Female	86	0	87.09	1.03	9.57	47.00	81.75	87.00	94.00
	Male	64	0	80.64	1.63	13.07	22.00	76.00	83.00	88.00

Variable	Gender	Maximum
Age		92.000
	Female	105.00
	Male	98.00

The mean age at death (87.09 years) of the 86 females is about six-and-one-half years greater than that for the 64 males (80.64 years). The t-test will determine if this difference is statistically significant, but it certainly seems striking and of considerable practical importance—for life insurance companies, for example. The standard deviations for the two samples of ages at death are quite different (9.57 years for females and 13.07 for males). There is one person who died at age 92, but whose gender could not be determined from the available records.

Minitab offers several options for testing whether one population mean is significantly greater than another. Because the males and females were assumed to be randomly and independently selected from all such affluent individuals, you can use the t-test for independent samples. Your hypotheses are:

Null Hypothesis, H_0: $\mu_F - \mu_M = 0$

Alternative Hypothesis, H_1: $\mu_F - \mu_M > 0$

where μ_F is the (unknown) mean age at death for all females living in affluent areas and μ_M is the corresponding mean for all males. You will use $\alpha = .05$.

In this case the sample values are stacked in one column (C5 Age) while the *subscripts* that identify which of the two samples the ages belong to are in another (C3-T Gender). Minitab uses the following notation for the two-sample t-test (where μ_1 is the mean for population 1 and μ_2 is the mean for population 2).

Hypothesis	Minitab Notation	
Null	H_0: $\mu_1 - \mu_2 = 0$	
Alternative	H_1: $\mu_1 - \mu_2 < 0$	less than
	H_1: $\mu1 - \mu_2 \neq 0$	not equal
	H_1: $\mu_1 - \mu_2 > 0$	greater than

When the samples are stacked and the column of subscripts is numeric, Minitab assumes that the smaller of the two numeric values the sample from population 1. When the column of subscripts contains text values with two names, Minitab assumes that the sample from population 1 corresponds to the name that begins with the letter that is closer to the beginning of the alphabet. Thus, in this case study on age of death, because F is closer to the beginning of the alphabet than M, Minitab assumes that Females are the sample from population 1 and Males are the sample from population 2. So, in Minitab notation, the appropriate choice of alternative hypothesis is "greater than".

Now, to obtain the test:

- Choose **Stat > Basic Statistics > 2-Sample t**
- Verify that the default "Samples in one column" option button is selected
- Double-click **C5 Age** as the "Samples" and **C3 Gender** as the "Subscripts"
- Click **Options**
- Verify that the default "Test difference" (hypothesized difference) is 0.0
- Click the "Alternative" **drop-down list arrow**, click **greater than**, and click **OK**

The completed 2-Sample t (Test and Confidence Interval) dialog box should look as shown in Figure 8-2.

FIGURE 8-2

The completed 2-Sample t (Test and Confidence Interval) dialog box for stacked data

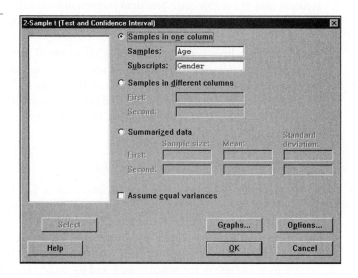

Notice that the "Assume equal variances" check box is not checked. In this case, the fact that the sample standard deviations for the two genders are quite different (9.57 and 13.07) supports the presumption of unequal population variances (remember, the variance is just the square of the standard deviation). The form of the two-sample t-test that assumes equal variances is simpler than the test that does not make this assumption, but it is sensitive to departures from the equal variances assumption. In general, unless there is a strong reason to make such an assumption, do not click the "Assume equal variances" check box.

■ Click **OK**

The results appear in the Session window and in Figure 8-3.

FIGURE 8-3

2-Sample t-test comparing mean female and male ages at death

```
MTB > TwoT 'Age' 'Gender';
SUBC>   Alternative 1.
```

Two-Sample T-Test and CI: Age, Gender

```
Two-sample T for Age

Gender   N   Mean  StDev  SE Mean
Female  86  87.09   9.57      1.0
Male    64   80.6   13.1      1.6


Difference = mu (Female) - mu (Male)
Estimate for difference:  6.45240
95% lower bound for difference:  3.24769
T-Test of difference = 0 (vs >): T-Value = 3.34  P-Value = 0.001  DF = 110
```

Minitab presents summarized information (sample size, mean, standard deviation, and standard error) for the females and males and then displays the t-statistic (T-Value = 3.34), p-value (0.001), and degrees of freedom for the t-statistic (DF = 110). Because the p-value of 0.001 is less than your level of significance (.05, in this case), you reject the null hypothesis. The data suggest that the mean age at death for females in affluent areas is significantly greater than that for males. You cannot attribute the differences in the sample means to sampling variation.

The difference between the sample means, 6.45240 years, is a point estimate for the corresponding difference in the population means, $\mu_F - \mu_M$. Minitab also provides a 95% upper one-sided confidence interval (3.24769 years to ∞) for $\mu_F - \mu_M$. The fact that the lower bound of the interval is above 0 suggests that, on average, females die at a greater age than males.

Two-sample t-tests, like their one-sample cousins that you examined in Tutorial 7, assume that the samples come from normal populations. Graphical procedures are often the best ways to examine this assumption. Minitab offers two built-in graphics options—boxplots and individual value plots—with the 2-Sample t-test. To see boxplots of the two samples of ages:

■ Click the **Edit Last Dialog** toolbar button 🖳

■ Click **Graphs** and click the **Boxplots of data** check box

■ Click **OK** twice

The resulting graph is as shown in Figure 8-4.

FIGURE 8-4

*Built-in boxplots for Age
by Gender*

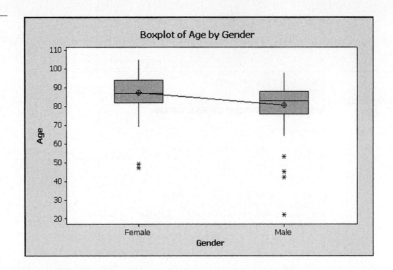

The boxplots are presented vertically. The mean age at death for each gender is indicated and connected. In each case, the distribution of the observations that are not outliers is fairly symmetric. However, the outlying values for both females and males suggest non-normal populations. This should not be a major problem in this case because of the large sample sizes, but in Tutorial 13 you will use a non-parametric test to compare the median ages of these two populations. This test does not depend upon an assumption of normality.

Histograms and stem-and-leaf displays are generally considered better than individual value plots or boxplots at representing the shape of the distribution of a data set. Minitab offers the latter two perhaps because they are easier to present on a single graph. In the next case study where the sample size is small, you will use a normal probability plot (NPP)—introduced in Tutorial 6—to check for normality.

Using Help to Find a Formula

You can find the formula for the test statistic for the two-sample t-test by using Minitab's Help, beginning either at the Help menu or with the Help button on the 2-Sample t (Test and Confidence Interval) dialog box. Following the latter path:

■ Click the **Edit Last Dialog** toolbar button ▣

This returns you to the 2-Sample t (Test and Confidence Interval) dialog box.

■ Click **Help**

You are taken to the 2-Sample t Help window.

■ Click on **see also** and then click on **Methods and formulas**

The new window shows the formulas used in the two-sample t-tests and t-confidence intervals. The first part of the window is as shown in Figure 8-5.

FIGURE 8-5

The Methods and Formulas - 2-Sample t Help window

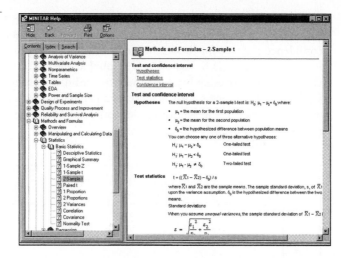

Notice that Minitab's test statistic is for testing the more general null hypothesis $H_0: \mu_1 - \mu_2 = \delta_0$ instead of the more standard $H_0: \mu_1 - \mu_2 = 0$.

- Click the **Close** button
- Click **Cancel** on the 2-Sample t (Test and Confidence Interval) dialog box

When you are comparing means, there is one Minitab plot that can provide considerable insight though it is not used for checking normality. This is an interval plot. To obtain such a plot for the ages at death for each gender:

- Choose **Graph > Interval Plot** and double-click *One Y, With Groups*
- Double-click **C5 Age** as the "Graph variables" and **C3 Gender** as the "Categorical variables for grouping (1-4, outermost first)"

The completed Interval Plot - One Y, With Groups dialog box is shown in Figure 8-6.

FIGURE 8-6

The completed Interval Plot - One Y, With Groups dialog box

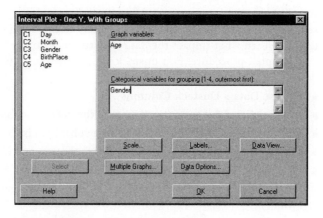

- Click **OK**

The *interval plot* is shown in Figure 8-7. It displays a 95% confidence interval for the mean age at death for all affluent females and the comparable confidence interval for the mean age at death for all affluent males. The sample mean age at death is indicated by a symbol at the center of each interval. The fact that there is no overlap between the two confidence intervals emphasizes the significant difference between these two population means. Recall that there is one 92-year-old person whose gender could not be determined. That person is represented on the left side of the plot.

FIGURE 8-7

Interval plot showing 95% confidence intervals for the population mean age at death by gender

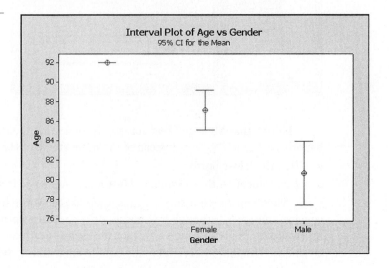

Now, return to the worksheet:

- Click the **Current Data Window** toolbar button ▦

Two-Sample t-Tests Using Unstacked Data

Another common way to store data from two samples is to place them in two separate columns. Recall that Minitab refers to this form of storage as unstacked data. To learn what impact the data structure has on the Minitab output for the two-sample t-procedure, first unstack the age data into two columns, the ages at death for the females and for the males.

- Choose **Data > Unstack Columns**
- Double-click **C5 Age** as the "Unstack the data in" column
- Double-click **C3 Gender** as the "Using subscripts in" column
- Click the **After last column in use** option button and click **OK**
- Click the **Current Data Window** toolbar button ▦, if necessary

There are now two new variables in your worksheet: C6 Age_Female and C7 Age_Male.

Now, use Minitab to test the same pair of hypotheses as before, but without the boxplots:

- Choose **Stat > Basic Statistics > 2-Sample t**

- Click the **Samples in different columns** option button
- Double-click **C6 Age_Female** as the "First" column
- Double-click **C7 Age_Male** as the "Second" column

    When the samples are in separate columns, Minitab assumes that the data you entered in the "First" text box are the sample from population 1.

- Click **Graphs** and clear the Boxplots of data check box
- Click **OK**

    The completed 2-Sample t (Test and Confidence Interval) dialog box looks as shown in Figure 8-8.

FIGURE 8-8

The completed 2-Sample t (Test and Confidence Interval) dialog box for unstacked data

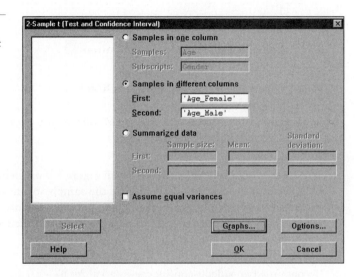

- Click **OK**

    The output is shown in Figure 8-9.

FIGURE 8-9

2-Sample t (Test and Confidence Interval) for unstacked data

```
MTB > TwoSample 'Age_Female' 'Age_Male';
SUBC>   Alternative 1.
```

Two-Sample T-Test and CI: Age_Female, Age_Male

```
Two-sample T for Age_Female vs Age_Male

            N    Mean   StDev   SE Mean
Age_Female  86   87.09   9.57     1.0
Age_Male    64   80.6   13.1      1.6


Difference = mu (Age_Female) - mu (Age_Male)
Estimate for difference:  6.45240
95% lower bound for difference:  3.24769
T-Test of difference = 0 (vs >): T-Value = 3.34  P-Value = 0.001  DF = 110
```

    The only major differences between the outputs for stacked and unstacked data are the names of the samples. For the stacked data the samples are named after the subscripts, Female and Male. Here they are the column names, Age_Female and Age_Male.

Two-Sample t-Tests Using Summarized Data

Minitab can perform a two-sample t-test when only summarized data are available. Suppose, for example, you had access to only the summarized sample data from the output in Figure 8-9.

```
              N    Mean   StDev
Age_Female   86   87.09    9.57
Age_Male     64   80.6    13.1
```

To obtain a two-sample t-test using these summarized results:

- Click the **Edit Last Dialog** toolbar button 🖻
- Click the **Summarized data** option button
- Type **86** as the "First: Sample size", **87.09** as the "First: Mean", and **9.57** as the "First: Standard deviation"
- Type **64** as the "Second: Sample size", **80.6** as the "Second: Mean", and **13.1** as the "Second: Standard deviation"
- Click **OK**

The output will differ from that in Figure 8-9 only slightly. When Minitab has access to the raw data, it computes the sample means and standard deviations more accurately than it prints them. These more accurate values are used to compute the t-statistic (and the p-value), and the confidence interval.

Obtaining a 95% Two-Sided Confidence Interval for $\mu_F - \mu_M$

To obtain a two-sided t-confidence interval for $\mu_F - \mu_M$ you need to change the alternative hypothesis to "not equal", as in the last tutorial. You can use these summarized data to obtain a 90% two-sided confidence interval for $\mu_F - \mu_M$.

- Click the **Edit Last Dialog** toolbar button 🖻
- Click **Options** and type **90** as the "Confidence level"
- Click the "Alternative" **drop-down list arrow**, click on **not equal**, and click **OK** twice

The output is shown in Figure 8-10.

FIGURE 8-10

90% two-sided confidence interval for $\mu_F - \mu_M$

```
MTB > TwoT 86 87.09 9.57 64 80.6 13.1;
SUBC>   Confidence 90.
```

Two-Sample T-Test and CI

```
Sample   N    Mean   StDev   SE Mean
1       86   87.09    9.57     1.0
2       64   80.6    13.1      1.6


Difference = mu (1) - mu (2)
Estimate for difference:  6.49000
90% CI for difference:  (3.27927, 9.70073)
T-Test of difference = 0 (vs not =): T-Value = 3.35  P-Value = 0.001  DF = 110
```

The output still includes the test of $H_0: \mu_F - \mu_M = 0$, but you can see the 90% two-sided confidence interval (3.27927 years to 9.70073 years). It is helpful to know that μ_F is significantly greater than μ_M, but it is equally helpful to know that you can be 90% confident that μ_F exceeds μ_M by anywhere from (approximately) 3.3 years to 9.7 years.

Before proceeding to the next case study save your project:

- Click the **Save Project** toolbar button 🖫

8.2   Inference on the Mean of Paired Data

In the previous case study in this tutorial, you presumed independent samples were drawn from each of two populations. Pairing subjects, such as identical twins or before-and-after measurements of a subject, can eliminate the effect of extraneous variables. Such data are called *paired*, or *dependent*, *samples*.

CASE STUDY	**HEALTH CARE—CEREAL AND CHOLESTEROL**

Recently 14 male patients suffering from high levels of cholesterol took part in an experiment to examine whether a diet that included oat bran would reduce cholesterol levels. Each was randomly assigned to a diet that included either corn flakes or oat bran. After two weeks, their low-density lipoprotein (LDL) cholesterol levels were recorded. Each man then repeated this process with the other cereal. The 14 pairs of LDL results are recorded in a data file called *Chol.mtw*. As a rookie analyst at the clinic where the research was conducted, you have been asked to analyze these data. (You may assume these men are a random sample of all men suffering from high levels of cholesterol.)

- Open the worksheet ***Chol.mtw***

In this data set, C1 CornFlke and C2 OatBran contain the LDL levels for each cereal, respectively, for each of the 14 patients in the study. Minitab enables you to obtain a paired t-test and t-confidence interval without computing the 14 (CornFlke – OatBran) differences. You form the hypotheses:

Null hypothesis, $H_0: \mu_{C-O} = 0$

Alternative hypothesis, $H_1: \mu_{C-O} > 0$

Here, μ_{C-O} is the population mean change in LDL that would result if every male with high cholesterol were included in the study. The alternative hypothesis is right-tailed, because your expectation is that if μ_{C-O} is not 0 it will be greater than 0. As a graphical check of the normality assumption, ask for a histogram of the 14 differences.

- Choose **Stat > Basic Statistics > Paired t**
- Double-click **C1 CornFlke** as the "First sample" and **C2 OatBran** as the "Second sample"

- Click **Options** and verify that the default "Confidence level" is 95.0 and that the default "Test mean" is 0.0
- Click the "Alternative" **drop-down list arrow**, click on **greater than**, and click **OK**
- Click **Graphs**, click the **Histogram of differences** check box, and click **OK**

The completed Paired t (Text and Confidence Interval) dialog box should look as shown in Figure 8-11.

FIGURE 8-11

The completed Paired t (Test and Confidence Interval) dialog box

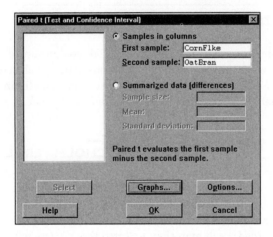

Notice that the Paired t (Text and Confidence Interval) dialog box indicates that inferences are based upon the First sample – Second sample differences.

- Click **OK**

The built-in histogram of the 14 differences is as shown in Figure 8-12.

FIGURE 8-12

Built-in histogram for differences

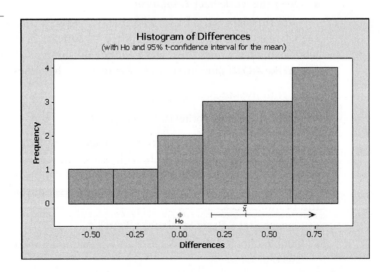

The null hypothesis value of 0 is highlighted, as is a 95% upper one-sided bound for the unknown value for μ_{C-O}. The fact that the one-sided interval does not contain the value 0 is a sign that you should reject the null hypothesis. The shape of the histogram is far from bell-shaped; this calls into question the assumption that the 14 differences come from a normal distribution. After reviewing the results you will create a normal probability plot (NPP) to investigate this matter further.

- Click the **Session Window** toolbar button 🔲

The output in the Session window is as shown in Figure 8-13.

```
MTB > Paired 'CornFlke' 'OatBran';
SUBC>    Alternative 1;
SUBC>    GHistogram.
```

Paired T-Test and CI: CornFlke, OatBran

```
Paired T for CornFlke - OatBran

              N      Mean      StDev    SE Mean
CornFlke     14    4.44357   0.96883   0.25893
OatBran      14    4.08071   1.05698   0.28249
Difference   14   0.362857   0.405964  0.108498


95% lower bound for mean difference: 0.170714
T-Test of mean difference = 0 (vs > 0): T-Value = 3.34   P-Value = 0.003
```

The mean of the differences is 0.362857, which corresponds to a t-statistic of 3.34. Because the p-value for this test is 0.003, you can reject H_0 at the 1% level of significance, for example. Your test provides significant evidence that the use of oat bran significantly reduces the mean LDL cholesterol level.

The paired t-procedures in this section are essentially identical to the one-sample t-procedures in the last tutorial with the hypothesized value equal to 0. You can review the test statistic in that tutorial.

When the sample size is as small as 14, it is particularly important to use a NPP to examine the normality in the data. To obtain a NPP of the 14 CornFlke – OatBran differences, you first have to compute these differences:

- Choose **Calc > Calculator**
- Type **Diff** as the "Store result in variable"
- Enter, by clicking or typing, **CornFlke – OatBran** as the "Expression" and click **OK**
- If necessary, click the **Current Data Window** toolbar button 🔲

The 14 differences will be in C3 Diff. To obtain the NNP:

- Choose **Stat > Basic Statistics > Normality Test**
- Double-click **Diff** as the "Variable" and click **OK**

The resulting plot is shown in Figure 8-14.

FIGURE 8-14

*NPP of the CornFlke –
OatBran differences*

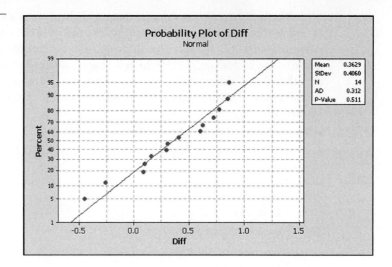

Recall that the closer the points on the NPP lie to the straight line, the more normal the data. By default, Minitab performs an Anderson-Darling test for normality. The null and alternative hypotheses are:

Null hypothesis, H_0: The data come from a normal distribution

Alternative hypothesis, H_1: The data come from a distribution that is not normal

You need not understand the details of how the test is performed because Minitab provides the p-value for the test as part of the output. This is one of those rare tests where the researcher does not want to reject the null hypothesis because to do this suggests that the normality assumption in the t-test is violated. In this case the p-value, shown to the right of the plot, is 0.511, which is far too high to reject the null hypothesis of normality. Minitab also provides the mean, standard deviation, and number of the observations in the plot. The value AD (0.312) is the value of the Anderson-Darling test statistic.

▶ **Note** Minitab's Stat > Basic Statistics > Paired t command is the simplest way of performing this test, but not the only way. Now that you have computed the CornFlke – OatBran differences you can use the 1-Sample t-test (covered in the last tutorial) on the differences. ■

▶ **Note** Minitab also offers the option of entering summarized data for a paired-t procedure. In this case you enter the sample size along with the mean and standard deviation of the differences. ■

Before proceeding to the next case study, save the current project:

■ Click the **Save Project** toolbar button 🖫

8.3 Sample Size and Power for Comparing the Means of Two Independent Samples

CASE STUDY	WELFARE REFORM— ERRORS IN GRANT DETERMINATION

Errors in determining how much welfare a family is entitled to can be expensive for both the recipient (if the grant is smaller than it should be) and for the state (if the grant is larger than it should be). Part of the difficulty in determining the grant amount is that the standard procedure for determining the amount is by reference to a manual consisting of complex state and federal regulations. To test a technique that might make this determination simpler, the federal government is underwriting an experiment at two welfare offices that serve two similar populations.

At one (Control) office, the standard procedure for determining the grant amount will be used. At the other (Experimental) office, a new procedure using Decision Logic Tables (DLTs) will be tried for six months. DLTs are logical pathways through the regulations that should make it easier to determine the appropriate grant in each case. After six months, auditors will select a random sample of cases from each center and determine the error, in dollars, for each case. As the statistician working for the agency responsible for this experiment, your first job is to decide upon the appropriate number of cases (n) to sample from each office.

First, create a new worksheet:

- Choose **File** > **New** and click **OK**

You formulate the appropriate hypotheses as follows:

Null hypothesis, H_0: $\mu_C - \mu_E = 0$

Alternative hypothesis, H_1: $\mu_C - \mu_E > 0$

Here, μ_C and μ_E are, respectively, the population mean dollar errors in the control and the experimental centers.

You decide to use a level of significance of $\alpha = .05$. Previous studies of errors in grant determination suggest that the standard deviation of error is approximately \$70. You have been informed that it is important to reject the null hypothesis if the true difference, $\mu_C - \mu_E$, is as small as \$20. This is the called the effect size. Minitab allows you to investigate the sample size associated with different values for the power of the test. Specifically, you intend to obtain the sample sizes needed to provide power values from .5 to .95 at intervals of .05. You will represent this sequence in the form .5:.95 / 05.

- Choose **Stat** > **Power and Sample Size** > **2-Sample t**
- Type **20** as the "Differences"
- Type **.5:.95/.05** as the "Power values"
- Type **70** as the "Standard deviation"

The completed Power and Sample Size for 2-Sample t dialog box should look as shown in Figure 8-15.

FIGURE 8-15

The completed Power and Sample Size for 2-Sample t dialog box

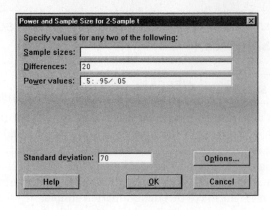

- Click **Options** and click the "Alternative Hypothesis" **Greater than** option button
- Verify that the "Significance level" is 0.05 and type **n** in the "Store sample sizes in" column
- Type **Power** as the "Store power values in" column
- Click **OK** twice

  The output is shown in the Session window and in Figure 8-16.

FIGURE 8-16

Sample size and power calculations

```
MTB > Name c1 "n" c2 "Power"
MTB > Power;
SUBC>    TTwo;
SUBC>       Difference 20;
SUBC>       Power .5:.95/.05;
SUBC>       Sigma 70;
SUBC>       Alternative 1;
SUBC>       SSample 'n';
SUBC>       SPower 'Power'.
```

Power and Sample Size

2-Sample t Test

Testing mean 1 = mean 2 (versus >)
Calculating power for mean 1 = mean 2 + difference
Alpha = 0.05 Assumed standard deviation = 70

	Sample	Target	
Difference	Size	Power	Actual Power
20	67	0.50	0.500142
20	78	0.55	0.552344
20	89	0.60	0.600161
20	102	0.65	0.651256
20	116	0.70	0.700095
20	133	0.75	0.751462
20	153	0.80	0.801937
20	177	0.85	0.850322
20	211	0.90	0.900617
20	266	0.95	0.950113

The sample size is for each group.

The sample sizes are given in the second column of the output. The third column contains the desired power values, which Minitab calls the "Target

Power". The exact, or "Actual Power" corresponding to the sample size is in the right-hand column. Note that the power of the test increases as the sample size increases. After consulting with your colleagues in the agency, you recommend that samples of n = 211 cases be selected and audited in each office. This will guarantee a power of 0.900617.

Now plot the power of the test against the sample size.

- Click the **Current Data Window** toolbar button ▦

  The values for n are stored in C1 n and the corresponding power values in C2 Power.

- Choose **Graph > Scatterplot** and double-click *With Connect Line*

- Double-click **C2 Power** as the "Y variables" column and then **C1 n** as the "X variables" column

- Click **OK**

  The resulting graph is shown in Figure 8-17.

FIGURE 8-17

Plot of Power against n

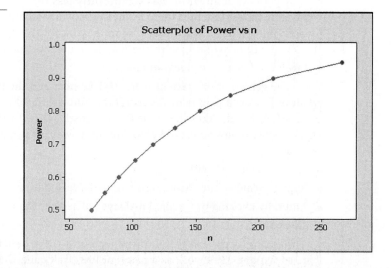

The plot shows that power increases with the value of n. It is also apparent that, because the power of the test cannot exceed 1, the graph begins to level off as n moves beyond 266 cases.

From Figure 8-15, note that the Stat > Power and Sample Size > 2-Sample t command can be used to calculate the sample size(s), the difference(s), or the power value(s). You need to specify acceptable values for two of these and Minitab will solve for the third.

Before proceeding to the next case study, save the current project:

- Click the **Save Project** toolbar button 🖫

8.4 Comparing Population Proportions from Two Independent Samples

| CASE STUDY | HEALTH CARE—WORK DAYS LOST TO PAIN (CONTINUED) |

As a researcher at a large Health Maintenance Organization (HMO), you have been asked to collaborate with a nurse who is completing her master's degree to select a random sample of 300 patients who have reported low-back pain in the past six months. Federal guidelines suggest that with the appropriate treatment, less than 60% of patients with low-back pain should lose workdays. You are to use the data to test whether this HMO has met this guideline. Because of incomplete records, you have an effective sample size of 279.

In the last tutorial, you tested whether the incidence of days lost to low-back pain by patients at an HMO was significantly less than .6. As a follow-up to this, you decide to test whether there is any statistically significant difference in the proportion of males and females who miss work for this health problem.

You will be working with the *Backpain* data, as you did in the last tutorial.

- Open the worksheet ***Backpain.mtw***

The gender of the patient is in C1-T Gender and the number of days lost as a result of low-back pain is in C3 LostDays. You will need to recreate the variable C5 LostInd, which takes the value 0 if C3 LostDays is 0 and the value 1 if C3 LostDays takes any non-zero value. You may want to review the procedure in Section 7.5.

- Choose **Calc > Calculator**
- Type **LostInd** as the "Store result in variable" column
- Enter, by clicking or typing, **LostDays > 0** as the "Expression"
- Click **OK**

Recall that the 1s in C5 LostInd represent those patients in the sample who lost at least one day's work as a result of low-back pain; the 0s represent those patients who did not. Minitab will assume that the larger of the two values (1) in C5 LostInd is the "success" and the smaller of the two (0) is the "failure".

Now formulate hypotheses to test whether the proportions of females (p_F) and males (p_M) (in the population) who miss work as a result of low-back pain are statistically, significantly different:

Null Hypothesis, H_0: $p_F - p_M = 0$

Alternative Hypothesis, H_1: $p_F - p_M \neq 0$

To obtain the test and the default 95% confidence interval for $p_F - p_M$:

- Choose **Stat > Basic Statistics > 2 Proportions**
- Verify that the "Samples in one column" option button is selected
- Double-click **C5 LostInd** as the "Samples" and **C1 Gender** as the "Subscripts"

- Click **Options** and verify that the default "Confidence level" is 95.0, that the default "Test difference" is 0.0, and that the default "Alternative" is not equal
- Click the **Use pooled estimate of p for test** check box (the option found in most textbooks)

  This option is often called the normal approximation method. It requires an assumption such as (sample size)*(sample proportion) ≥ 5 and (sample size)*(1 – sample proportion) ≥ 5 for each sample.

  The completed 2 Proportions - Options dialog box should look as shown in Figure 8-18.

FIGURE 8-18

The completed
2 Proportions - Options
dialog box

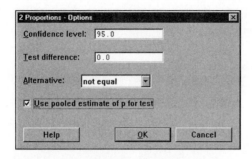

- Click **OK** twice

  The output in the Session window is as shown in Figure 8-19.

FIGURE 8-19

2 Proportions Z-inference
for the proportion of days
lost by gender

```
MTB > PTwo 'LostInd' 'Gender';
SUBC>    Stacked;
SUBC>    Pooled.
```

Test and CI for Two Proportions: LostInd, Gender

```
Event = 1


Gender   X     N   Sample p
Female   61   106   0.575472
Male     93   173   0.537572


Difference = p (Female) - p (Male)
Estimate for difference:  0.0378994
95% CI for difference:  (-0.0819902, 0.157789)
Test for difference = 0 (vs not = 0):  Z = 0.62   P-Value = 0.537
```

The sample proportions are quite close. Almost 58% of the 61 females and almost 54% of the 93 males lost workdays. The difference in the sample proportions (0.0378994) translates into a Z-statistic of only 0.62. The corresponding p-value is 0.537. Because this exceeds any standard level of significance, you cannot reject the null hypothesis. The data suggest that there is no significant difference between the proportion of females and the proportion of males who lose work time due to low-back pain. This lack of statistical significance is reflected in the 95% confidence interval for $p_F - p_M$, which runs from approximately –082 to .158 and contains 0.

Comparing Two Proportions Using Summarized Data

It is quite common when dealing with proportions, to know only the sample size and number of successes but not have the raw data. Minitab allows you to compare proportions by entering only the number of observations ("Trials") and the number of successes ("Events") for each sample. Try this option using the results in Figure 8-18:

- Click the **Edit Last Dialog** toolbar button 🖳
- Click the **Summarized data** option button
- Type **106** as the "First: Trials" value and **61** as the "First: Events" value
- Type **173** as the "Second: Trials" value and **93** as the "Second: Events" value

The completed 2 Proportions (Test and Confidence Interval) dialog box is shown in Figure 8-20.

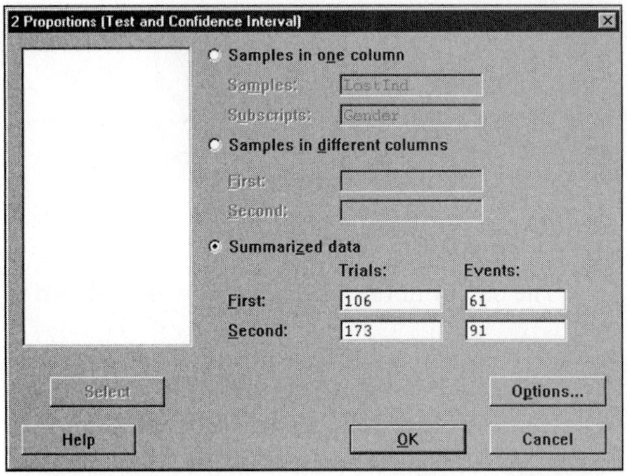

FIGURE 8-20

*The completed
2 Proportions (Test and
Confidence Interval)
dialog box with
summarized data*

- Click **OK**

The output is similar to that in Figure 8-19; the generic 1 and 2 replace the text values Female and Male.

The test statistic used in the 2 Proportions test with pooled proportions is

$$Z_0 = \frac{\hat{p}_1 - \hat{p}_2}{\sqrt{\hat{p}(1-\hat{p})\left(\frac{1}{n_1} + \frac{1}{n_2}\right)}}$$

where $\hat{p}_1$ is the proportion of successes in sample 1, $\hat{p}_2$ is the proportion of successes in sample 2, and

$$\hat{p} \text{ is the } pooled\ proportion = \frac{n_1\hat{p}_1 + n_2\hat{p}_2}{n_1 + n_2}$$

A $(1 - \alpha)100\%$ two-sided confidence interval for $p_1 - p_2$ is

$$\hat{p}_1 - \hat{p}_2 \pm Z \sqrt{\frac{\hat{p}_1(1 - \hat{p}_1)}{n_1} + \frac{\hat{p}_2(1 - \hat{p}_2)}{n_2}}$$

where Z is the $(1 - \alpha/2)100$th percentile of the standard normal distribution.

8.5 Sample Size and Power for Comparing Two Independent Proportions

In the previous analysis, you could not reject the null hypothesis of no difference in the population proportion of all male and all female sufferers of low-back pain who miss work. You fear that this may be because the power of the test is inadequate to detect an important difference. Minitab does not allow you to compute the power of this test when the sample sizes are not the same. However, you can compute how many patients of each gender you would need to survey in order to detect a specified difference that has a specified power. For example, suppose you want to be 80% sure of detecting a difference between p_F and p_M of .10 (the effect size), where the level of significance is kept at the standard .05. To do this, you need to specify provisional values for p_F and p_M. From your statistics class, you know that sample sizes are likely to be conservatively high if you select values for these proportions that are close to .5, so you will use $p_F = .6$ and $p_M = .5$. The sample data you do have support these approximations.

To perform the power calculation:

- Choose **Stat > Power and Sample Size > 2 Proportions**
- Type **.6** as "Proportion 1 values"
- Type **.8** as the "Power values"

▶ **Note** Minitab does not allow you to enter a percentage for the power value. You must enter a decimal value. This is different from the two possible entries for the confidence level. ■

- Verify that the default "Proportion 2" is 0.5
- Click **Options** and verify that the default "Alternative Hypothesis" option is Not equal and that the default "Significance level" is 0.05
- Click **OK**

The completed Power and Sample Size for 2 Proportions dialog box should look as shown in Figure 8-21.

FIGURE 8-21

The competed Power and Sample Size for 2 Proportions dialog box

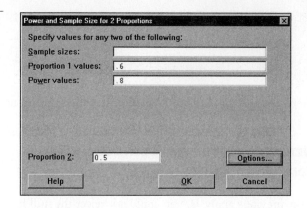

■ Click **OK**

The results of the calculations in the Session window are shown in Figure 8-22.

FIGURE 8-22

2-proportions sample-size calculations

```
MTB > Power;
SUBC>   PTwo;
SUBC>     PrOne .60;
SUBC>     Power .80;
SUBC>     PrTwo 0.5.
```

Power and Sample Size

Test for Two Proportions

Testing proportion 1 = proportion 2 (versus not =)
Calculating power for proportion 2 = 0.5
Alpha = 0.05

```
                Sample  Target
Proportion 1     Size   Power  Actual Power
        0.6       388    0.8      0.800672
```

The sample size is for each group.

The output suggests that you would need to select 388 patients with low-back pain of each gender in order to be 80% sure of rejecting the null hypothesis of no difference if, in fact, a difference of 0.1 exists.

■ Click the **Save Project** toolbar button 🖫

■ Exit Minitab

◆ **Note** Minitab contains procedures for testing for the equality of two variances. These tests are sensitive to departures from the normality assumption, particularly when the sample sizes are different. You will find these tests under Stat > ANOVA > Test for Equal Variance. ■

Congratulations on finishing Tutorial 8! You have seen how Minitab performs the most widely used two-sample procedures. In the next tutorial you will examine the standard procedures for comparing three or more samples.

MINITAB *at work*

SCIENTIFIC RESEARCH

A supernova, the explosion of a star, is a rarely observed phenomenon. Researchers often study it by examining the remnants it leaves behind. Astronomers have charted 29 supernova remnants (SNRs) in a nearby galaxy named the Large Magellanic Cloud. They predicted that SNRs produced by stars with relatively short lives would be close to areas of recent star formation, called H II regions. A physicist, a mathematician, and an undergraduate at a Colorado university teamed up to study the Large Magellanic Cloud supernovas by testing whether the remnants are indeed close to H II regions.

The team calculated the distance between each of the 29 known SNRs in the cloud and its nearest H II region, and then found the mean distance. They used Minitab to simulate what the mean distance would have been had the SNRs been randomly distributed throughout the galaxy. A two-sample t-test showed that the actual mean distance was significantly less than Minitab's randomly generated SNR locations had led them to expect.

This suggested that supernova remnants in the Cloud are not randomly distributed. The SNRs cluster around areas of recent star formation and are likely to have resulted from the explosions of short-lived stars.

Minitab Command Summary

This section describes the commands introduced in, or related to, this tutorial. Use Minitab's online Help for a complete explanation of all commands.

Minitab Menu Commands

Menu	Command	Description
Stat ➤	Basic Statistics	
	➤ 2-Sample t	Performs inferences (tests and confidence intervals) on the differences between two population means using independent samples.
	➤ Paired t	Performs inferences (tests and confidence intervals) on the differences between two population means using paired (dependent) data.
	➤ 2 Proportions	Performs inferences (tests and confidence intervals) on the differences between two population proportions using independent samples.
	Power and Sample Size	
	➤2-Sample t	Computes power for different sample sizes or for different values for $\mu_1 - \mu_2$ based upon a t distribution. Also computes sample sizes for different power values.
	➤ 2 Proportions	Computes power for different sample sizes or for different values for $p_1 - p_2$. Also computes sample sizes for different power values.
Graph ➤	Interval Plot	
	One Y, With Groups	Produces plots of means and either confidence intervals or error bars for one or more group variables.

Review and Practice

Matching Problems

Match the following terms to their definitions by placing the correct letter next to the term it describes.

_____ 0.5

_____ Samples in one column

_____ Assume equal variances

_____ Samples in different columns

_____ not equal

_____ Subscripts

_____ Basic Statistics

_____ 2-Sample t

_____ 0.05

_____ Paired t

a. The default alternative hypothesis in Minitab hypothesis test commands

b. The Minitab option that lets you perform a two-sample t-test, assuming the standard deviations of the populations are equal

c. The Minitab menu that includes all of the commands for two-sample inferences about means and proportions

d. The Minitab command that performs inference on the differences between two population means using independent samples

e. The Minitab term for the column containing the values identifying each sample in two-sample inferences

f. The default value for Proportion 2 in Minitab's command to determine power and sample size for two proportions

g. The Minitab two-sample test option that performs inference on means from two different populations using stacked data

h. The Minitab two-sample test option that performs inference on means from two different populations using unstacked data

i. The default value for significance level in Minitab's command to determine power and sample size for 2-Sample t

j. The Minitab command that performs a t-test for two dependent samples

True/False Problems

Mark the following statements with a *T* or an *F*.

a. ____ Minitab can perform a two-sample t-test if the data from the populations are stored in either one or two columns.

b. ____ Minitab cannot obtain the power of a test for two proportions when the sample sizes are different.

c. ____ The 2-Sample t command performs a two-sample test for the difference between two population proportions.

d. ____ The output of the test for a significant difference between two population means includes a confidence interval for the difference between the two population means.

e. ___ Minitab can perform two-sample t-tests, assuming the population variances are equal or assuming they are unequal.

f. ____ There is no need to specify the standard deviation when computing the power of a two-sample t-test.

g. ____ The default confidence level for two-sample confidence intervals is 90%.

h. ____ In a two-sample test Minitab requires a subscripts column when the two-sample data are stored in a single column.

i. ____ Minitab can perform a two-sample t-test only on raw data.

j. ____ In a two-sample test for proportions, Minitab assumes the smaller of the two numeric subscript values is the success.

Practice Problems

The practice problems instruct you to save your worksheets with a filename prefixed by P, followed by the tutorial number. In this tutorial, for example, use *P8* as the prefix. If necessary, use Help or Appendix A Data Sets, to review the contents of the data sets referred to in the following problems. Interpretations should use the language of the subject matter of the question. If you are using the Student version, be sure to close worksheets when you have completed a problem.

1. Open *Assess.mtw*.

   a. Regard these 79 homes as a random sample. It is reasonable to expect that, in general, the presence of a garage would tend to increase the assessed value of a home. Formulate the null and alternative hypotheses appropriate for testing this presumption.

   b. Perform a two-sample t-test of your null hypothesis in part a using the numeric values in Total$ as your "Samples" and the text values in Garage? as the "Subscripts". What is the difference in the sample means? What is the p-value?

   c. Carefully state your conclusion.

   d. Interpret the 95% one-sided confidence interval that is a portion of your output.

   e. Use the Graph > Boxplots > *With Groups* command to obtain separate boxplots for the values of Total$ for homes with and without garages. What do these plots suggest about the validity of the normality assumptions? Explain briefly.

   f. Obtain an interval plot for these data. Does your plot suggest your conclusion in part c? Explain briefly.

2. Open *Textbooks.mtw*.

   a. Regard the 17 general education texts and the 19 business/economics texts as random samples of their respective types of texts. Obtain a 90% two-sided confidence interval for the difference between the mean number of pages for business/economics texts and the corresponding mean for general education texts.

   b. Interpret your confidence interval in part a for a group of publishing executives.

   c. Obtain boxplots of the number of pages by type of textbook. What do your plots suggest about the normality assumptions?

   d. The output you obtained in part a included a p-value. What were the implicit hypotheses in this case? What conclusion does your p-value suggest?

   e. Obtain an interval plot for these data. Does your plot suggest your conclusion in part d? Explain briefly.

3. Open *RandomIntegers.mtw*. Assume that these students can be considered a random sample of undergraduate and graduate students.

   a. Test whether there is a significant difference between the mean random number selected by male and female graduate students.

   b. Obtain a two-sided confidence interval for the difference between the two means. Explain how this interval can be used to test the hypothesis in part a.

   c. Obtain NPPs of these two samples. Based upon these plots, did you use an appropriate test in part a? Explain briefly.

4. Open *Murderu.mtw*. Assume the state murder rates are a random sample of the murder rates over time.

   a. Test whether a southern state has a higher mean murder rate than a northeastern state. Use a level of significance of .05. Perform this test with both individual value plots and boxplots of the sample data.

   b. Which of the two graphs in part a best displays the results of your test in part a? Explain your choice.

   c. Are the graphs you obtained in part a consistent with the normality assumptions? Explain briefly.

5. The data below are for the 30 teams that played in the National Hockey League (NHL) during the 2002–2003 season. The NHL is divided into an East and a West division, each having 15 teams. The values below include the mean and the standard deviation of the total number of goals scored per team. For each division, regard the 15 numbers of goals per team as a random sample over time.

Division	Number of Teams	Mean Number of Goals	Standard Deviation of Number of Goals
East	15	213.73	25.07
West	15	221.60	28.27

   a. Use these data to test whether there is any significant difference in the mean number of goals per team for the two divisions.

   b. Obtain a 95% confidence interval for the difference in the average goal production per team in the two divisions.

   c. Write a brief explanation of your results in parts a and b.

6. Open *DrugMarkup.mtw*. Assume that these drug prices are a random sample of all available drugs.

   a. Test whether the price of Xanax is, on average, higher than the price of its generic equivalent. Include in your analysis a built-in histogram of differences. Explain how the histogram illustrates your answer to the last query. Based upon this test, what would you recommend to an individual who needs the benefits from such a drug?

   b. Use the same data set to perform a two-tailed test of whether the price of Xanax is, on the average, ten dollars higher than the price of its generic equivalent. The test should still be significant at any reasonable value of α. At what value of the alternative test mean will the test be significant when α is .05?

   c. Create a new variable that contains the differences between the price of Xanax and its generic equivalent. Name this new variable Difference. Test whether the difference is greater than 0 dollars. Again, include a built-in histogram with your analysis. Discuss the similarities and differences between the result of this test and the test in part a. What are the differences in the Minitab output between the two tests?

7. Open *Force.mtw*. Assume that the 12 undergraduate women are a random sample.

   a. Use these data to test whether the average force produced at 5° is significantly greater than the average force produced at 45°.

   b. You suspected that the average force produced at 5° would be substantially greater than that produced at 45° and want to construct a 90% two-sided confidence interval for the difference in average force for the two situations. Obtain this interval. Explain your interval to your physical therapist.

8. Refer to the first case study in this tutorial.

   a. Compute the power of the test that the mean age at death from an affluent suburb for men is the same as that for women versus the alternative that such women live .5, 1.0, 1.5, ..., 10.0 years longer (the effect size). Assume samples of 100 men and 100 women and that the standard deviation of age at death is 15 years for both genders.

   b. Plot power against the effect size and comment on what your plot shows.

9. Open *PulseA.mtw*. Assume these 92 students can be considered a random sample of 18- to 21-year-olds in college.

   a. Test whether there is any significant difference between the proportions of males (p_M) and females (p_F) that smoke. Use a level of significance of .05 and the normal approximation method. What conclusion seems appropriate?

   b. Obtain an 80% two-sided confidence interval for $p_M - p_F$. Interpret your interval.

   c. Use Minitab's Help to explore the "Exact" method used by Minitab when comparing two proportions.

   d. Repeat parts a and b using Minitab's "Exact" method. Do the results differ very much from those in parts a and b? Explain briefly.

10. Open *Backpain.mtw*.

   a. Create a new variable that contains the numeric value 0 when LostDays is 0 and the numeric value 1 otherwise.

   b. Create a new variable that contains the text value "younger" when the patient is 39 years old or younger and the text value "older" otherwise.

   c. Test whether the proportion of older patients with lost work days is significantly greater than the corresponding proportion of younger patients. Use the pooled estimate of the proportion for this inference.

   d. Obtain a 90% one-sided confidence interval for the difference in the population proportions referred to in part c.

   e. Interpret your results in parts c and d for the medical director of the health plan.

   f. What assumptions are implicit in the procedures you used in parts c and d? Do what you can to check the validity of these assumptions.

11. In a cloud-seeding experiment 50 clouds were seeded with silver nitrate pellets in the hope that the pellets would increase the chance of the clouds producing rain. A control group of 50 similar clouds was not seeded. Both sets of clouds were randomly selected. Forty-two of the seeded clouds, and 36 of the unseeded clouds produced measurable amounts of rain. Do these data suggest that clouds that are seeded are more likely to produce rain than unseeded ones? Use the pooled estimate of the proportion for this problem.

   a. State the appropriate hypotheses in this case.

   b. What conclusion is appropriate?

   c. Interpret the default confidence interval in this case.

12. Open *PulseA.mtw*. Assume that these 92 students can be considered a random sample of 18- to 21-year-olds in college.

   a. It is believed that the proportion of such females who smoke is less than the proportion of such males who smoke. Use these data to perform an appropriate test at the .05 level of significance. Include your hypotheses. Use the normal approximation method.

   b. How powerful is your test in part a versus the alternative that the true difference in the proportion that smoke is .10? Explain briefly.

   c. Answer part b for the following alternative values: .2, .18, .16, ..., .02. Produce a plot of power against the value of the difference in proportions and comment on what your plot shows.

   d. Answer part b for the following levels of significance: .1, .01, and .005. Explain how the power of the test varies with the level of significance.

On Your Own

At the end of each year magazines and almanacs list famous men and women who died in that year. Obtain such a list and enter the genders and ages of the deceased. On average, do famous women live longer than famous men do? Use both a hypothesis test and a confidence interval.

Answers to Matching Problems

(f) 0.5
(g) Samples in one column
(b) Assume equal variances
(h) Samples in different columns
(a) not equal
(e) Subscripts
(c) Basic Statistics
(d) 2-Sample t
(i) 0.05
(j) Paired t

Answers to True/False Problems

(a) T, (b) T, (c) F, (d) T, (e) T, (f) F, (g) F, (h) T, (i) F, (j) F

Tutorial 9

Comparing Population Means: Analysis of Variance

In Tutorial 8, you used Minitab to perform t-tests to compare the means of two populations. You can extend this capability to compare the means of three or more populations by using a statistical tool called *Analysis of Variance (ANOVA)*. In this tutorial, you will perform *one-way ANOVAs*, in which you test for the equality of population means when the populations are defined by the different values for a single *explanatory variable*. Then, in a *two-way ANOVA*, you will assess the effect of two explanatory variables on a *response variable*.

OBJECTIVES

In this tutorial, you learn how to:

- Compare the means of several populations when the observations for all of the samples are in one column and all of the subscripts that identify the samples are in another
- Check the validity of the assumptions in a one-way ANOVA
- Perform Tukey's multiple comparison test for significant differences between pairs of population means
- Compare the means of several populations when the samples from each population are placed in different columns
- Assess the effects of two factors on a response variable
- Examine the interaction between two factors as they affect a response variable
- Check the validity of the assumptions in a two-way ANOVA

9.1 Comparing the Means of Several Populations

CASE STUDY — **CHILD DEVELOPMENT— INFANT ATTENTION SPANS**

You are a psychology student taking a course in child development. As part of your project for the course, you have been examining how different designs vary in their ability to capture an infant's attention. You devised an experiment to see whether an infant's attention span varies with the type of design on a mobile. You randomly divided a group of 30 three-month-old infants into five groups of six. Each group was shown a mobile with one of five multicolored designs: A, B, C, D, or E. Because of time constraints, each infant could be shown only one design. You recorded the median time (in seconds) that the infant spent looking at the design. In a Minitab worksheet, you recorded the times in C1 Times and the designs that each infant saw in C2-T Design. The worksheet is saved in the file *Baby.mtw*.

Begin by opening Minitab:

- Open **Minitab** and maximize the Session window
- If necessary, **Enable Commands** and clear Output Editable from the Editor menu
- Click the **Save Project** toolbar button 🖫 and, in the location where you are saving your work, save this project as **T9.mpj**
- Open the worksheet **Baby.mtw**
- If necessary, click the **Current Data Window** toolbar button ▦

One value in the data set was recorded by a friend in your class who admits to not paying close attention to the infant being studied. The value 5.6 for mobile E is the one your friend recorded. You are concerned about its accuracy, so you replace it with an asterisk (*) to indicate a missing numeric value.

- Type * in **row 27** of **C1Time** as the value to replace 5.6

You are advised by the teaching assistant for the course that a one-way ANOVA should be used to compare the mean times for each mobile. However, before doing an ANOVA you should examine graphical and numerical summaries of your data. Begin by obtaining a description of the times for each design with boxplots:

- Choose **Stat > Basic Statistics > Display Descriptive Statistics**
- Double-click **C1 Time** as the "Variables"
- Double-click **C2 Design** as "By variables (optional)"
- Click **Graphs** and click the **Boxplot of data** check box
- Click **OK** twice

The boxplots are shown in Figure 9-1.

FIGURE 9-I

Boxplots for attention times by design

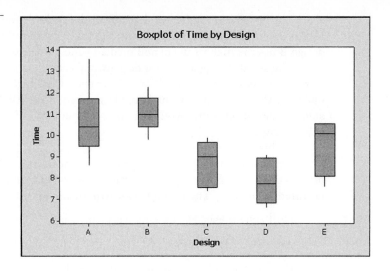

The boxplots suggest that there is considerable spread in the median attention times, with design B having the highest median and design D the lowest. The times for design A seem to be a little more variable than those for the other designs. The numerical summaries appear in the Session window as shown in Figure 9-2. To see them:

■ Click the **Session Window** toolbar button

FIGURE 9-2

Descriptive statistics for attention times by Design

```
MTB > Describe 'Time';
SUBC>   By 'Design';
SUBC>   Mean;
SUBC>   SEMean;
SUBC>   StDeviation;
SUBC>   QOne;
SUBC>   Median;
SUBC>   QThree;
SUBC>   Minimum;
SUBC>   Maximum;
SUBC>   N;
SUBC>   NMissing;
SUBC>   GBoxplot.
```

Descriptive Statistics: Time

Variable	Design	N	N*	Mean	SE Mean	StDev	Minimum	Q1	Median
Time	A	6	0	10.650	0.683	1.672	8.600	9.500	10.400
	B	6	0	11.050	0.349	0.855	9.800	10.400	11.000
	C	6	0	8.750	0.440	1.078	7.400	7.550	9.000
	D	6	0	7.833	0.416	1.019	6.600	6.825	7.750
	E	5	1	9.480	0.591	1.322	7.600	8.100	10.100

Variable	Design	Q3	Maximum
Time	AA	11.725	13.600
	B	11.775	12.300
	C	9.675	9.900
	D	8.950	9.100
	E	10.550	10.600

The N* column indicates that one of the times for design E is missing. The results are, in general, consistent with the findings from the boxplot. That is, design B has the highest mean and median times, while design D has the lowest of both. The standard deviations for designs B, C, and D, are quite similar while the standard deviation for designs A and E are somewhat higher. You know that the one-way ANOVA assumes equal population variances. Although these five sample standard deviations differ, you judge that they do not differ enough to invalidate this assumption.

In this context, the *F-test* in a one-way ANOVA allows you to use the five sample means to test whether the population mean times associated with the five designs differ significantly (in the statistical sense of being unlikely to have occurred by chance). The two relevant hypotheses for the F-test are:

Null Hypothesis, H_0: $\mu_A = \mu_B = \mu_C = \mu_D = \mu_E$

Alternative Hypothesis, H_1: These five population means are not all equal.

Here, μ_A, μ_B, μ_C, μ_D, and μ_E are the population mean attention times for the five designs.

You intend to use a level of significance (α) of .05. To perform the test:

■ Choose **Stat > ANOVA > One-Way**

■ Double-click **C1 Time** as the "Response" and **C2 Design** as the "Factor"

In this context, the term *factor* refers to the explanatory variable, Design. When ANOVA is applied in the context of an experiment, it is quite common to speak of the explanatory variable (or variables, in more complicated designs) as factors. The completed dialog box should look as shown in Figure 9-3.

FIGURE 9-3

The completed One-Way Analysis of Variance dialog box

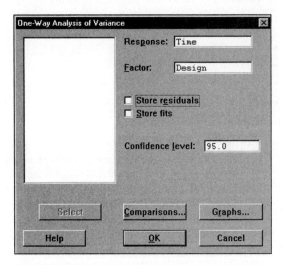

■ Click **OK**

The output in the Session window is shown in Figure 9-4.

FIGURE 9-4

One-way ANOVA of attention times

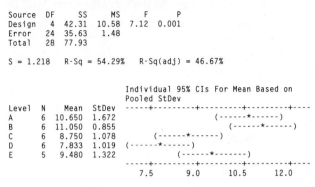

```
MTB > Oneway 'Time' 'Design'.

One-way ANOVA: Time versus Design

Source  DF    SS     MS     F     P
Design   4  42.31  10.58  7.12  0.001
Error   24  35.63   1.48
Total   28  77.93

S = 1.218   R-Sq = 54.29%   R-Sq(adj) = 46.67%

                              Individual 95% CIs For Mean Based on
                              Pooled StDev
Level  N    Mean   StDev   -----+---------+---------+---------+----
A      6  10.650   1.672                      (------*------)
B      6  11.050   0.855                      (------*------)
C      6   8.750   1.078           (------*------)
D      6   7.833   1.019   (------*------)
E      5   9.480   1.322              (------*-------)
                          -----+---------+---------+---------+----
                              7.5       9.0      10.5      12.0

Pooled StDev = 1.218
```

The Session window display includes five components:

1. An ANOVA table that includes the value of the F-statistic (7.12) and corresponding p-value (0.001). Because this p-value < α = .05, you would reject the null hypothesis at the .05 level of significance.

   The formula for the F-statistic is:

$$F_{k-1,n-k} = \frac{\dfrac{1}{k-1} \sum_{\text{all designs}} n_i \left(\overline{X}_i - \overline{X}\right)^2}{\dfrac{1}{n-k} \sum_{\text{all designs}} \sum_{\text{all times within a design}} \left(X_{ij} - \overline{X}_i\right)^2}$$

   where n_i is the number of observations (attention times) for sample (design) i, n is the total number of observations, k is the number of samples (designs), X_{ij} is the jth attention time for the ith design, $\overline{X}_i$ is the mean attention time for the ith design, and $\overline{X}$ is the mean attention time over all babies.

2. The quantities S, R-Sq, and R-Sq(adj) are comparable to their counterparts in linear regression. In this context, S = 1.218 seconds is an estimate of the standard deviation of attention time for each design. It is a pooled estimate based upon the times in each of the five designs. The value for R-Sq = 54.29% indicates that more than 54% of the variation in the 29 times can be associated with the design factor. You will learn more about interpreting the value for R-Sq(adj) in subsequent tutorials.

3. A descriptive summary of the attention times for each design that includes the sample size, mean, and standard deviation.

4. A diagram of the individual 95% confidence interval for the mean attention time for each design based upon the pooled standard deviation (the asterisk represents the sample mean and the parentheses indicate the 95% confidence bounds).

5. The pooled standard deviation used in the confidence intervals, in this case, 1.218 seconds.

The data suggest that the population mean attention times for the five designs are not all equal. From the diagram of the five individual confidence intervals, you notice that designs A and B together seem to elicit a longer mean attention span than do designs C, D, and E. You will look into this more closely in Section 9.3.

Interval plots are frequently used in conjunction with an ANOVA. To obtain an interval plot for the five designs:

- Choose **Graph > Interval Plot** and double-click **One Y, With Groups**
- Double-click **C1 Time** as the "Graph variables" and **C2 Design** as the "Categorical variables for grouping (1-4, outermost first)"
- Click **OK**

The interval plot, with Minitab's default title, is shown in Figure 9-5. The plot is a more sophisticated and vertical version of the five confidence intervals in Figure 9-4. One small difference between the intervals in the two plots is that in Figure 9-4 the intervals are computed using a common, pooled standard deviation. By contrast, in Figure 9-5 each interval is computed using the standard deviation for the times in that sample.

FIGURE 9-5

Interval plot showing 95% confidence intervals for the population mean attention times by design

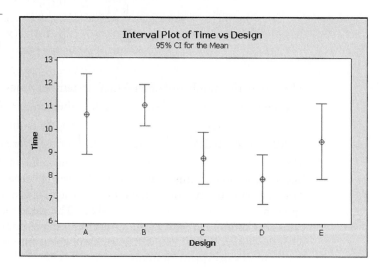

▶ **Note** You can obtain exactly the same interval plot as that in Figure 9-5 with the command Stat > ANOVA > Interval Plot. ■

To obtain a brief review of the F-test, examine the term *F-test* in Minitab's glossary:

- Choose **Help > Help**
- In the right-hand side of the window, hold your cursor over Reference, and click on **Glossary**
- Click on the letter **F** at the top of the Glossary window and then click on **F-test**

  The resulting definition is shown in Figure 9-6. The definition gives you the brief review for this important test.

FIGURE 9-6

The Glossary entry for F-test

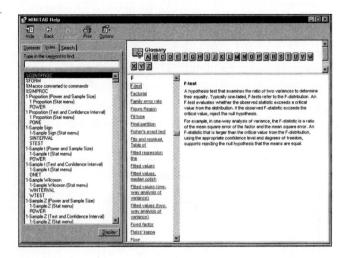

- Click on the **Close** button

9.2 | *Checking the Assumptions for a One-Way ANOVA*

The ANOVA test requires three assumptions about the measurements in the study:

1. The observations must be randomly selected.
2. The populations from which the observations for each sample are taken must all be approximately normally distributed.
3. The observations in each sample must come from populations that have equal variances.

You are satisfied that the infants you included in the study can be regarded as a random sample of such children. You can check the normality assumption using the residuals. To store the fitted values and the residual values in the next available columns:

- Choose **Stat > ANOVA > One-Way**
- Click the **Store residuals** and **Store fits** check boxes
- Click **OK**

  To see the residuals and the fitted values:

- Click the **Current Data Window** toolbar button

The Data window displays columns containing the residuals (C3 RESI1) and the fitted values (C4 FITS1).

In a one-way ANOVA, the *fitted value* associated with any actual value is simply the sample mean associated with that value. The *residual value* is the difference between the actual and fitted values; that is, actual value minus fitted value. Compare the fitted values to the sample means shown in Figure 9-4. Note that the fitted values—10.6500, 11.0500, 8.7500, 7.8333, and 9.4800—are the sample means for each design. Minitab obtains the residuals, by subtracting the corresponding fitted value (design mean) from each median time. The residuals can be used to check the normality of the pooled samples. They should "look" as though they come from a normal distribution. You could check for normality by examining a histogram of the residuals. However, a more sensitive procedure is a normal probability plot (NPP) of the residuals.

■ Choose **Stat > Basic Statistics > Normality Test**

■ Double-click **C3 RESI1** as the "Variable" and click **OK**

The resulting NPP, with Minitab's default title, is shown in Figure 9-7.

FIGURE 9-7

NPP of residuals

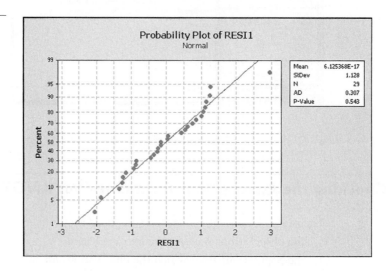

Of the 29 residuals, 28 lie close to the straight line, so the NPP suggests one large outlier. This outlier corresponds to the largest value in the data set.

■ Hold the cursor over this outlier on the far right of the plot to obtain information about this point

The data tip indicates that this point corresponds to row 3 and that the corresponding value for RESI1 is 2.95000 seconds. The value on the vertical axis, 97.619, is the result of a rather complicated calculation. Tutorial 6 included an explanation of how NPPs are constructed.

One approach to the presence of such an outlier is to remove the corresponding observation and examine the impact that its removal has on the ANOVA. In fact, removing this value does not change the significant differences you discovered earlier.

The legend on the right of the NPP in Figure 9-6 contains important information. The mean of the 29 residuals is essentially 0 (recall that 6.125368E-17 is exponential notation for the value 0.00000000000000006125368); the standard deviation of these residuals is 1.128 seconds. There are N = 29 residuals in all. AD = 0.307 is the value for the test statistic for the Anderson-Darling test of the null hypothesis that the residuals come from a normal distribution. Because the corresponding p-value is 0.543 you cannot reject this hypothesis of normality (despite the outlier).

After examining the boxplots of times in Figure 9-1 and the numerical summaries in Figure 9-2, you decide that the constant variance assumption was not violated.

▶ **Note** There are methods for testing for the equality of several population variances, but they are extremely sensitive to departures from the normality assumption and so are rarely used. Minitab offers a number of such tests via the Stat > ANOVA > Test for Equal Variances command. ■

9.3 *Performing Tukey's Multiple Comparisons Test*

Recall that earlier you rejected the null hypothesis of equality of attention time means at the .05 level of significance. You can use Minitab's *multiple comparison test* capabilities to determine which means differ from each other. Minitab provides four multiple comparison methods—named after their respective creators: *Tukey, Fisher, Dunnett*, and *Hsu*. Each method takes a different approach to assessing where the significant differences occur. You are advised to use Tukey's, which involves comparing each pair of means.

In this case Tukey's comparisons test involves the ten sets of hypotheses (one set for each pair of designs) listed below.

$H_0: \mu_A = \mu_B$ $H_0: \mu_A = \mu_C$ $H_0: \mu_A = \mu_D$ $H_0: \mu_A = \mu_E$

$H_1: \mu_A \neq \mu_B$ $H_1: \mu_A \neq \mu_C$ $H_1: \mu_A \neq \mu_D$ $H_1: \mu_A \neq \mu_E$

$H_0: \mu_B = \mu_C$ $H_0: \mu_B = \mu_D$ $H_0: \mu_B = \mu_E$

$H_1: \mu_B \neq \mu_C$ $H_1: \mu_B \neq \mu_D$ $H_1: \mu_B \neq \mu_E$

$H_0: \mu_C = \mu_D$ $H_0: \mu_C = \mu_E$

$H_1: \mu C \neq \mu_D$ $H_1: \mu C \neq \mu_E$

$H_0: \mu_D = \mu_E$

$H_1: \mu_D \neq \mu_E$

The Tukey's test forms confidence intervals for the differences between each pair of population means. To perform Tukey's multiple comparisons test:

- Choose **Stat > ANOVA > One-Way**
- Verify the "Response" is Time and the "Factor" is Design
- Clear the Store residuals and Store fits check boxes
- Click **Comparisons**
- Click the **Tukey's, family error rate** check box

At this point, the completed One-Way Multiple Comparisons dialog box looks as shown in Figure 9-8.

FIGURE 9-8

The completed One-Way Multiple Comparisons dialog box

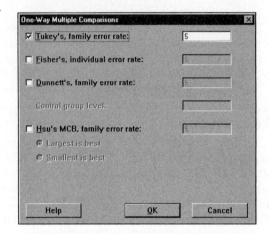

Because it is not apparent what the number 5 in the text box to the right of the check box refers to, seek help from Minitab:

- Click **Help**

The resulting Comparisons - One-Way Multiple Comparisons with Stacked Data Help window is shown in Figure 9-9.

FIGURE 9-9

The Comparisons - One-Way Multiple Comparisons with Stacked Data Help window

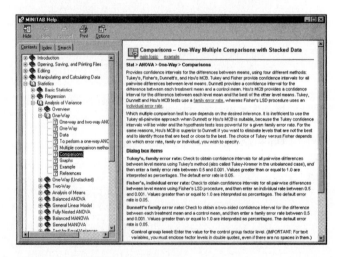

Under "Tukey's, family error rate" you note that "Values greater than or equal to 1.0 are interpreted as percentages. The default error rate is 0.05." So, the 5 refers to a family error rate of 5%.

To investigate the meaning of "family error rate":

- Click **family error rate** in the first paragraph

The pop-up window contains the definition, "Maximum probability of obtaining one or more confidence intervals that do not contain the true difference between level means."

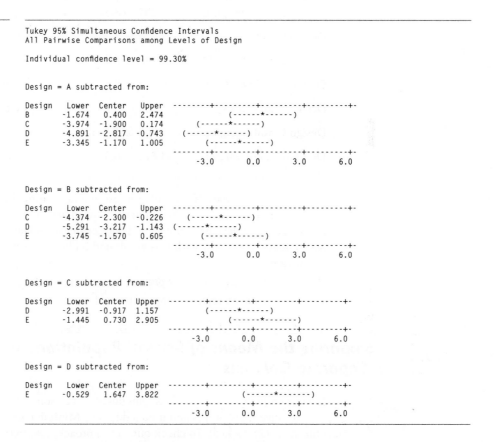

- Click the **Close** button
- Click **OK** twice

Figure 9-10 shows the Stat >ANOVA > One-Way command output that contains the Tukey pairwise comparisons.

FIGURE 9-10
Tukey pairwise comparisons

```
Tukey 95% Simultaneous Confidence Intervals
All Pairwise Comparisons among Levels of Design

Individual confidence level = 99.30%


Design = A subtracted from:

Design   Lower  Center   Upper  --------+---------+---------+---------+-
B       -1.674   0.400   2.474              (------*------)
C       -3.974  -1.900   0.174       (------*------)
D       -4.891  -2.817  -0.743    (------*------)
E       -3.345  -1.170   1.005        (------*------)
                                --------+---------+---------+---------+-
                                     -3.0       0.0       3.0       6.0


Design = B subtracted from:

Design   Lower  Center   Upper  --------+---------+---------+---------+-
C       -4.374  -2.300  -0.226     (------*------)
D       -5.291  -3.217  -1.143  (------*------)
E       -3.745  -1.570   0.605       (------*------)
                                --------+---------+---------+---------+-
                                     -3.0       0.0       3.0       6.0


Design = C subtracted from:

Design   Lower  Center  Upper  --------+---------+---------+---------+-
D       -2.991  -0.917  1.157        (------*------)
E       -1.445   0.730  2.905             (------*-------)
                               --------+---------+---------+---------+-
                                    -3.0       0.0       3.0       6.0


Design = D subtracted from:

Design   Lower  Center  Upper  --------+---------+---------+---------+-
E       -0.529   1.647  3.822                 (------*-------)
                               --------+---------+---------+---------+-
                                    -3.0       0.0       3.0       6.0
```

For each comparison, the output contains the lower bound, the center, and the upper bound of the interval, and a pictorial representation of the interval.

According to Tukey's test, you should reject the null hypothesis that the two population means are equal whenever the confidence interval for the difference in the means does not contain 0. The family error rate (sometimes called the *overall error rate*) is the probability of making at least one Type I error (falsely rejecting H_0) in the ten tests. A summary of the results follows:

Difference Between	Confidence Interval 95% Family Confidence	Decision
Design A and Design B	–2.474 to 1.674	Cannot reject H_0: $\mu_A = \mu_B$
Design A and Design C	–0.174 to 3.974	Cannot reject H_0: $\mu_A = \mu_C$
Design A and Design D	0.743 to 4.891	Reject H_0: $\mu_A = \mu_D$
Design A and Design E	–1.005 to 3.345	Cannot reject H_0: $\mu_A = \mu_E$
Design B and Design C	0.226 to 4.374	Reject H_0: $\mu_B = \mu_C$
Design B and Design D	1.143 to 5.291	Reject H_0: $\mu_B = \mu_D$
Design B and Design E	–0.605 to 3.745	Cannot reject H_0: $\mu_B = \mu_E$
Design C and Design D	–1.157 to 2.991	Cannot reject H_0: $\mu_C = \mu_D$
Design C and Design E	–2.905 to 1.445	Cannot reject H_0: $\mu_C = \mu_E$
Design D and Design E	–3.822 to 0.529	Cannot reject H_0: $\mu_D = \mu_E$

From the results of the Tukey's test you determine that there is significant evidence of a difference in the population mean attention spans between designs A and D, between designs B and C, and between designs B and D. You conclude your psychology experiment with this result and include it in your report about infant perception.

At this point, save your project.

■ Click the **Save Project** toolbar button 🖫

9.4 Comparing the Means of Several Populations with Responses in Separate Columns

In your child development class, you observe that some students entered their data in separate columns and used a different Minitab command, Stat > ANOVA > One-Way (Unstacked). To check out this approach, restructure your data by placing each of the five groups of attention spans for each design into separate columns:

■ Click the **Current Data Window** toolbar button 🖩
■ Choose **Data > Unstack Columns**
■ Double-click **C1 Time** as the "Unstack the data in" column
■ Double-click **C2 Design** as the "Using subscripts in" column

- Click the **After last column in use** option button and click **OK**

Minitab has placed the five sets of attention times in the five columns C5 to C9 and named them Time_A, Time_B, Time_C, Time_D, and Time_E.

Now, repeat the ANOVA procedure by using the Stat > ANOVA > One-Way (Unstacked) command that is appropriate when the data for the different samples are in separate columns.

- Choose **Stat > ANOVA > One-Way (Unstacked)**
- Double-click **C5 Time_A, C6 Time_B, C7 Time_C, C8 Time_D**, and **C9 Time_E** as the "Responses (in separate columns)"
- Click **OK**

The results appear in the Session window, as shown in Figure 9-11.

FIGURE 9-11

One-way (unstacked) ANOVA of attention times

```
MTB > AOVOneway 'Time_A' 'Time_B' 'Time_C' 'Time_D' 'Time_E'.

One-way ANOVA: Time_A, Time_B, Time_C, Time_D, Time_E

Source  DF    SS     MS     F     P
Factor   4  42.31  10.58  7.12  0.001
Error   24  35.63   1.48
Total   28  77.93

S = 1.218   R-Sq = 54.29%   R-Sq(adj) = 46.67%


                              Individual 95% CIs For Mean Based on
                              Pooled StDev
Level    N    Mean   StDev   -----+---------+---------+---------+----
Time_A   6  10.650   1.672                     (------*------)
Time_B   6  11.050   0.855                      (------*------)
Time_C   6   8.750   1.078        (------*------)
Time_D   6   7.833   1.019   (------*------)
Time_E   5   9.480   1.322            (------*------)
                             -----+---------+---------+---------+----
                              7.5       9.0      10.5      12.0

Pooled StDev = 1.218
```

Minitab produces the same display in this case as it did when all the observations were stacked in one column. There are only two differences: the analysis of variance table no longer identifies the response and explanatory variables and the levels are now identified by the column names Time_A, Time_B, and so on, rather than by the text values A, B, C, D, and E.

◆ **Note** The exact appearance of this ANOVA in your session window will depend on whether you select the columns C5 Time_A, C6 Time_B, C7 Time_C, C8 Time_D, and C9 Time_E one at a time or together. ■

◆ **Note** For unstacked data, Minitab uses the AOVOneway Session command. For stacked data, it uses the Oneway Session command. ■

Before proceeding to the next case study, save your project:

- Click the **Save Project** toolbar button 🖫

9.5 | Performing a Two-Factor Analysis of Variance

Thus far in this tutorial, you have analyzed the effect of a single factor with five levels (the five mobile designs) on a response (infant attention times). However, many responses are affected by more than one factor. Statisticians frequently design studies to take this fact into account. In this section you will analyze the results of a study designed to measure the impact of two factors on a response.

<table>
<tr><td>CASE STUDY</td><td>PSYCHOLOGY—MEASURING DEPTH PERCEPTION</td></tr>
</table>

You have been hired to assist in the analysis of a study of depth perception under different lighting conditions involving 36 randomly selected people. The people were divided into three groups on the basis of age. Each of the 12 members of an age group was randomly assigned to one of three "treatment" groups. All 36 people were asked to judge how far they were from a number of different objects. An average "error" in judgment (in feet) was recorded for each person. One treatment group was shown the objects in bright sunshine, another under cloudy conditions, and the third at twilight.

You receive the data in the form shown in Figure 9-12.

FIGURE 9-12

Errors in depth perception arranged by age groups and light conditions

Age Group	Lighting		
	Sun	**Clouds**	**Twilight**
Young	5.2	7.1	8.9
	4.3	6.4	10.4
	5.8	7.9	8.4
	6.0	5.6	7.9
MiddleAge	5.2	7.0	9.5
	4.2	7.0	9.0
	6.0	8.2	7.6
	5.4	6.0	10.7
Older	6.2	8.0	10.6
	7.1	6.7	12.1
	6.7	9.1	13.1
	6.5	9.2	12.4

Each of the 36 values in the table is an average "error" in feet. The two factors are age group and lighting condition. Each factor has three levels. (It is not necessary that each factor have the same number of levels.) There are four observations for each combination of levels of the two factors. You learn that these data are stored in a Minitab data file *Depth.mtw*.

- Open the worksheet ***Depth.mtw***

The Data window after opening the Depth worksheet is as shown in Figure 9-13. The text values for the age groups are in C1-T AgeGroup, the text values for lighting conditions are listed in C2-T Light, and the errors themselves are in C3 Error.

FIGURE 9-13

The Depth worksheet window

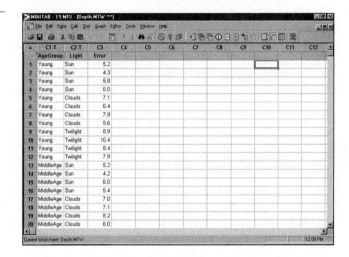

You are interested in the impact of the factors age group and lighting condition on the errors. You know you must also consider the *interaction* between these two factors. Recall that two factors interact if the effect of one of the two factors on the response variable depends on the value of the other factor. In your case study, this might indicate, for example, that the effect of lighting conditions on the errors differs by age group. It is customary to test for significant interaction before testing for the effects of each of the two factors. The relevance of the results of testing for each factor depends on whether there is significant interaction between them.

The Two-Way ANOVA

For this two-way ANOVA, you want to test three sets of hypotheses:

Null Hypothesis, H_0: There is no interaction between age group and lighting condition in the population.

Alternative Hypothesis, H_1: There is an interaction between age group and lighting condition in the population.

Null Hypothesis, H_0: There is no difference in the population mean error for different age groups.

Alternative Hypothesis, H_1: There is a difference in the population mean error for different age groups.

Null Hypothesis, H_0: There is no difference in the population mean error for different lighting conditions.

Alternative Hypothesis, H_1: There is a difference in the population mean error for different lighting conditions.

You perform a two-way ANOVA by using the Stat > ANOVA > Two-Way command. This command requires that there be an equal number of observations for all combinations of factor levels. That is, the data must be *balanced*. In your example, each combination of age group and light has four values so that the data are balanced.

The ANOVA > Two-Way command can produce not only the two-way ANOVA table, but also various residual and other helpful plots. You decide to construct graphical displays of confidence intervals for both factors and an NPP of the residuals from the two-way ANOVA model. This latter plot is helpful because the F-tests in the two-way ANOVA assume that the observations are drawn from normal populations.

▶ **Note** Minitab refers to the two factors as the *row factor* and the *column factor*. These terms are derived from the traditional way of setting out the data. Figure 9-12 showed that the errors in depth perception are organized with the AgeGroup as the row factor and Light as the column factor. ■

To obtain the analyses:
- Choose **Stat > ANOVA > Two-Way**
- Double-click **C3 Error** as the "Response," **C1 AgeGroup** as the "Row factor," and **C2 Light** as the "Column factor"
- Click the **Display means** check boxes to the right of both factors
- Click **Graphs** and click the "Residual Plots" **Normal plot of residuals** check box
- Click **OK**

The completed Two-Way Analysis of Variance dialog box is shown in Figure 9-14.

FIGURE 9-14

*The completed Two-Way
Analysis of Variance
dialog box*

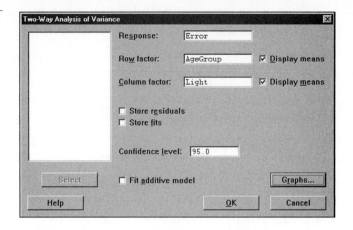

- Click **OK**

    The built-in NPP of the residuals graph is shown in Figure 9-15. The points on the plot appear to lie close to a straight line so that you can be assured that the normality assumption has not been violated. This built-in NPP does not include a formal test for normality.

FIGURE 9-15

*Built-in NPP of residuals
in a two-way ANOVA*

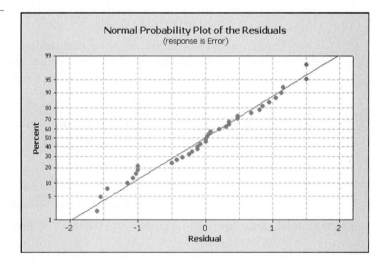

    To view the actual two-way ANOVA output:

- Click the **Session Window** toolbar button 🔳

    The results in the Session window are shown in Figure 9-16. (You may need to scroll up to see them all.)

FIGURE 9-16

*Two-Way ANOVA
of Errors*

```
MTB > Twoway 'Error' 'AgeGroup' 'Light';
SUBC>   Means 'AgeGroup' 'Light';
SUBC>   GNormalplot;
SUBC>   NoDGraphs.
```

Two-way ANOVA: Error versus AgeGroup, Light

```
Source        DF      SS       MS       F      P
AgeGroup       2   29.047  14.5233   15.66  0.000
Light          2  114.872  57.4358   61.95  0.000
Interaction    4    5.097   1.2742    1.37  0.269
Error         27   25.033   0.9271
Total         35  174.047

S = 0.9629   R-Sq = 85.62%   R-Sq(adj) = 81.36%


                        Individual 95% CIs For Mean Based on
                        Pooled StDev
AgeGroup     Mean     +---------+---------+---------+---------
Young      6.99167    (------*-------)
MiddleAge  7.15833     (------*-------)
Older      8.97500                           (------*------)
                      +---------+---------+---------+---------
                      6.40      7.20      8.00      8.80


                        Individual 95% CIs For Mean Based on
                        Pooled StDev
Light        Mean     ------+---------+---------+---------+---
Sun        5.7167     (---*---)
Clouds     7.3583                (---*---)
Twilight  10.0500                            (---*---)
                      ------+---------+---------+---------+---
                      6.0       7.5       9.0       10.5
```

The first component in the output is the standard ANOVA table. It lists the sources of variation (AgeGroup, Light, Interaction, and Error), the values for the F-statistics for each test, and the corresponding p-values. The results of the three hypothesis tests are summarized in the following table.

Test	Effect	F-statistics	p-value	Decision
1	AgeGroup * Light Interaction	1.37	0.269	Do not reject H_0; the data suggest no significant interaction between AgeGroup and Light.
2	AgeGroup	15.66	0.000	Reject H_0; the data suggest that the population mean errors are not the same for the different age groups.
3	Light	61.95	0.000	Reject H_0; the data suggest that the population mean errors are not the same for the different light conditions.

The order of the results here differs from that in Minitab's display. The test for the interaction effect is listed first here because the main effect tests are usually performed only if you do not reject the null hypothesis associated with the interaction test.

The results indicate no significant interaction between age group and lighting condition (p-value = 0.269). However, the mean errors for the three age groups do differ significantly (p-value < .00005) and the mean errors for the three lighting conditions also differ significantly (p-value < .00005).

Immediately below the two-way ANOVA table in Figure 9-16 are values for S, R-Sq, and R-Sq(adj). In this context, S = 0.9629 feet is an estimate of the standard deviation of error for each combination of age group and lighting condition. It is a pooled estimate based upon the errors in each of the nine combinations of factors. The value R-Sq = 85.62% indicates that almost 86% of the variability in the 36 errors can be associated with age group, lighting condition, and their interaction. Finally, R-Sq(adj) = 81.36% is the R-Sq value adjusted for the number of factors (and their interactions) included in the study.

The last two components of the output consist of the sample mean errors for each age group and for each lighting condition. Graphical representations of the 95% confidence intervals for the corresponding population mean errors are also included. The statistically significant F-test for differences in age group means is due to the fact that the mean error for older subjects (8.97500 feet) is substantially greater than those for the middle-aged (7.15833 feet) and the young (6.99167 feet).

The mean error in twilight (10.0500 feet) is considerably greater than that for clouds (7.3583 feet), which, in turn, is much greater than the mean error under sunny conditions (5.7167 feet).

Checking the Assumptions for the F-tests in a Two-Way ANOVA

The assumptions for the F-tests in the two-way ANOVA are similar to those in the one-way ANOVA. They are:

1. The observations must be randomly selected.
2. The populations from which the observations for each factor combination are selected must be approximately normally distributed.
3. The populations from which the observations for each factor combination are selected must have equal variances.

In this case you are prepared to regard the four observations for each combination of age group and lighting condition as random samples from the nine populations. You have checked the normality assumptions using the built-in NPP of the residuals. You can informally check the assumption of equal variances by obtaining the sample variance of Error for each of the nine combinations of age group and lighting condition. To do this:

- Choose **Stat > Basic Statistics > Display Descriptive Statistics**
- Press F3 to clear the previous selections
- Double-click **C3 Error** as the "Variables"
- Double-click **C1 AgeGroup** and **C2 Light** as the "By variables (optional)"
- Click **Statistics**

- Clear the default check boxes, click the **Variance** check box, and click **OK**

The completed Display Descriptive Statistics dialog box should look as shown in Figure 9-17.

FIGURE 9-17

The completed Display Descriptive Statistics dialog box

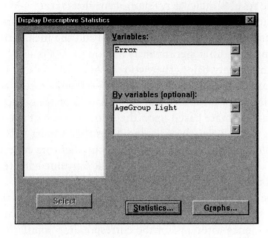

- Click **OK**

The table of variances is shown in Figure 9-18. It is ordered according to the occurrence of each column's text values in the worksheet.

FIGURE 9-18

Variances for error by age groups and light conditions

```
MTB > Describe 'Error';
SUBC>   By 'AgeGroup' 'Light';
SUBC>   Variance.
```

Descriptive Statistics: Error

Results for AgeGroup = Young

```
Variable  Light      Variance
Error     Sun           0.583
          Clouds        0.963
          Twilight      1.167
```

Results for AgeGroup = MiddleAge

```
Variable  Light      Variance
Error     Sun           0.560
          Clouds        0.809
          Twilight      1.647
```

Results for AgeGroup = Older

```
Variable  Light      Variance
Error     Sun           0.142
          Clouds        1.363
          Twilight      1.110
```

The variances range in value from 0.142 squared feet to 1.647 squared feet. This large range of variances is of some concern. However, it is important to keep in mind two points. First, even if the nine population variances were equal, in nine samples you would expect some variation in the values for the sample variances. Second, some variation among the variances is tolerable if the design is balanced, as it is in this case. You decide that the assumption of equal variances is not violated.

▶ **Note** When you have a balanced ANOVA with two or more factors, you can use the more general Minitab command, Stat > ANOVA > Balanced ANOVA. This command performs a multi-way ANOVA and is more complex to use than the Stat > ANOVA > Two-Way command. If you need to use this command, you may want to seek help or refer to the steps outlined in Minitab's Help. ■

▶ **Note** The Professional version contains a command to perform unbalanced ANOVAs: STAT > ANOVA > General Linear Model. ■

The Interactions Plot

The two-way ANOVA above suggests no significant interaction between the factors age group and lighting conditions in their effect on depth perception. Minitab can also provide a nice graphical display of the extent of interaction in a two-way ANOVA, the *interactions plot*. To see this plot:

■ Choose **Stat > ANOVA > Interactions Plot**
■ Double-click **C3 Error** as the "Responses" and **C1 AgeGroup** and **C2 Light** as the "Factors"
■ Click **OK**

The resulting plot, with Minitab's default title, is shown in Figure 9-19.

FIGURE 9-19

Interactions plot

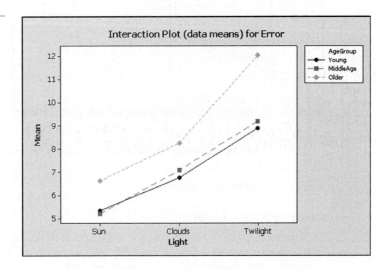

The plot shows the mean error for each of the nine combinations of factor levels. Different symbols are used for the different age groups and, for each age group, the lighting condition means are connected. A complete lack of interaction between the two factors would appear as completely parallel sets of connected lines. The fact that the three sets of connected lines are not too far from parallel supports your earlier finding that the two factors, age group and lighting conditions, interact very little in their effects on depth perception.

▶ **Note** The F-tests in the two-way ANOVA require balanced data. However, the interactions plot is based only upon the means for each combination of factor levels and can be obtained whether or not the data are balanced. ■

To find out more about the two-way ANOVA output, go to Minitab's StatGuide.

■ Click the **Session Window** toolbar button 🖼

■ Scroll up to the Two-Way ANOVA output, click on the output, and click the **StatGuide** toolbar button 🖼

Minitab opens two windows, as shown in Figure 9-20. On the right-hand side is the StatGuide's Two-Way ANOVA Summary window that contains an overview of the subject. The left window is the MiniGuide to Two-Way ANOVA: Topics.

FIGURE 9-20

StatGuide's Two-Way ANOVA Summary window and the MiniGuide window with Two-Way ANOVA topics

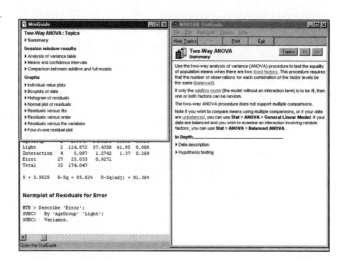

The MiniGuide window provides a pathway to more detailed information about all aspects of the output.

■ Close the StatGuide windows

Creating Factor Levels

Examining Figure 9-12 once again, it is tempting to ignore the factor levels and conclude that the only data in the table are the 36 errors. It is not unusual, especially with larger-scale experiments, to enter only the data for the response variable and then "create" the corresponding factor levels. Practice creating factor levels with the Depth data.

First, return to the Depth data.

■ Click the **Current Data Window** toolbar button 🖼

Now, erase the variables C1-T AgeGroup and C2-T Light:

■ Choose **Data > Erase Variables**

■ Double-click **C1 AgeGroup** and **C2 Light** as the "Columns, constants, and matrices to erase" and click **OK**

Only the column C3 Error remains. You will begin the process of creating (or, more properly, recreating) the factor levels by constructing the age group text values in C1. By comparing the appearance of the data in Figure 9-11 with the data in C3 Error, note that the first 12 entries in C1 should be Young, the next 12 should be MiddleAge, and the last 12 should be Older. You could type in these 36 entries or you could use Autofill as you did earlier. Here you will use Minitab's Calc > Make Patterned Data > Text Values command.

■ Name C1 **AgeGroup**

■ Choose **Calc > Make Patterned Data > Text Values**

■ Double-click **C1 AgeGroup** as the "Store patterned data in" column

■ Type **Young MiddleAge Older** as the "Text values (eg, red "light blue")"

■ Type **12** as the "List each value [ ] times" value

The Text Values dialog box should look as shown in Figure 9-21.

FIGURE 9-21

The completed Text Values dialog box

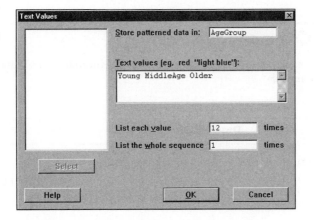

■ Click **OK**

Minitab has entered the correct text values into C1-T AgeGroup.

Now you will enter the lighting condition text values in C2. Again, comparing the organization of the data in Figure 9-11 with the 36 errors in C3, note that the first 12 entries in C2 will have to be four "Sun" entries, followed by four "Clouds" entries, and then four "Twilight" entries. This pattern has to appear three times.

■ Name C2 **Light**

■ Click the **Edit Last Dialog** toolbar button 🔲

■ Press F3 to clear the previous selections

■ Double-click **C2 Light** as the "Store patterned data in" column

- Type **Sun Clouds Twilight** as the "Text values"
- Type **4** as the "List each value [ ] times" value and type **3** as the "List the whole sequence [ ] times" value
- Click **OK**

  The columns C1-T and C2-T will now contain the factor levels.

  It's worth noting that you would have had to use Autofill four times to create the column C2 as you did above. Cutting and pasting would also have involved multiple steps.

  ◆ **Note** In this process of creating factor levels, the patterned data have been text data. You can also enter patterned numeric data with the command Calc > Make Patterned Data > Simple Set of Numbers (or with Autofill). ∎

- In the location where you are saving your work, save the current project **T9** and exit Minitab

  Congratulations! You have now completed Tutorial 9. In the next two tutorials you will explore how to use Minitab to make inferences in simple and multiple regressions.

Minitab Command Summary

This section describes the commands introduced in, or related to, this tutorial. Use Minitab's online Help for a complete explanation of all commands.

Minitab Menu Commands

Menu	Command	Description
Calc ➤	Make Patterned Data	
	➤ Text Values	Generates sequences or repetitive patterns of text values and stores the results in a column.
Stat ➤	ANOVA	
	➤ One-Way	Performs a one-way analysis of variance on stacked data.
	➤ One-Way (Unstacked)	Performs a one-way analysis of variance on unstacked data.
	➤ Two-Way	Performs a two-way analysis of variance for stacked and balanced data.
	➤ Balanced ANOVA	Performs a multi-way analysis of variance for stacked and balanced data.
	➤ Test for Equal Variances	Performs hypothesis tests for equality or homogeneity of variance using Bartlett's test and Levene's test.
	➤ Interval Plot *One Y, With Groups*	Produces plots of means and either confidence intervals or error bars for one or more group variables.
	➤ Interactions Plot	Produces an interaction plot for two factors or a matrix of interaction plots for three to nine factors.

Review and Practice

Matching Problems

Match the following terms to their definitions by placing the correct letter next to the term it describes.

_____ Text Values

_____ Simple Set of Numbers

_____ Response variable

_____ Balanced Analysis of Variance

_____ Balanced

_____ Comparisons

_____ Factor

_____ Unstacked

_____ Two-Way

_____ Interaction

a. The command that can perform a multi-way ANOVA

b. The command to create a text variable made up of patterned data

c. The command that performs a two-factor ANOVA

d. Data structured so that the observations for each level of the factor are in separate columns

e. A term that indicates that there are equal numbers of observations for each factor combination

f. The Minitab option that performs multiple contrasts between the means in a one-way ANOVA

g. The term used in ANOVA to denote a variable presumed to affect a quantitative outcome

h. The term used in ANOVA to denote the variable presumed to be affected by the treatment(s)

i. An effect in a two-way ANOVA other than one of the main effects

j. The command to create a numeric variable made up of patterned data

True/False Problems

Mark the following statements with a *T* or an *F*.

a. _____ Minitab's Two-Way ANOVA command allows unbalanced data, whereas the Balanced ANOVA command must have balanced data.

b. _____ The commands for both one-way and two-way ANOVAs allow you to store residuals and/or fitted values.

c. _____ Minitab can perform multiple comparisons only for stacked data.

d. _____ For the One-Way ANOVA command, there must be the same number of observations in each group.

e. _____ The user specifies the columns in which the ANOVA commands store the residuals and/or fitted values.

f. _____ You can perform multi-way ANOVAs by using the Balanced ANOVA command.

g. _____ 95% confidence intervals for each population mean are part of the default output for the One-Way ANOVA command.

h. _____ Three multiple comparison procedures are available with the One-Way ANOVA command.

i. _____ Both one-way ANOVA procedures offer the option of obtaining a histogram of the responses for each sample.

j. _____ Minitab cannot perform an ANOVA if any of the factors are in text columns.

Practice Problems

The practice problems instruct you to save your worksheets with a filename prefixed by P, followed by the tutorial number. In this tutorial, for example, use *P9* as the prefix. If necessary, use Help or Appendix A Data Sets, to review the contents of the data sets referred to in the following problems. Interpretations should use the language of the subject matter of the question. If you are using the Student version, be sure to close worksheets when you have completed a problem.

1. Open *Baby.mtw*. Replace both the value 5.6 and the outlier, 13.6, with missing values. Show that removing this outlier has little impact on the result of the one-way ANOVA of Times by Design compared to the analysis in Section 9.1. How does removing the outlier change the NPP?

2. Open *Temco.mtw*.

   a. Determine if the population average salary differs for the four departments at Temco. If there are significant differences among departments, explain which is the most highly paid and the least highly paid.

   b. Is it reasonable to assume that the members of these departments are a random sample? How does your answer affect your conclusions in part a?

3. Open *YogurtData.mtw*.

   a. You want to examine how the cost per ounce and the calories per serving vary with nutritional rating. Notice, however, that there is only one yogurt with a rating of excellent. You can merge that yogurt with those that are rated as very good. Create a new variable, in C4, which accomplishes this merge. (When you use the Data > Code > Text to Text command put the text value Very Good in double quotes. This will ensure that Minitab does not treat Very Good as two text values, Very and Good.) Name this new variable.

   b. Use a one-way ANOVA twice to determine if the population mean cost per ounce and the population mean calories per serving differ by nutritional rating. Assume random samples. Examine the validity of the other assumptions. Write a brief paragraph summarizing your conclusions.

4. Open *Infants.mtw*.

   a. Use a one-way ANOVA to test whether there are significant differences among the mean ages for the three ethnic groups. Use $\alpha = .10$ in this case.

   b. If you find significant differences use Tukey's multiple comparisons test to assess which pairwise differences are statistically significant.

   c. Check the normality and the equal variance assumptions in this case. You may assume that the women in each ethnic group are a random sample.

5. Open *PulseA.mtw*. Assume random samples.

   a. Perform a one-way ANOVA to investigate how initial pulse rate varies with level of activity. (i) What are the appropriate hypotheses in this case? (ii) Summarize the differences in means for the three groups and explain if there is a significant difference among the population means. (iii) Do you think that the equal variance assumption is valid in this case? Explain briefly. (iv) Check whether the normality assumption is valid for these data.

b. How much of the variability in pulse rate can be explained by differences in activity levels?

c. If there are significant differences among the activity groups, use Tukey's multiple comparisons test to compare the three pairs of means.

d. Obtain an interval plot of pulse rate by activity level. How does your plot differ from the text plots of confidence intervals that form part of the one-way ANOVA output?

6. Open *Cotinine.mtw.*

a. Perform a one-way ANOVA on these data to test for significant differences in the mean cotinine serum level for the three populations. State the appropriate hypotheses and summarize your conclusions.

b. You should find a significant F-statistic value in part a. By examining the three confidence intervals in the output, can you explain this large value for the F-statistic?

c. Perform Tukey's multiple comparisons test. Do the results of this test support your conclusions in part b? Explain briefly.

d. Check the validity of (i) the equal variances and (ii) the normality assumptions. What do you conclude? Assume random samples.

7. Open *Murderu.mtw.* You should treat each of the four sets of murder rates as random samples of such rates over time.

a. Perform a one-way ANOVA on these data to test for significant differences in the mean murder rates for the four regions. State the appropriate hypotheses and summarize your conclusions.

b. You should find a significant value for the F-statistic in part a. By examining the three confidence intervals in the output, can you explain this large value for the F-statistic?

c. Perform Tukey's multiple comparisons test. Do the results of this test support your conclusions in part b? Explain briefly.

d. Check the validity of (i) the equal variances and (ii) the normality assumptions. What do you conclude?

8. Open *AgeDeath.mtw.*

a. Perform a two-tailed two-sample t-test of whether the mean age at death is the same for males and females. Assume random samples and equal variances. What is: (i) the p-value?, (ii) the pooled standard deviation?, and (iii) the 95% confidence interval for the difference in the mean age at death? Write a brief conclusion.

b. Perform a one-way ANOVA with age at death as the response variable and gender as the factor. What is: (i) the p-value?, (ii) the pooled standard deviation?, and (iii) the 95% confidence interval for the difference in the mean age at death? Write a brief conclusion.

c. The two tests in parts a and b are equivalent so your three answers in part a should match those in part b. Verify that the value for F-statistic in the ANOVA is the square of the value for t-statistic in part a.

d. In what sense does the t-test offer the user more flexibility than the ANOVA in the case where there are only two groups?

9. The data below are the pulse rates for 36 students. They are organized by gender and level of exercise. (These data are a subset of the data in *Survey.mtw* that can be assumed to be a random sample.)

	Exercise			No Exercise		
Female	76	90	68	60	71	85
	73	65	88	60	78	75
	70	68	67	80	86	100
Male	54	72	67	72	70	66
	47	40	64	67	69	60
	84	60	80	70	75	72

a. Enter the 36 pulse rates into C1. Enter the 18 female rates first with the rates for those who exercise before those who do not exercise. Then enter the males' rates in the same order.

b. Use the Calc > Make Patterned Data > Text Values command to create the factor Gender in C2 and the factor Exercise in C3.

c. Perform a two-way ANOVA on the pulse rates. Check first for a significant interaction effect and only then for significant main effects. Write a brief account of your conclusions.

d. What fraction of the variability in pulse rates can be associated with the two factors and their interaction?

e. Obtain an interactions plot for these data. Does the plot support your earlier analysis? Explain briefly.

10. Open *Pancake.mtw*. (This worksheet is in the Data folder so you will have to navigate to that folder.) Assume random samples.

a. Perform a two-way ANOVA on the quality ratings. Check first for a significant interaction effect and only then for significant main effects. Be sure to obtain a NPP of the residuals.

b. Obtain an interactions plot for these data. Does the plot support your earlier analysis? Explain briefly.

c. Use the Stat > Basic Statistics > Display Descriptive Statistics command to obtain the variances of quality ratings for each Supplement–Whey combination. What do the results suggest about the assumption of equal variances?

d. Write a brief account of your conclusions. Your account should deal with (i) the validity of the assumptions underlying your analysis and (ii) your advice on the impact of the two factors on the quality rating of pancakes.

11. Open *Compliance.mtw*. Assume that the 34 patients were randomly sampled.

a. Use the Calc > Row Statistics command to create, in C11, a measure of diet compliance that is the sum of the five seven-point scale scores. Name this variable Compliance.

b. Create a new variable, in C12, which has two text values: High, if the patient had at least some college education, and Low, otherwise. Name this new variable Education.

c. You are to explore the impact of two variables, Education and HistHD on Compliance. Because there are not the same numbers of patients in each of the four combinations of levels of these variables, the data are not balanced and so you cannot use the Stat > ANOVA > Two-Way command. However, you can do a descriptive analysis, looking only at mean compliance scores. First, use the Stat > Tables > Descriptive Statistics command to obtain the mean compliance score for each combination of levels of HistHD and Education. What is the difference (if any) in scores between those patients who have and do not have, a history of heart disease? What is the comparable difference if you look at only those with a low level of education? What is the comparable difference if you look at only those with a high level of education? Does there seem to be any interaction between HistHD and Education in their effects on compliance scores?

d. Obtain an interactions plot of these data. Does the plot support your conclusion in part c? Explain briefly.

On Your Own

The data set *Academe.mtw* contains information about the distribution of faculty by rank and gender at a variety of institutions of higher education. It also has data on faculty salaries similarly distributed by rank and gender. The institutions are balanced on the basis of geography and type of college. You are to create two response variables: (i) the percentage of each rank that are female and (ii) for each rank, the percentage that female salaries are of male salaries. The major task in this assignment is to analyze, summarize, and report on, how these two quantities vary by geography, by type of institution, and by rank. You are also to investigate interactions among these variables. For each response variable, you should perform three one-way ANOVAs, three two-way ANOVAs, and, if you are feeling particularly ambitious, you can perform a balanced ANOVA to explore the impact of these three factors on the two quantities. Be sure to test the normality and the equal variance assumptions. You may assume that this is a random sample over time.

Answers to Matching Problems

(b) Text Values
(j) Simple Set of Numbers
(h) Response variable
(a) Balanced Analysis of Variance
(e) Balanced
(f) Comparisons
(g) Factor
(d) Unstacked
(c) Two-Way
(i) Interaction

Answers to True/False Problems

(a) F, (b) T, (c) F, (d) F, (e) F, (f) T, (g) T, (h) F, (i) F, (j) F

Fundamentals of Linear Regression

You explored some of the descriptive aspects of simple linear regression in Tutorial 5. In this tutorial, you will focus on the inferential aspects of linear regression. Regression allows you to investigate and model the relationship between a *quantitative response (dependent)* variable and one or more *predictors (independent or exploratory variables)*. A regression model can be used to estimate a future response based upon the value of the predictor variables. You will begin with simple linear regression, building a model of the effect of a single variable and testing its predictive ability. Then you will examine how to build a multiple linear regression with three predictor variables. In each case, you will use Minitab's regression features to perform hypothesis tests and generate confidence intervals to evaluate the model.

OBJECTIVES

In this tutorial, you learn how to:

- Test for a zero population slope in a linear relationship between a response variable and a predictor variable
- Obtain a confidence interval for the population slope
- Obtain a confidence interval for the mean value of the response variable for a given predictor value
- Obtain a prediction interval for an individual value of the response variable for a given predictor value
- Generate a quadratic regression model
- Use multiple linear regression to investigate the form of the linear relationship between a response variable and several predictor variables

10.1 Fitting a Straight Line to Data: Simple Linear Regression

CASE STUDY	INSTITUTIONAL RESEARCH— TUITION MODELING

In this and the next tutorial, you will play the role of an institutional researcher. You have been asked by the administration of the Massachusetts college at which you work to generate a model for tuition based upon data collected from other Massachusetts colleges. For each of 60 colleges, ten variables were identified that could be related to the tuition charged there. These variables, all measured for 2002, are stored in the worksheet *CollMass.mtw*. You are to use these data to build a model that the administration can use to determine a reasonable tuition for the college.

Begin by opening Minitab:

- Open **Minitab** and maximize the Session window
- If necessary, **Enable Commands** and clear Output Editable from the Editor menu
- Click the **Save Project** toolbar button 🖫 and, in the location where you are saving your work, save this project as *T10.mpj*
- Open the worksheet *CollMass.mtw*
- If necessary, click the **Current Data Window** toolbar button 🖽

   The variables in the data set are listed below.

Column	Name	Description
C1-T	College	Name of the college
C2-T	PubPriv	Type of ownership; Private or Public
C3	SFRatio	Ratio of faculty to students
C4	%FacPhD	Percentage of faculty with Ph.D.
C5	%Accept	Percentage of applicants that are accepted
C6	%Enroll	Percentage of accepted students that enroll
C7	FrClass	Size of the Freshman class
C8	SATV	Average SAT Verbal score
C9	SATM	Average SAT Math score
C10	%Graduate	Percentage of Freshman class who graduate in six years or fewer
C11	%Top20%	Percentage of Freshman class who were in the top 20% of their High School class
C12	Tuition	Annual tuition

The dean of admissions is particularly interested in the relationship between tuition and average verbal SAT scores for Massachusetts colleges. So, begin your analysis of these data by investigating a *simple linear regression* with Tuition as the response variable and SATV as the predictor. First, examine a scatterplot of Tuition against SATV.

■ Choose **Graph > Scatterplot** and double-click *Simple*

■ Double-click **C12 Tuition** as the "Y variables" column for Graph 1

■ Double-click **C8 SATV** as the "X variables" column for Graph 1

■ Click **Labels**, type **Scatterplot of Tuition Against Average Verbal SAT Score** as the "Title", and click **OK** twice

The scatterplot should resemble Figure 10-1.

FIGURE 10-1

Scatterplot of Tuition against SATV

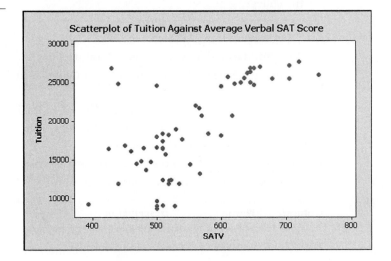

The scatterplot suggests a positive relationship between these two variables; that is, the higher the average verbal SAT score, the higher the tuition tends to be. There is a slight curvature to the data, but the relationship is approximately linear, although there are two or three outlying observations in the upper left-hand corner. You can measure the strength of the linear relationship between the two variables by computing the correlation coefficient.

■ Choose **Stat > Basic Statistics > Correlation**

■ Double-click **C12 Tuition** and **C8 SATV** as the "Variables"

■ Click **OK**

The Session window reports a correlation coefficient of 0.693. This value suggests a moderately strong positive linear relationship between tuition and average verbal SAT scores among Massachusetts colleges in 2002. This is verified by the corresponding p-value of 0.000. This p-value is less than any reasonable level of significance, so you reject the null hypothesis that the population correlation (ρ) between these two variables is 0. The small p-value supports the graphical evidence of a moderately strong positive linear relationship. This suggests that an equation for this relationship is

$$\text{Tuition} = \beta_0 + \beta_1 \text{SATV} + \varepsilon$$

where β_0 is the population y-intercept, β_1 is the population slope, and ε is the error. That is, ε is how far the observed tuition is from the modeled tuition. You are ready to obtain the linear equation that relates tuition to average verbal SAT score based upon the method of least squares using these data. This estimated regression equation, or simply the regression equation, will have the form

$$\text{Tuition} = b_0 + b_1 \text{SATV}$$

where b_0 is the estimated y-intercept and b_1 is the estimated slope.

- Choose **Stat > Regression > Regression**
- Double-click **C12 Tuition** as the "Response" variable
- Double-click **C8 SATV** as the "Predictors" variable
- Click **OK**

The Session window display is shown in Figure 10-2.

FIGURE 10-2

Simple linear regression of Tuition against SATV

```
MTB > Regress 'Tuition' 1 'SATV';
SUBC>   Constant;
SUBC>   Brief 2.
```

Regression Analysis: Tuition versus SATV

```
The regression equation is
Tuition = - 9146 + 50.6 SATV ──────────────── Regression equation

58 cases used, 2 cases contain missing values
```

Predictor table

```
Predictor     Coef   SE Coef      T      P
Constant      -9146     3926   -2.33  0.023
SATV        50.635     7.033    7.20  0.000

S = 4374.66   R-Sq = 48.1%   R-Sq(adj) = 47.1% ──── Summary statistics
```

ANOVA table

```
Analysis of Variance

Source          DF         SS          MS      F      P
Regression       1   991864700   991864700  51.83  0.000
Residual Error  56  1071708394    19137650
Total           57  2063573094
```

Unusual observations table

```
Unusual Observations

Obs   SATV  Tuition    Fit  SE Fit  Residual  St Resid
  5    430    26858  12627    1034     14231      3.35R
 15    440    24830  13133     976     11697      2.74R
 25    750    26019  28830    1505     -2811     -0.68 X

R denotes an observation with a large standardized residual.
X denotes an observation whose X value gives it large influence.
```

The equation for the least squares/regression line is Tuition = –9146 + 50.6 SATV. It is based upon 58 out of the 60 colleges because two colleges do not have either Tuition or SATV values. Predicted tuition is obtained by subtracting $9,146 from the product of 50.6 and the average verbal SAT score for the college. For example, the predicted tuition for a college with an average verbal SAT score of

530 is –9,146 + 50.6*530 = $17,672. The slope of the line, 50.6, indicates that for every one-point increase in average verbal SAT score, predicted tuition increases by $50.60.

The slope of the line is recorded with more decimal places (50.635) in the *predictor table* immediately beneath the equation. With this value, the predicted tuition for a college with average verbal SAT score of 530 is –9,146 + 50.635 * 530 = $17,690.55. Later in this tutorial, you will use Minitab to obtain this prediction.

The predictor table contains the sample regression coefficients (–9,146 and 50.635), their standard errors, as well as the t-statistics and p-values for testing the hypothesis that a population coefficient is zero. The t-test of whether the slope of the population line (β_1) is 0 is of particular importance. In this context, the appropriate hypotheses for this test might be:

Null Hypothesis, H_0: $\beta_1 = 0$

Alternative Hypothesis, H_1: $\beta_1 \neq 0$

where you might think of β_1 as the change in mean Tuition for each additional average verbal SAT score for all Massachusetts colleges.

The statistic for this test is $t_{n-2} = (b_1 - 0)/SE(b_1)$, where $SE(b_1)$ is the standard error of $b_1 = (50.635 - 0)/7.033 = 7.20$. The test's p-value of 0.000 for SATV in the predictor table indicates that you should reject the null hypothesis. The data suggest that there is a non-zero, probably positive, population slope. This p-value is the same as the p-value you obtained when computing the sample correlation previously. In fact, this test for a non-zero slope is equivalent to the test that the population correlation equals zero ($\rho = 0$).

▶ **Note** The p-value computed by Minitab is for a two-tailed test. For a one-tailed test with an alternative of $\beta_1 > 0$, you must modify Minitab's p-value depending upon the sign of b_1: (i) if $b_1 \geq 0$ the appropriate p-value is Minitab's p-value divided by 2 and (ii) if $b_1 < 0$ the appropriate p-value is 1 minus Minitab's p-value divided by 2. For a one-tailed test with an alternative of $\beta_1 < 0$, you must modify Minitab's p-value depending upon the sign of b_1: (i) if $b_1 \leq 0$ the appropriate p-value is Minitab's p-value divided by 2 and (ii) if $b_1 > 0$ the appropriate p-value is 1 minus Minitab's p-value divided by 2. ■

The output shown in Figure 10-2 contains information that helps you assess the usefulness of the linear model. Many of the statistics were discussed in Tutorial 5.

The summary statistics row includes S, the estimated standard deviation around the regression line. This statistic is often called the standard error of the estimate; in this case, its value is $4,374.66. R-sq, the coefficient of determination, indicates that 48.1% of the variation in tuition can be explained by its linear relationship to the average verbal SAT scores at the Massachusetts colleges. Recall that the higher the R^2 value, the better the estimated regression line fits the data. (R-sq(adj) is the R^2 value adjusted for the number of predictor variables in the regression. It is not something that the beginning student needs to worry about in simple linear regression.)

The F-test in the ANOVA table is exactly equivalent to the previous two-sided t-test. The hypotheses are the same and so are the p-values. The value of the F-statistic 51.83 is the square of the value of the t-statistic, 7.20.

In the last table in the output, Minitab lists two types of *unusual observations*. It is good statistical practice to carefully examine such observations. The first type of unusual observation is based upon residuals. A *residual* is the difference between an observed value for y (in this case, Tuition) and a predicted value (shown in the Fit column) of an observation. A *standardized residual* is the residual divided by an estimate of its variation. Minitab lists outliers that have a standardized residual of more than 2 or less than –2. These values are marked with an "R". For example, Minitab reports that unusual standardized residuals exist for the colleges in rows 5 (Atlantic Union College with an average verbal SAT score of 430 and a tuition of $26,828) and 34 (Curry College with an average verbal SAT score of 440 and a tuition of $24,830). Both Atlantic Union, with a standardized residual of 3.35, and Curry College, with a standardized residual of 2.74, have tuition rates substantially higher than would be predicted from their average verbal SAT score.

You can brush your scatterplot to see if the outliers in the scatterplot correspond to these two colleges.

- Choose **Window > Scatterplot of Tuition vs SATV**
- Hold your cursor over the scatterplot, click the **right-hand mouse button**, and choose **Brush**

The default brushing palette appears. You want it to identify the name of the college.

- Hold your cursor over the scatterplot, click the **right-hand mouse button**, and choose **Set ID Variables**
- Double-click **C1 College** as the "Variables" and click **OK**
- Drag your cursor to draw a box around the two points in the upper left-hand corner of the scatterplot.

The brushing palette indicates that the two schools are indeed Atlantic Union College and Curry College.

- Click the **Session Window** toolbar button 🖻

The last section of the output in Figure 10-2 also lists any values for the X variable (SATV in this case) that differ substantially from the other predictor values in the data. These *influential points*, the second type of unusual observations, are marked with an "X". Harvard University, with an average verbal SAT score of 750 and a tuition of $26,019, has the largest (or the greatest) influence upon the regression. Another name for influential points are *leverage points*. Minitab uses this latter name. You will look in more detail at unusual observations in Tutorial 11, but if you want to learn more about how Minitab marks such observations you need only consult the unusual observations entry in its glossary.

A Confidence Interval for the Population Slope

As you can see from Figure 10-2, Minitab provides you with a lot of valuable information in the regression output. However, it does not provide confidence intervals for the unknown population regression coefficients β_0 and β_1—respectively, the intercept and the slope of the population line. You can, however, use the corresponding sample regression coefficient and its standard error to compute such an interval. For example, you could compute a 95% two-sided confidence interval for the unknown population slope, β_1, by using the sample slope 50.635—which is a point estimate for β_1—and the related standard error of 7.033 in the predictor table. The form of the confidence interval is $\beta_1 \pm t_{n-2} SE(\beta_1)$ or 50.635 $\pm t_{56} * 7.033$, where the value t_{56} is the 97.5th percentile of the t distribution with 56 DF (degrees of freedom). (You need the 97.5th percentile because 95% of the distribution lies between the 2.5th percentile and the 97.5th percentile.) You can find t by using Calc > Probability Distributions (as you did in Tutorial 6):

- Choose **Calc > Probability Distributions > t**
- Click the **Inverse cumulative probability** option button and type **56** as the "Degrees of freedom"
- Click the **Input constant** option button and type **.975** as the "Input constant"
- Click **OK**

The value for t is 2.00324, so the 95% two-sided confidence interval for b_1 is 50.635 ± 2.00324*7.033, or (50.635 ± 14.0888), or ($36.55 for each average verbal SAT score to $64.72). You can be 95% confident that the slope of the population line lies in this interval. (For these hypothesis tests and confidence intervals to be valid, certain assumptions must be satisfied. These are discussed in Tutorial 11.)

Obtaining Residuals

Minitab can compute (and store for later use) the residuals and fitted or predicted values for each observation.

- Choose **Stat > Regression > Regression**
- Click **Storage**
- Click the "Diagnostic Measures" **Residuals** and **Standardized residuals** and the "Characteristics of Estimated Equation" **Fits** check boxes

The completed Regression - Storage dialog box should look as shown in Figure 10-3.

FIGURE 10-3

The completed Regression - Storage dialog box

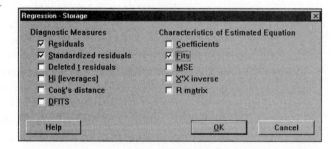

- Click **OK** twice
- Click the **Current Data Windo**w toolbar button 🖩
- If necessary, scroll right to display columns C13, C14, and C15

Minitab computes the residuals, standardized residuals, and fitted values for each observation in the data set. In addition, Minitab automatically names the columns. Your Data window should resemble that shown in Figure 10-4.

FIGURE 10-4

The Data window showing stored statistics

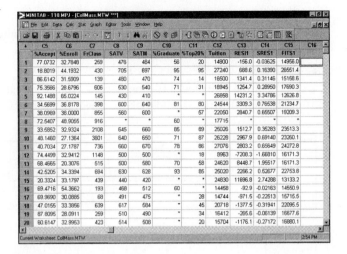

The new columns contain the information in the following chart.

Column	Name	Description
C13	RESI1	Contains the residual for each college. This is the difference between each college's actual tuition and the fitted (predicted) tuition.
C14	SRES1	Contains each residual divided by its standard deviation. Extreme standardized residuals (with an absolute value greater than 2) determine the unusual observations marked by Rs.
C15	FITS1	Contains each college's predicted tuition, as computed from the regression equation.

For instance, for Assumption College in row 4:
- Average verbal SAT score is 530.
- Tuition is $18,945.
- Fitted (predicted) tuition is, using the regression equation, $-9,146 + 50.635*530 = \$17,690.3$, which is the value for C15 FITS1.
- The residual, in C13 RESI1, is $18,945 - 17,690.3 = \$1254.7$.

In Tutorial 11 you will use the fitted values and residuals to check assumptions and investigate possible tuition inequities

The Fitted Line Plot

Minitab's fitted line plot, introduced in Tutorial 5, draws the least squares/regression line with the data.

To create a fitted-line plot for these data:

- Choose **Stat > Regression > Fitted Line Plot**
- Double-click **C12 Tuition** as the "Response (Y)" variable
- Double-click **C8 SATV** as the "Predictor (X)" variable
- Click **OK**

The fitted line plot of tuition against average verbal SAT score, with Minitab's default title, should look as shown in Figure 10-5.

FIGURE 10-5

Fitted line plot of Tuition against SATV

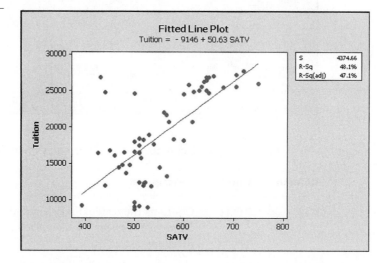

The scatterplot shows the fitted line. This is obtained by plotting the connected fitted values in C15 FITS1 against SATV. You can review the contents of this display by utilizing StatGuide.

- Click your cursor on the display and click the **StatGuide** toolbar button

Figure 10.6 shows the right-hand window of the first of five Fitted Line Plot Graphs StatGuide windows.

FIGURE 10-6

*The first Fitted Line Plot
Graphs StatGuide window*

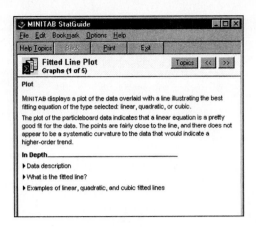

You should close the StatGuide windows after this review.

■ Click **Exit** in the right-hand window

10.2   *Computing Response Variable Estimates*

Now you will use Minitab to obtain point and interval estimates for Tuition
for a given value of SATV. For a college with an average verbal SAT score of
530, for example, the predicted tuition is, from the regression equation,
–9,146 + 50.635*530 = $17,690.55. This predicted value is a point estimate
for two quantities:

1. The mean tuition for the population of all Massachusetts colleges with an
   average verbal SAT score of 530
2. The tuition for a specific college with an average verbal SAT score of 530

    Minitab not only can obtain a point estimate for these two quantities, but
can also compute an interval estimate for each. When the mean tuition is being
estimated, the interval is called, as usual, a confidence interval. But when the tuition
at a specific college is being estimated, the interval is called a *prediction interval*.

    As an example, compute a 95% two-sided confidence interval for the mean
tuition over all colleges with average verbal SAT scores of 530 and a 95% two-
sided prediction interval for the tuition for a specific college with an average
verbal SAT score of 530.

■ Choose **Stat > Regression > Regression**

■ Press F3 to clear the previous selections

■ Double-click **C12 Tuition** as the "Response" variable and **C8 SATV** as the
  "Predictors" variable

■ Click **Options**

■ Type **530** as the "Prediction intervals for new observations" (530 is the value
  for the predictor variable)

    The completed Regression - Options dialog box should look as shown in
Figure 10-7.

FIGURE 10-7
The completed Regression - Options dialog box

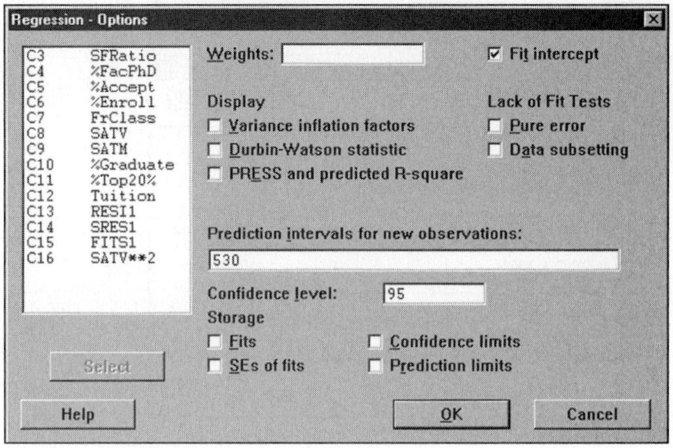

- Click **OK** twice

The last two blocks in the Session window display are shown in Figure 10-8.

FIGURE 10-8
Point and interval estimates for tuition when the average verbal SAT score is 530

```
Predicted Values for New Observations

New
Obs     Fit  SE Fit       95% CI            95% PI
  1   17690     595  (16498, 18883)  (8846, 26535)


Values of Predictors for New Observations

New
Obs   SATV
  1    530
```

The predicted tuition for a Massachusetts college with an average verbal SAT score of 530 is $17,690. You are 95% confident that the mean tuition for all colleges with this score will be between $16,498 and $18,883. For an individual college with a score of 530, you can predict with 95% confidence that its tuition will be between $8,846 and $26,535.

Note As you can see in Figure 10.7, you can change the default 95% confidence level if you wish. ■

You can display all the 58 fitted values, the corresponding 95% two-sided confidence interval values, and the corresponding 95% two-sided prediction interval values on a scatterplot of data for a simple linear regression.

- Choose **Stat > Regression > Fitted Line Plot**
- Click **Options**
- Click the "Display options" **Display confidence interval** and **Display prediction interval** check boxes

The completed Fitted Line Plot - Options dialog box should look as shown in Figure 10-9.

FIGURE 10-9

The completed Fitted Line Plot - Options dialog box

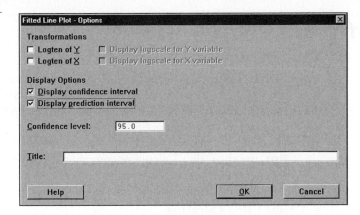

■ Click **OK** twice

The resulting graphs, with Minitab's default title, should resemble Figure 10-10.

FIGURE 10-10

Fitted Line Plot of Tuition against SATV with 95% confidence band and 95% prediction band

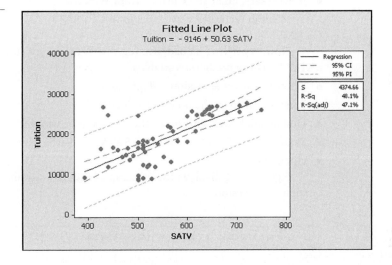

In addition to the fitted line, the graph shows the *95% confidence bands* and *95% prediction bands* for values between the minimum and maximum scores of SATV. Note that two colleges are above the upper prediction interval band. You suspect that these two values are the unusually large standardized residuals referred to previously.

Before going on, save your project:

■ Click the **Save Project** toolbar button 🖫

10.3 Performing a Quadratic Regression

In the previous section, you examined the simplest case, in which tuition is expressed as a linear function of the SATV. Perhaps this relationship can be better described by a curve given the slight curvilinear relationship observed in Figure 10-1. You can create a second-degree variable, such as $SATV^2$, and use it as a variable for a *quadratic regression* to explain Tuition. A quadratic equation is a special case of a polynomial equation.

First, use the Calc > Calculator command to create a second-degree variable, $SATV^2$. (You will use the variable name SATV**2. Recall that ** is the Minitab symbol for exponentiation.)

- Choose **Calc > Calculator**
- Type **SATV**2** as the "Store result in variable"
- Enter, by clicking or typing, **SATV**2**, as the "Expression"
- Click **OK**
- If necessary, click the **Current Data Window** toolbar button 🖩

Minitab records a new variable in C16 SATV**2, the square of each term in C8 SATV.

Now, try fitting a quadratic regression model for tuition using SATV and SATV**2 as the predictor variables.

- Choose **Stat > Regression > Regression**
- Press F3 to clear the previous selections
- Double-click **C12 Tuition** as the "Response" variable
- Double-click **C8 SATV** and **C16 SATV**2** as the "Predictors"
- Click **OK**

Figure 10-11 displays the creation of the new variable and the results of the quadratic regression.

FIGURE 10-11

Quadratic regression of Tuition against average verbal SAT score

```
MTB > Regress 'Tuition' 3 '%Enroll' 'SATV' '%Top20%';
SUBC>    Constant;
SUBC>    Brief 2.
```

Regression Analysis: Tuition versus %Enroll, SATV, %Top20%

```
The regression equation is
Tuition = 5296 - 104 %Enroll + 21.8 SATV + 119 %Top20%


41 cases used, 19 cases contain missing values


Predictor     Coef   SE Coef      T      P
Constant      5296      6378    0.83  0.412
%Enroll    -104.49     43.91   -2.38  0.023
SATV         21.77     15.13    1.44  0.159
%Top20%     119.33     37.73    3.16  0.003


S = 2694.87    R-Sq = 78.2%    R-Sq(adj) = 76.4%
```

```
Analysis of Variance

Source              DF          SS          MS      F      P
Regression           3   962726810   320908937  44.19  0.000
Residual Error      37   268706450     7262336
Total               40  1231433260


Source      DF      Seq SS
%Enroll      1     3020298
SATV         1   887080416
%Top20%      1    72626096

Unusual Observations

Obs  %Enroll  Tuition     Fit  SE Fit  Residual  St Resid
 12     32.9     8963   14885     614     -5922     -2.26R
 22     29.3     8936   15755     889     -6819     -2.68R
 25     77.6    26019   25208    1705       811      0.39 X
 45     27.7    18360   24092     937     -5732     -2.27R

R denotes an observation with a large standardized residual.
X denotes an observation whose X value gives it large influence.
```

The quadratic (or second-degree) equation, which appears at the top of the output, is Tuition = 36245 − 112 SATV + 0.143 SATV**2. For a college with an average verbal SAT score of 530, for example, predicted tuition is 36245 − 112*530 + .143*530^2 = $17,053.70.

Minitab displays the t-statistics for testing that each of the population coefficients is 0 and their corresponding p-values. The table reports a p-value of 0.201 for the SATV coefficient; this value is not significant in this model. The quadratic term, SATV**2, has a p-value of 0.065, for example, which also is not significant for any value for α less than .065, for example, .05.

The standard error of the estimate of the quadratic regression equals $4,278.43. This value is only slightly less than the corresponding simple linear regression value of $4,374.66. The regression explains 51.2% of the variation—this isn't much better than the value R-Sq = 48.1% for simple linear regression model used previously.

The ANOVA table reports an F-statistic of 28.87. The corresponding p-value of 0.000 suggests that at least one of the coefficients in the model is not zero. This result suggests that Tuition depends upon SATV and/or the quadratic term SATV**2. But this appears to contradict the results of the t-tests in the predictor table that appeared to suggest that tuition was linearly unrelated to both SATV and SATV**2. This apparent contradiction is probably due to the strong positive, linear correlation (r = .997) between the two predictors. In Tutorial 11, you will explore this issue further. Based upon these results, you decide not to recommend the quadratic model to the administration.

▶ **Note** The content and interpretation of the Seq SS table following the ANOVA table is rather involved. You can find information on this table by following this sequence: (i) clicking Help in the Stat > Regression > Regression command dialog box, (ii) clicking on see also, (iii) clicking on Methods and formulas, and (iv) clicking on Sequential sum of squares. This Help window also provides information on all of Minitab's regression output. You may need to seek further assistance from a statistician. ■

The last table in Figure 10-11 identifies six colleges as unusual observations. Three have large standardized residuals, identified by Rs, and three have predictor variable values that greatly influence the regression equation, identified by Xs.

▶ **Note** You may also obtain a quadratic regression by choosing Stat > Regression > Fitted Line Plot and clicking the "Type of Regression Model" Quadratic option button. ■

Using Transformations

Minitab can include not only higher powers of a predictor variable, but also other transformations of both the response and predictor variables. Among the most common transformations of the response variable are square roots, logarithms, and negative reciprocals. You can use products of predictor variables such as X_1X_2 to create complex regression models. In addition, more advanced transformations are sometimes appropriate. For example, if a variable is cyclic in nature, you may consider including the cosine of that variable.

Next, save your work.

- Click the **Save Project** toolbar button 🖫

10.4 | *Performing Multiple Linear Regression*

The plot in Figure 10-5 showed that there is quite a bit of variation around the regression line. Tuition at Massachusetts colleges in 2002 is not based solely upon their average verbal SAT scores—there are other factors at work. To investigate further, add two predictor variables to the model with the average verbal SAT predictor:

1. %Enroll—percentage of accepted students that enroll
2. %Top20%—percentage of the Freshman class who were in the top 20% of their High School graduating class

This model, with more than one predictor variable, is called a *multiple linear regression* model.

- Click the **Edit Last Dialog** toolbar button 🔲
- Press F3 to clear the previous selections
- Double-click **C12 Tuition** as the "Response" variable
- Double-click **C6 %Enroll**, **C8 SATV**, and **C11 %Top20%** as the "Predictors"
- Click **OK**

The Session window shows the results in Figure 10-12.

FIGURE 10-12

*Multiple linear regression
with three predictors*

```
MTB > Regress 'Tuition' 3 '%Enroll' 'SATV' '%Top20%';
SUBC>    Constant;
SUBC>    Brief 2.
```

Regression Analysis: Tuition versus %Enroll, SATV, %Top20%

```
The regression equation is
Tuition = 5296 - 104 %Enroll + 21.8 SATV + 119 %Top20%


41 cases used, 19 cases contain missing values


Predictor    Coef   SE Coef      T      P
Constant     5296      6378    0.83  0.412
%Enroll   -104.49     43.91   -2.38  0.023
SATV        21.77     15.13    1.44  0.159
%Top20%    119.33     37.73    3.16  0.003


S = 2694.87   R-Sq = 78.2%   R-Sq(adj) = 76.4%


Analysis of Variance

Source           DF          SS         MS      F      P
Regression        3   962726810  320908937  44.19  0.000
Residual Error   37   268706450    7262336
Total            40  1231433260


Source    DF      Seq SS
%Enroll    1     3020298
SATV       1   887080416
%Top20%    1    72626096


Unusual Observations

Obs  %Enroll  Tuition    Fit  SE Fit  Residual  St Resid
 12     32.9     8963  14885     614     -5922    -2.26R
 22     29.3     8936  15755     889     -6819    -2.68R
 25     77.6    26019  25208    1705       811     0.39 X
 45     27.7    18360  24092     937     -5732    -2.27R

R denotes an observation with a large standardized residual.
X denotes an observation whose X value gives it large influence.
```

The multiple least squares/regression equation is Tuition = 5296 – 104 %Enroll + 21.8 SATV + 119 %Top20%. It is based upon 41 out of the 60 colleges because 19 colleges do not have Tuition, %Enroll, SATV, or %Top20% values. The coefficient of %Top20% (119), for example, indicates that for every percentage increase in the percentage of the Freshman class that was in the top 20% of their class predicted tuition increases by $119 among those colleges with the same average verbal SAT score and the same percentage of accepted students that enroll. Note that the interpretation of this coefficient and the coefficients for SATV and %Enroll is different from the interpretation of the coefficients of the same predictor in simple linear regression because here you have to hold the other predictors constant.

Minitab displays the t-statistic for each coefficient and the corresponding p-value. In multiple regression the t-tests measure the benefit of adding that predictor variable given the other variables in the model. A small p-value suggests that the corresponding population regression coefficient is not zero and therefore, suggests that the corresponding variable is significant in predicting the response variable.

The p-values for %Enroll (.023) and for %Top20% (.003) suggest that these two variables are significant in predicting tuition. The high p-value (0.159) for SATV suggests that, after the inclusion of the other two variables, this variable is not significant in predicting tuition.

▶ **Note** The adjusted R-Sq value plays a more important role in multiple linear regression. If you include non-significant predictor variables in the model, R-Sq (78% in Figure 10-12) can be artificially high. The R-Sq(adj) (76.4%) is adjusted to reflect the number of predictor variables in the model. ■

With a standard error of the estimate of $2,694.87, this model explains 78.2% of the variation in tuition at Massachusetts colleges. The adjusted percentage of variation is only slightly less at 76.4%. The addition of other variables substantially improved your model over the linear and quadratic models considered previously.

The F-statistic in the ANOVA table (which tests the hypothesis that all predictor population coefficients are zero) has a p-value of 0.000. This suggests that at least one of the population coefficients is different from zero at any reasonable level of significance.

Minitab has noted several unusual observations in the data set. There are three observations with large standardized residuals (in rows 12, 22, and 45) and one observation with large influence (in row 25). You will investigate these values in Tutorial 11.

▶ **Note** Recall that in multiple linear regression, unlike in simple linear regression, the overall F-test is not equivalent to the two-sided t-test. ■

▶ **Note** As with the corresponding section of the quadratic regression output, the content and interpretation of the Seq SS table following the ANOVA table is rather involved. You may again want to examine the Minitab Help or seek further assistance from a statistician. ■

▶ **Note** The quadratic regression you developed in the last section is a special case of a multiple regression model with two (related) predictor variables. ■

10.5 Obtaining Multiple Linear Regression Response Variable Estimates

Based upon the output in Figure 10-12, obtain a regression equation without C8 SATV. Using the new regression equation, you can generate response estimates for this multiple linear regression. For example, find tuition estimates for a college with 45% of its accepted Freshman class enrolling at the college and with 35% of its Freshman class in the top 20% of their High School graduating class (these are values that your college aspires to). Generate 99% interval estimates, instead of the default 95% interval estimates.

■ Click the **Edit Last Dialog** toolbar button 🖳

- Clear SATV from the "Predictors"
- Click **Options**
- Type **45 35** as the "Prediction intervals for new observations" (be sure to put a space between the two values)
- Type **99** as the "Confidence level"
- Click **OK** twice

Minitab reports a regression equation of Tuition = 14218 – 76.2 %Enroll + 169 %Top20%, as shown in Figure 10-13.

FIGURE 10-13

Multiple linear regression with two predictors and response variable estimates

```
MTB > Regress 'Tuition' 2 '%Enroll' '%Top20%';
SUBC>    Constant;
SUBC>    Predict 45 35;
SUBC>      Confidence 99;
SUBC>    Brief 2.
```

Regression Analysis: Tuition versus %Enroll, %Top20%

```
The regression equation is
Tuition = 14218 - 76.2 %Enroll + 169 %Top20%


41 cases used, 19 cases contain missing values


Predictor     Coef   SE Coef      T      P
Constant     14218      1511   9.41  0.000
%Enroll     -76.21     39.81  -1.91  0.063
%Top20%     169.25     15.05  11.25  0.000


S = 2732.55   R-Sq = 77.0%   R-Sq(adj) = 75.7%


Analysis of Variance

Source          DF         SS         MS      F      P
Regression       2  947693175  473846588  63.46  0.000
Residual Error  38  283740084    7466844
Total           40 1231433260


Source    DF     Seq SS
%Enroll    1    3020298
%Top20%    1  944672878


Unusual Observations

Obs  %Enroll  Tuition     Fit  SE Fit  Residual  St Resid
 12     32.9     8963   14754     616     -5791     -2.18R
 22     29.3     8936   14862     645     -5926     -2.23R
 24     30.7    26871   21021     491      5850      2.18R
 25     77.6    26019   24891    1714      1128      0.53 X
 29     44.9    13198   18584     563     -5386     -2.01R
 45     27.7    18360   24967     723     -6607     -2.51R

R denotes an observation with a large standardized residual.
X denotes an observation whose X value gives it large influence.


Predicted Values for New Observations

New
Obs    Fit  SE Fit       99% CI           99% PI
  1  16712     622  (15026, 18398)  (9113, 24311)


Values of Predictors for New Observations

New
Obs  %Enroll  %Top20%
  1     45.0     35.0
```

The coefficient of %Top20%, 169, indicates that for every percentage increase in the percentage of the Freshman class that was in the top 20% of their High School class, predicted tuition increases by $169, among those colleges with the same percentage of accepted students that enroll.

The Unusual Observations table (near the bottom of Figure 10-13) identifies six colleges. Five (those in rows 12, 22, 24, 29, and 45) have large standardized residuals, while one (in row 25) has predictor-variable values that greatly influence the regression equation. The last section of the output indicates that the predicted tuition for a college with %Enroll = 45 and %Top20% = 35 is $16,712. A 99% two-sided confidence interval for the mean tuition for colleges with this combination of values is $15,026 to $18,398. This looks like a good model. With just two predictor variables, it explains 77.0% of the variation in tuition, only slightly less than the 78.2% for the model with three predictor variables that you investigated earlier.

You have completed your initial regression analysis of tuition at Massachusetts colleges and are ready to submit your preliminary recommendations. At this point, you will recommend to the administration the model that relates tuition to the percentage of the college's Freshman class that was in the top 20% of their High School graduating class and to the percentage of accepted students that enrolls. You will show how the regression can provide tuition estimates for different types of colleges and thereby assist the administration in setting future tuition levels. You will also indicate that another possible model would be one without the percentage of accepted students that enrolls because of the negative sign of its coefficient and the size of its p-value being greater than .05.

In the next tutorial, you will continue your analysis of the CollMass data using some of Minitab's advanced regression features to obtain a better model to explain the variation in tuition. Next, however, save your data.

- Click the **Save Project** toolbar button 💾 and exit Minitab

▶ **Note** If you have completed this tutorial, you have a sound understanding of how to make effective use of Minitab to perform a regression analysis. As mentioned in Tutorial 5 Minitab comes with a series of tutorials—not to be confused with the tutorials in this book—that provide excellent overviews of different aspects of the software. To reinforce your understanding of the regression, work through Session Two, which is called Entering and Exploring Data. You may access this Minitab tutorial by choosing the Help > Tutorials command. ■

MINITAB at work

HUMAN RESOURCES

A faculty member, as a member of her College Compensation Committee, was asked to analyze the current salaries of faculty members with the aim of uncovering any inequities. Different multiple regression models were used to predict salaries for professors, associate professors, and assistant professors.

For professors, almost 80% of the variability in salary from one individual to another could be explained by a linear model consisting of just three variables: years at that rank, whether the individual held an endowed chair, and the length, in months, of the individual's annual contract. A similar linear model based upon these same three variables explained 70% of the variability in the salaries of associate professors. There were no endowed chairs among the assistant professors, but the other two variables explained almost 60% of the variability in their salaries. Surprisingly, neither gender nor department was a significant predictor of salary at any rank.

When these three models were used, it was possible to obtain a predicted salary and a corresponding residual for each instructor based upon their values for these variables. When the data set for each rank was ordered according to the magnitude of the residual, it was easy to identify those instructors earning substantially less than their predicted salaries (that is, those with large negative residuals). Those faculty members were given salary equity adjustments before receiving the basic annual increase enjoyed by every faculty member.

Minitab Command Summary

This section describes the commands introduced in, or related to, this tutorial. Use Minitab's online Help for a complete explanation of all commands.

Minitab Menu Commands

Menu	Command	Description
Stat ➤	Basic Statistics	
	➤ Correlation	Computes the correlation, and, optionally, the corresponding p-value, between pairs of variables.
	Regression	
	➤ Regression	Performs simple or multiple linear regression, allowing for both descriptive and inferential analyses.
	➤ Fitted Line Plot	Graphs a regression line and a scatterplot of the regression data, and optionally, confidence bands and prediction bands around the line.

Review and Practice

Matching Problems

Match the following terms to their definitions by placing the correct letter next to the term it describes.

_____ SRES1

_____ FITS1

_____ X**2

_____ X

_____ R

_____ X1*X2

_____ PI

_____ R-Sq(adj)

_____ CI

_____ R-Sq

a. The Minitab symbol that denotes an influential point

b. A variable in a quadratic model

c. A name given to the column of standardized residuals

d. A name assigned to the column of predicted values

e. Minitab's notation for the coefficient of determination that accounts for the number of predictors

f. Minitab's notation for an interval estimate of an individual response for a given predictor value

g. An example of a complex multiple linear regression variable

h. Minitab's notation for the coefficient of determination

i. Minitab's notation for an interval estimate of a mean response for a given predictor value

j. The symbol that Minitab uses to identify an unusual observation that shows a large absolute standardized residual

True/False Problems

Mark the following statements with a T or an F.

a. ____ You use the Regression command to perform simple linear regression and multiple linear regression.

b. ____ The Fitted Line Plot command can graph 90% confidence bands and prediction bands for a simple linear regression line.

c. ____ The number of entries in the "Prediction intervals for new observations" text box is equal to the number of predictors plus one.

d. ____ Among the most used transformations of the response variable are cube roots, logarithms, and negative reciprocals.

e. ____ Minitab regression output includes an ANOVA table.

f. ____ A quadratic model is an example of a polynomial model.

g. ____ The Regression command can print a 95% confidence interval for the slope of a simple linear regression.

h. ____ You can store fitted values in the worksheet only if you also store the residual values.

i. ____ Even if you don't store the standardized residual and fitted values in the worksheet, Minitab displays these values for unusual observations.

j. ____ Minitab uses "S" to denote the standard deviation of the response variable in its regression output.

Practice Problems

The practice problems instruct you to save your worksheets with a filename prefixed by P, followed by the tutorial number. In this tutorial, for example, use *P10* as the prefix. If necessary, use Help or Appendix A Data Sets, to review the contents of the data sets referred to in the following problems. Interpretations should use the language of the subject matter of the question. If you are using the Student version, be sure to close worksheets when you have completed a problem.

1. Open *Homes.mtw*. Perform a simple linear regression with Price as the response variable and Area as the predictor variable, as follows:

   a. Construct a scatterplot with Price on the vertical axis and Area on the horizontal axis. Based upon this plot, do you believe there is a linear relationship between these two variables? Explain briefly.

   b. Compute the correlation coefficient between these two variables and the corresponding p-value. Do these values surprise you? Explain briefly.

   c. What is the regression equation of the line relating price to area?

   d. Interpret the slope of the line in this case.

   e. Interpret the values of the standard error of the estimate and the coefficient of determination.

   f. Test the hypotheses that the population slope is equal to 0 versus the alternative hypothesis that it is not equal to 0. Explain your conclusion.

   g. Construct and interpret a 95% two-sided confidence interval for the population slope.

2. Open *PulseA.mtw*. This problem concerns the possible benefit of simple linear regressions on subsets of a data set.

   a. Plot Weight against Height. Based upon this plot, do you believe there is a linear relationship between these two variables? Explain briefly.

   b. What is the regression equation of the line relating Weight to Height?

   c. Interpret the slope of this line.

   d. Interpret the values of the standard error of the estimate and the coefficient of determination?

   e. Test the hypotheses that the population slope is equal to 0 versus the alternative hypothesis that it is not equal to 0. Explain your conclusion. Be sure to mention the p-value in this case.

   f. Construct a 95% two-sided confidence interval for the population slope.

   g. For the mean height of the 92 students, construct a point estimate for the weight based upon this model and its corresponding 95% two-sided confidence and prediction intervals. Identify these intervals on a Fitted Line Plot that displays 95% confidence and prediction bands.

   h. Plot Weight against Height with Gender as the categorical variable for grouping. Based upon this display, are there linear relationships for each gender? Explain briefly.

i. Use the Data > Unstack Columns command to separate the heights and the weights by gender. For the male students only obtain the regression equation of the line relating weight to height. Interpret the slope of the line. What fraction of the variability in weight can be associated with differences in height? Test the hypotheses that the population slope is equal to 0 versus the alternative hypothesis that it is not equal to 0.

j. For the female students only obtain the regression equation of the line relating weight to height. Interpret the slope of the line. What fraction of the variability in weight can be associated with differences in height? Test the hypotheses that the population slope is equal to 0 versus the alternative hypothesis that it is not equal to 0.

k. Reviewing your previous answers, should you report the simple regression analysis results for the entire sample or for each gender in the sample, separately? Explain briefly.

3. Open *MnWage2.mtw*. This problem considers transformations of a response variable.

a. Plot minimum wage against year. Describe the appearance of the plot. Perform a simple linear regression relating minimum wage to year.

b. Compute the square root of minimum wage. Name this new variable SQRTMW. Plot this new variable against year. Perform a related simple linear regression relating SQRTMW to year.

c. Compute the natural logarithm of minimum wage. Name this new variable LOGEMW. Plot this new variable against year. Perform a related simple linear regression relating LOGEMW to year.

d. Compute the negative reciprocal of minimum wage. Name this new variable NRMW. Plot this new variable against year. Perform a related simple linear regression relating NRMW to year.

e. Which of the four models in parts a, b, c, and d do you prefer? Briefly explain your answer using the plots and the corresponding regression output.

4. Open *DrivingCosts.mtw*.

a. Use simple linear regression to obtain a point estimate of the cost of driving a car in 2003. In 2008. In 2013. Which of these estimates do you trust the most? Explain briefly.

b. Obtain a 95% two-sided prediction interval estimate of the cost of driving a car in 2003. In 2008. In 2013. Which of these estimates do you trust the most based upon the widths of each interval? Explain how your answer strengthens the advantage of using an interval estimate.

5. Open *Temco.mtw*.

a. Perform a simple linear regression to explain salary based upon the number of years employed, YrsEm. Does the output suggest that this is a useful model? Explain briefly.

b. Perform a quadratic regression to explain the salary based upon YrsEm and YrsEm**2. Does the output suggest that this is a useful model? Explain briefly.

c. Which of the two regression models do you consider to be the best? Briefly explain your answer by referring to the regression output.

6. Open *SpeedCom.mtw*.

   a. Plot speed (SPDGigaflops) against year. Based upon this plot is the relationship between speed and year closer to linear or quadratic? Explain briefly.

   b. Perform a related simple linear regression relating speed to year and then a quadratic regression relating speed to year. Based upon these regression analyses, is the relationship between speed and year linear or quadratic? Explain briefly.

   c. Use the Stat > Regression > Fitted Line Plot command to obtain a display and related regression output for the simple linear regression model. Be sure to include 99% confidence and prediction bands in your display.

   d. Use the Stat > Regression > Fitted Line Plot command to obtain a display and related quadratic regression output for the quadratic regression model. Be sure to include 99% confidence and prediction bands in your display.

   e. How do the results of parts a and b differ from the results of parts c and d?

   f. Based upon your answer to part e, do you prefer to use the Graph > Scatterplot and Stat > Regression > Regression or the Stat > Regression > Fitted Line Plot command for this analysis? Explain briefly.

7. Open *Temco.mtw*.

   a. Perform a multiple linear regression with Salary as the response variable and the three variables (YrsEm, PriorYr, and Educ) as predictors. What is the regression equation?

   b. Interpret the coefficients of each predictor.

   c. Interpret the values of the standard error of the estimate and the coefficients of determination.

   d. Test the hypotheses that all predictor population coefficients are equal to 0 versus the alternative hypothesis that at least one of them is not equal to 0. Explain your conclusion.

   e. Test the hypotheses that the YrsEm population coefficient is equal to 0 versus the alternative hypothesis that it is not equal to 0. Perform similar tests for the PriorYr and Educ population coefficients. Explain each conclusion.

   f. Based upon your analysis in part e, do you believe that any predictor(s) should be dropped from the model? Explain briefly. If you believe any predictor(s) should be dropped, perform a multiple or simple linear regression analysis with the appropriate subset of predictors. Interpret your results.

8. Open *OpenHouse.mtw*.

   a. Perform a multiple linear regression with Price as the response variable and the other two quantitative variables as predictors. What is the resulting regression equation? Are you pleased with this multiple linear regression model? Explain briefly.

   b. For the median number of bedrooms and baths, what is the 90% two-sided confidence interval for an average price and what is the 90% two-sided prediction interval for an individual price?

c. For the median number of bedrooms and baths, what is the 95% two-sided confidence interval for an average price and what is the 95% two-sided prediction interval for an individual price?

d. For the median number of bedrooms and baths, what is the 99% two-sided confidence interval for an average price and what is the 99% two-sided prediction interval for an individual price?

e. Discuss the differences, if any, between each of the three pairs of interval estimates.

f. Discuss the differences, if any, among the 90%, 95%, and 99% interval estimates.

9. Open *Textbooks.mtw*. This problem concerns finding a "best" subset of predictors based upon a multiple linear regression.

a. Perform a multiple linear regression with the list price of a textbook as the response variable and the other quantitative variables, except the price of a textbook at Amazon, as the predictors. What is the resulting regression equation? Are you pleased with this multiple linear regression model? Explain briefly.

b. Based upon your answer in part a, do you believe that any of these predictors should be included in a model? Explain briefly. If you believe that any of these predictors should be included, perform a simple or multiple linear regression analysis with the appropriate subset of predictors. Interpret your results.

On Your Own

There are 93 data files in the Studnt14 folder. Use Windows Explorer to determine the size of each file. Based upon the data file information given in Appendix A Data Sets, obtain a linear regression equation to explain the size of a data file based upon the number of observations in the file, the number of numeric variables, and the number of text variables. Write a paragraph summarizing your results.

Answers to Matching Problems

(c) SRES1
(d) FITS1
(b) X**2
(a) X
(j) R
(g) X1*X2
(f) PI
(e) R-Sq(adj)
(i) CI
(h) R-Sq

Answers to True/False Problems

(a) T, (b) T, (c) F, (d) F, (e) T, (f) T, (g) F, (h) F, (i) T, (j) F

Building Regression Models

In this tutorial, you will use some of Minitab's advanced regression commands. First, you will see an excellent example of why it is important to use graphs in regression analysis. Then, you will examine the problems caused by predictor variables that are highly correlated and check the validity of the assumptions underlying regression analysis. You will identify unusual observations and employ a technique that allows you to include qualitative variables in a regression model. In addition, you will use a technique for selecting an optimal subset of predictor variables from a large number of possible predictors. Finally, you will be introduced to binary logistic regression, an increasingly popular technique for modeling dichotomous variables.

OBJECTIVES

In this tutorial, you learn how to:

- Appreciate the importance of graphs in a regression analysis
- Identify the presence of collinearity in a multiple linear regression
- Perform a residual analysis to check the assumptions of a linear regression
- Identify unusual observations (outlying points and points of influence) in a linear regression
- Create indicator (dummy) variables so that you can use qualitative variables as predictors in a linear regression
- Use best subsets regression to explore the relative importance of predictor variables
- Use binary logistic regression to model a dichotomous response variable

The Importance of Graphs in Regression

CASE STUDY	CASE STUDY: DATA ANALYSIS— IMPORTANCE OF GRAPHS

The statistician Frank J. Anscombe created a data set to illustrate the importance of inspecting scatterplots before examining the standard computer regression output. Four pairs of variables with very different relationships all result in the same basic regression output. These data are saved in the file *Fja.mtw*. There are 11 values for each of the following six variables: C1 X, C2 Y1, C3 Y2, C4 Y3, C5 X4, and C6 Y4. (X represents predictors, and Y represents response variables.) The instructor in your introductory statistics class gives these data to your group and asks you to summarize the output from four simple linear regressions: Y1 on X, Y2 on X, Y3 on X, and Y4 on X4.

Begin by opening Minitab:

- Open **Minitab** and maximize the Session window
- If necessary, **Enable Commands** and clear Output Editable from the Editor menu
- Click the **Save Project** toolbar button 🖫 and, in the location where you are saving your work, save this project as **T11.mpj**

You'll begin working with the data file *Fja.mtw*:

- Open the Worksheet **Fja.mtw**

Your group is skeptical about doing a regression analysis before examining plots of the data, but you proceed as you did in the last tutorial:

- Choose **Stat > Regression > Regression**
- Double-click **C2 Y1** as the "Response " variable and **C1 X** as the "Predictors" variable

To focus on the essential aspects of the output, limit the amount of regression output. In the Results dialog box, Minitab allows you some control over the amount of output produced with the Stat > Regression > Regression command. In this example, with one predictor variable, your choice will eliminate the printing of any unusual observations.

- Click **Results**
- Click the **Regression equation, table of coefficients, s, R-squared, and basic analysis of variance** option button
- Click **OK** twice

Minitab produces the first regression. Notice in the Session commands printed at the top of the output that your selection in the Results dialog box is equivalent to entering a brief 1 Session subcommand. (Brief 1 produces minimal output, Brief 2 is the default, and Brief 3 produces the most output.)

- Click the **Edit Last Dialog** toolbar button 🖾, and enter, in turn, the remaining three pairs in the following chart

Response	Predictor
C3 Y2	C1 X
C4 Y3	C1 X
C6 Y4	C5 X4

When you have finished the four regressions, the Session window contains the output shown in Figure 11-1 (you will have to scroll to see all of this output).

FIGURE 11-1

Four Brief 1 regression analyses

```
MTB > Regress 'Y1' 1 'X';
SUBC>    Constant;
SUBC>    Brief 1.
```

Regression Analysis: Y1 versus X

```
The regression equation is
Y1 = 3.00 + 0.500 X
```

First regression equation

```
Predictor     Coef  SE Coef     T      P
Constant     3.000    1.125   2.67  0.026
X           0.5001   0.1179   4.24  0.002


S = 1.23660   R-Sq = 66.7%   R-Sq(adj) = 62.9%


Analysis of Variance

Source         DF      SS      MS      F      P
Regression      1  27.510  27.510  17.99  0.002
Residual Error  9  13.763   1.529
Total          10  41.273

MTB > Regress 'Y2' 1 'X';
SUBC>    Constant;
SUBC>    Brief 1.
```

Regression Analysis: Y2 versus X

```
The regression equation is
Y2 = 3.00 + 0.500 X
```

Second regression equation

```
Predictor     Coef  SE Coef     T      P
Constant     3.001    1.125   2.67  0.026
X           0.5000   0.1180   4.24  0.002


S = 1.23721   R-Sq = 66.6%   R-Sq(adj) = 62.9%


Analysis of Variance

Source         DF      SS      MS      F      P
Regression      1  27.500  27.500  17.97  0.002
Residual Error  9  13.776   1.531
Total          10  41.276

MTB > Regress 'Y3' 1 'X';
SUBC>    Constant;
SUBC>    Brief 1.
```

Regression Analysis: Y3 versus X

```
The regression equation is
Y3 = 3.00 + 0.500 X
```

Third regression equation

```
Predictor     Coef   SE Coef    T      P
Constant     3.002    1.124   2.67  0.026
X           0.4997   0.1179   4.24  0.002


S = 1.23631   R-Sq = 66.6%   R-Sq(adj) = 62.9%


Analysis of Variance

Source          DF      SS       MS      F      P
Regression       1   27.470   27.470  17.97  0.002
Residual Error   9   13.756    1.528
Total           10   41.226

MTB > Regress 'Y4' 1 'X4';
SUBC>   Constant;
SUBC>   Brief 1.
```

Regression Analysis: Y4 versus X4

```
The regression equation is
Y4 = 3.00 + 0.500 X4
```

Fourth regression equation

```
Predictor     Coef   SE Coef    T      P
Constant     3.002    1.124   2.67  0.026
X4          0.4999   0.1178   4.24  0.002


S = 1.23570   R-Sq = 66.7%   R-Sq(adj) = 63.0%


Analysis of Variance

Source          DF      SS       MS      F      P
Regression       1   27.490   27.490  18.00  0.002
Residual Error   9   13.742    1.527
Total           10   41.232
```

Note that the four regression equations are the same. They also all have a t-statistic of 4.24, with a corresponding p-value of 0.002, which Minitab uses to test whether the population slope is zero. Except for rounding, the standard error of the estimates and the coefficients of determinations are also the same. It's tempting to assume that the response and predictor variables are the same in the four cases. But if you return to the Data window, you will see that this is not true.

■ Click the **Current Data Window** toolbar button 🔢

The data are shown in Figure 11-2.

FIGURE 11-2

The Data window showing the Anscombe data set

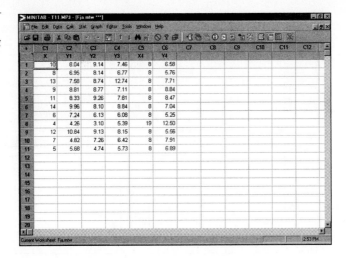

	C1	C2	C3	C4	C5	C6	C7	C8	C9	C10	C11	C12
	X	Y1	Y2	Y3	X4	Y4						
1	10	8.04	9.14	7.46	8	6.58						
2	8	6.95	8.14	6.77	8	5.76						
3	13	7.58	8.74	12.74	8	7.71						
4	9	8.81	8.77	7.11	8	8.84						
5	11	8.33	9.26	7.81	8	8.47						
6	14	9.96	8.10	8.84	8	7.04						
7	6	7.24	6.13	6.08	8	5.25						
8	4	4.26	3.10	5.39	19	12.50						
9	12	10.84	9.13	8.15	8	5.56						
10	7	4.82	7.26	6.42	8	7.91						
11	5	5.68	4.74	5.73	8	6.89						

The four pairs of variables are very different. Now, produce scatterplots for each pair of variables:

- Choose **Graph > Scatterplot** and double-click *Simple*
- Double-click **C2 Y1** as the "Y variables" column for Graph 1
- Double-click **C1 X** as the "X variables" column for Graph 1
- Double-click **C3 Y2** as the "Y variables" column for Graph 2
- Double-click **C1 X** as the "X variables" column for Graph 2
- Double-click **C4 Y3** as the "Y variables" column for Graph 3
- Double-click **C1 X** as the "X variables" column for Graph 3
- Double-click **C6 Y4** as the "Y variables" column for Graph 4
- Double-click **C5 X4** as the "X variables" column for Graph 4

To see all four scatterplots in separate panels but in the same display:

- Click **Multiple Graphs**
- Click the "Show Pairs of Graph Variables" **In separate panels of the same graph** option button
- Click **OK** twice

The four resulting scatterplots, with Minitab's default title, are shown in Figure 11-3.

FIGURE 11-3

*Four Anscombe
data set scatterplots*

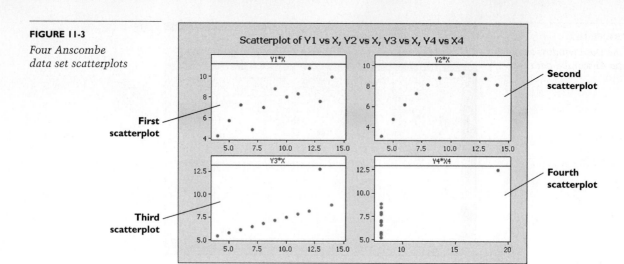

First
scatterplot

Second
scatterplot

Third
scatterplot

Fourth
scatterplot

These plots demonstrate why Anscombe constructed this data set. The cloud of points in the first scatterplot suggests a straightforward linear relationship. The second clearly calls for a quadratic model to describe the relationship. An outlying observation—the third observation with coordinates (13, 12.74)—is apparent in the third scatterplot. Finally, the fourth scatterplot shows an unusual and influential variable value: ten out of the 11 observations have an X4 value of 8, while X4 is 19 for the other observation. This eighth observation greatly influences the regression analysis.

Your group reports on these four very different situations. Your instructor asks you to repeat the four regressions with the amount of output restored to the default option. This default output can be helpful in detecting some of these situations. You will do this by copying your previous commands from the History window to the Command Line Editor and then editing the four Brief subcommands. This time, Minitab will produce more output.

- Click the **Show History** toolbar button ☑
- Highlight the **12 regression command lines**
- Choose **Edit > Command Line Editor**
- Change each Brief 1 to **Brief 2**

The completed Command Line Editor dialog box should resemble Figure 11-4.

FIGURE 11-4

*The completed Command
Line Editor with pasted
and modified Regression
commands*

```
Command Line Editor
Regress 'Y1' 1 'X';
  Constant;
  Brief 2.
Regress 'Y2' 1 'X';
  Constant;
  Brief 2.
Regress 'Y3' 1 'X';
  Constant;
  Brief 2.
Regress 'Y4' 1  'X4';
  Constant;

  Submit Commands                    Cancel
```

- Click **Submit Commands**

The Session window displays the output shown in Figure 11-5. Minitab notes the unusual observations you saw in the last two scatterplots. For the data in the third plot, it flags the third observation with an R (an observation with a large absolute standardized residual); in the fourth, it marks the eighth observation with an X (an observation whose predictor value give it large influence). Notice though that Minitab doesn't notify you of the need for a quadratic model for the second data set. Although Minitab's unusual observation regression output table is valuable, you should still plot your data whenever you perform a regression.

FIGURE 11-5

Four Brief 2 regression analyses

```
MTB > Regress 'Y1' 1 'X';
SUBC>    Constant;
SUBC>    Brief 2.
```

Regression Analysis: Y1 versus X

```
The regression equation is
Y1 = 3.00 + 0.500 X


Predictor    Coef   SE Coef     T      P
Constant    3.000     1.125   2.67  0.026
X           0.5001   0.1179   4.24  0.002


S = 1.23660   R-Sq = 66.7%   R-Sq(adj) = 62.9%


Analysis of Variance

Source          DF      SS      MS      F      P
Regression       1  27.510  27.510  17.99  0.002
Residual Error   9  13.763   1.529
Total           10  41.273

MTB > Regress 'Y2' 1 'X';
SUBC>    Constant;
SUBC>    Brief 2.
```

Regression Analysis: Y2 versus X

```
The regression equation is
Y2 = 3.00 + 0.500 X


Predictor    Coef   SE Coef     T      P
Constant    3.001     1.125   2.67  0.026
X           0.5000   0.1180   4.24  0.002


S = 1.23721   R-Sq = 66.6%   R-Sq(adj) = 62.9%


Analysis of Variance

Source          DF      SS      MS      F      P
Regression       1  27.500  27.500  17.97  0.002
Residual Error   9  13.776   1.531
Total           10  41.276

MTB > Regress 'Y3' 1 'X';
SUBC>    Constant;
SUBC>    Brief 2.
```

Regression Analysis: Y3 versus X

```
The regression equation is
Y3 = 3.00 + 0.500 X


Predictor    Coef  SE Coef     T      P
Constant    3.002    1.124  2.67  0.026
X           0.4997   0.1179  4.24  0.002


S = 1.23631   R-Sq = 66.6%   R-Sq(adj) = 62.9%

Analysis of Variance

Source          DF      SS      MS      F      P
Regression       1  27.470  27.470  17.97  0.002
Residual Error   9  13.756   1.528
Total           10  41.226


Unusual Observations

Obs    X      Y3    Fit  SE Fit  Residual  St Resid
  3  13.0  12.740  9.499   0.601     3.241      3.00R

R denotes an observation with a large standardized residual.

MTB > Regress 'Y4' 1 'X4';
SUBC>   Constant;
SUBC>   Brief 2.
```

Regression Analysis: Y4 versus X4

```
The regression equation is
Y4 = 3.00 + 0.500 X4


Predictor    Coef  SE Coef     T      P
Constant    3.002    1.124  2.67  0.026
X4          0.4999   0.1178  4.24  0.002


S = 1.23570   R-Sq = 66.7%   R-Sq(adj) = 63.0%

Analysis of Variance

Source          DF      SS      MS      F      P
Regression       1  27.490  27.490  18.00  0.002
Residual Error   9  13.742   1.527
Total           10  41.232


Unusual Observations

                                            St
Obs    X4     Y4     Fit  SE Fit  Residual  Resid
  8  19.0  12.500  12.500   1.236     0.000    * X

X denotes an observation whose X value gives it large influence.
```

Before moving on to the next case study, save your work:

■ Click the **Save Project** toolbar button 💾

In this tutorial, you will again play the role of an institutional researcher as you did in Tutorial 10. You have been asked by the administration of the Massachusetts college at which you work to generate a model for tuition based upon data collected from Massachusetts colleges. For each of 60 colleges, ten variables were identified that could be related to the tuition charged there. These variables, all measured for 2002, are stored in the worksheet *CollMass2.mtw*, along with four additional variables created in the last tutorial. You are to use these data to build an advanced model that the administration can use to determine a reasonable tuition for the college.

■ Open the worksheet ***CollMass2.mtw***

You may wish to review the contents of this data set by referencing section 10.1, Appendix A Data Sets, or Minitab's Help.

In Tutorial 10, you performed a quadratic regression analysis, that is, a multiple linear regression analysis with C12 Tuition as the response variable and C8 SATV and C16 SATV**2 as the predictor variables. If you look back at Figure 10-11, you can see the regression equation: Tuition = 36245 − 112 SATV + 0.143 SATV**2. According to the predictor table, each coefficient is not significant at the .05 level of significance, in the presence of the other. (The p-value is 0.201 for the SATV coefficient and 0.065 for the SATV**2 coefficient.) In the simple linear regression the p-value for SATV was 0.000, as shown in Figure 10-2. The reason for the change in the significance of SATV is the increase in the standard error of the slope for SATV from 7.033 to 86.68. Moreover, the ANOVA table in Figure 10-11 reports that at least one of the two coefficients is not zero because the p-value is 0.000.

This apparent contradiction is due to the high degree of linear association between the two predictors. The value of their correlation coefficient is 0.997. (You could use the Stat > Basic Statistics > Correlation command to verify this value.) This is an example of *collinearity* (or *multicollinearity*) that occurs in multiple linear regression when there is a high correlation among the predictors.

Minitab's Variance inflation factors option helps identify collinearity. The *variance inflation factor (VIF)* is a measure of how much the variance of an estimated regression coefficient increases if your predictors are linearly correlated (collinear) relative to the situation were they independent. The higher the VIF value, the greater the degree of collinearity. VIF values above five are considered evidence that collinearity is affecting the variability of the regression coefficients. VIF values above ten provide even more evidence.

▶ **Note** If there are p predictor variables, the VIF for the jth predictor is $1/(1-R_j^2)$, where R_j^2 is the coefficient of determination from regressing the jth predictor on the remaining p − 1 predictors. ■

To redo the quadratic regression analysis from Tutorial 10 using the VIF option:

- Choose **Stat > Regression > Regression**
- Press F3 to clear the previous selections
- Double-click **C12 Tuition** as the "Response" variable and **C8 SATV** and **C16 SATV**2** as the "Predictors"
- Click **Options**
- Click the "Display" **Variance inflation factors** check box
- Click **OK** twice

Figure 11-6 displays the results of the quadratic regression with the VIFs.

FIGURE 11-6

Quadratic regression with VIFs

```
MTB > Regress 'Tuition' 2 'SATV' 'SATV**2';
SUBC>   Constant;
SUBC>   VIF;
SUBC>   Brief 2.
```

Regression Analysis: Tuition versus SATV, SATV**2

```
The regression equation is
Tuition = 36245 - 112 SATV + 0.143 SATV**2


58 cases used, 2 cases contain missing values
```

——— **Variance inflation factor**

```
Predictor      Coef   SE Coef      T      P     VIF
Constant      36245     24404   1.49  0.143
SATV        -112.12     86.68  -1.29  0.201   158.8
SATV**2     0.14275   0.07579   1.88  0.065   158.8


S = 4278.43   R-Sq = 51.2%   R-Sq(adj) = 49.4%

Analysis of Variance

Source          DF          SS          MS      F      P
Regression       2  1056798679   528399339  28.87  0.000
Residual Error  55  1006774416    18304989
Total           57  2063573094


Source     DF     Seq SS
SATV        1  991864700
SATV**2     1   64933978


Unusual Observations

Obs  SATV  Tuition    Fit  SE Fit  Residual  St Resid
  5   430    26858  14429    1392     12429      3.07R
 13   500    24620  15874     685      8746      2.07R
 15   440    24830  14550    1215     10280      2.51R
 25   750    26019  32453    2423     -6434     -1.82 X
 30   720    27728  29522    1740     -1794     -0.46 X
 40   393     9175  14231    2219     -5056     -1.38 X

R denotes an observation with a large standardized residual.
X denotes an observation whose X value gives it large influence.
```

The predictor table contains the VIFs. Each coefficient's VIF is 158.8—with only two predictor variables, the VIFs are always the same. Because these values indicate a high degree of collinearity, there may be statistical and computational difficulties with this model.

Now that you have identified a problem with this quadratic regression, try to remedy it by creating and using two new, less correlated predictor variables. The way to do this is to create a *centered variable* by subtracting the mean SATV from each SATV score. First, return to the CollMass2 worksheet.

- Click the **Current Data Window** toolbar button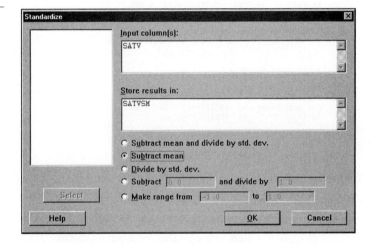

    To create the centered variable,

- Choose **Calc > Standardize**
- Double-click **C8 SATV** as the "Input column(s)"
- Type **SATVSM** as the "Store results in" column
- Click the **Subtract mean** option button

    The completed Standardize dialog box should look as shown in Figure 11-7.

FIGURE 11-7

The completed Standardize dialog box

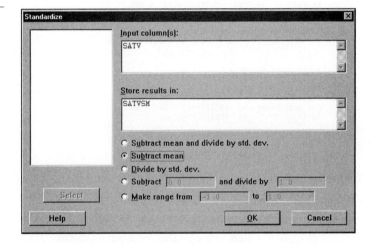

- Click **OK**

    Scroll over to see the new variable in C17 SATVSM. Next, create the square of this new variable:

- Choose **Calc > Calculator**
- Type **SATVSM**2** as the "Store result in variable"
- Enter **SATVSM**2**, by either clicking or typing, as the "Expression" and click **OK**

    By using the Stat > Basic Statistics > Correlation command you can verify that the correlation between C17 SATVSM and C18 SATVSM**2 is only 0.367.

    Now, as your solution to collinearity in polynomial regression, regress tuition on these two new variables:

- Choose **Stat > Regression > Regression**
- Verify that the "Response" variable is Tuition
- Double-click **C17 SATVSM** and **C18 SATVSM**2** to replace the previous "Predictors"
- Click **OK**

Figure 11-8 displays the results of this new regression. Note that the VIF column now contains VIFs of 1.2. You no longer have a problem with collinearity.

FIGURE 11-8

Quadratic regression with centered predictors

```
MTB > Regress 'Tuition' 2  'SATVSM' 'SATVSM**2';
SUBC>    Constant;
SUBC>    VIF;
SUBC>    Brief 2.
```

Regression Analysis: Tuition versus SATVSM, SATVSM2**

```
The regression equation is
Tuition = 17860 + 45.5 SATVSM + 0.143 SATVSM**2


58 cases used, 2 cases contain missing values


Predictor      Coef  SE Coef       T      P  VIF
Constant    17860.0    755.7   23.63  0.000
SATVSM       45.520    7.395    6.16  0.000  1.2
SATVSM**2   0.14275  0.07579    1.88  0.065  1.2


S = 4278.43   R-Sq = 51.2%   R-Sq(adj) = 49.4%


Analysis of Variance

Source          DF          SS         MS      F      P
Regression       2  1056798679  528399339  28.87  0.000
Residual Error  55  1006774416   18304989
Total           57  2063573094


Source      DF    Seq SS
SATVSM       1  991864700
SATVSM**2 1    64933978


Unusual Observations

Obs  SATVSM  Tuition    Fit  SE Fit  Residual  St Resid
  5    -122    26858  14429    1392     12429      3.07R
 13     -52    24620  15874     685      8746      2.07R
 15    -112    24830  14550    1215     10280      2.51R
 25     198    26019  32453    2423     -6434     -1.82 X
 30     168    27728  29522    1740     -1794     -0.46 X
 40    -159     9175  14231    2219     -5056     -1.38 X

R denotes an observation with a large standardized residual.
X denotes an observation whose X value gives it large influence.
```

Note, also, that there are many similarities between the outputs in Figures 11-6 and 11-8. The values of S, R-Sq, F, and the p-value for F are identical. In addition, Minitab flags the same unusual observations. However, there is a difference in the predictor table: The first predictor coefficient changed. Minitab indicates that this coefficient is significant by its p-value of 0.000. You no longer have a contradiction between the t-tests and F-test. Because the quadratic term is not significant at the .05 significance level, you need not include this term when developing your tuition model for Massachusetts colleges.

11.3 | *Verifying Linear Regression Assumptions*

To use the t-test, F-tests, and interval estimates introduced in Tutorial 10, you assumed that the data met certain conditions. Most of those conditions concerned the error term, or unexplained component in a model. Here are the three major assumptions of linear regression:

1. *Normality Assumption:* The error term is normally distributed.
2. *Constant Variation (Homoscedasity) Assumption:* The error terms have a constant variation.
3. *Independence Assumption:* The error terms for two observations are independently distributed.

Serious departures from these assumptions affect the validity of inferences in both simple and multiple linear regression. You can check the validity of these assumptions by using the residuals and measures related to the residuals.

Storing the Residuals

The differences between the values for the response variable and the corresponding fitted values are called residuals. In fact, residuals are estimates of the error terms in a linear regression model. When you perform a linear regression, Minitab allows you to store many columns of useful information, including the residuals. For example, in Tutorial 10 you stored the residuals, the standardized residuals, and the fitted values for a simple linear regression.

Now, you will store this same information, along with two other statistics related to the residuals, for the multiple linear regression model you recommended to the administration at the end of Tutorial 10 (with C12 Tuition as the response and C6 %Enroll and C11 %Top20% as the predictors). You can use the results to check the three major assumptions of linear regression.

- Choose **Stat > Regression > Regression**
- Verify that the "Response" variable is Tuition
- Double-click **C6 %Enroll** and **C11 %Top20%** to replace the previous "Predictors"
- Click **Storage**, and click the "Diagnostic Measures" **Residuals**, **Standardized residuals**, **Hi (leverages)**, **Cook's distance**, and "Characteristics of Estimated Equations" **Fits** check boxes

The completed Regression - Storage dialog box is shown in Figure 11-9.

FIGURE 11-9

The completed Regression - Storage dialog box

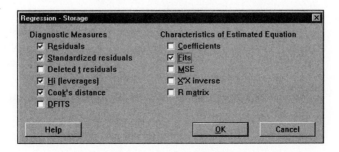

- Click **OK** twice

Figure 11-10 shows the results.

FIGURE 11-10

Multiple linear regression with storage

```
MTB > Name c19 "RESI2" c20 "SRES2" c21 "HI2" c22 "COOK2" c23 "FITS2"
MTB > Regress 'Tuition' 2  '%Enroll' '%Top20%';
SUBC>    Residuals 'RESI2';
SUBC>    SResiduals 'SRES2';
SUBC>    Hi 'HI2';
SUBC>    Cookd 'COOK2';
SUBC>    Fits 'FITS2';
SUBC>    Constant;
SUBC>    VIF;
SUBC>    Brief 2.
```

Regression Analysis: Tuition versus %Enroll, %Top20%

```
The regression equation is
Tuition = 14218 - 76.2 %Enroll + 169 %Top20%


41 cases used, 19 cases contain missing values


Predictor    Coef   SE Coef      T      P   VIF
Constant    14218      1511   9.41  0.000
%Enroll    -76.21     39.81  -1.91  0.063   1.1
%Top20%    169.25     15.05  11.25  0.000   1.1


S = 2732.55   R-Sq = 77.0%   R-Sq(adj) = 75.7%


Analysis of Variance

Source           DF          SS          MS      F      P
Regression        2   947693175   473846588  63.46  0.000
Residual Error   38   283740084     7466844
Total            40  1231433260


Source     DF     Seq SS
%Enroll     1    3020298
%Top20%     1  944672878


Unusual Observations

Obs  %Enroll  Tuition    Fit  SE Fit  Residual  St Resid
 12     32.9     8963  14754     616     -5791     -2.18R
 22     29.3     8936  14862     645     -5926     -2.23R
 24     30.7    26871  21021     491      5850      2.18R
 25     77.6    26019  24891    1714      1128      0.53 X
 29     44.9    13198  18584     563     -5386     -2.01R
 45     27.7    18360  24967     723     -6607     -2.51R

R denotes an observation with a large standardized residual.
X denotes an observation whose X value gives it large influence.
```

Notice, from the first line of Figure 11-10, that Minitab has created five new variables in the Name Session command line. In this case, there doesn't appear to be a problem with collinearity based upon the two low VIF values. The rest of the output is identical to Figure 10-13, except for the confidence and prediction intervals.

To see the five new variables:

- Click the **Current Data Window** toolbar button ▦ and scroll over to see C19–C23

You can use either the residuals or the standardized residuals for a residual analysis. Some statisticians prefer residuals (with a mean of 0); others prefer standardized residuals (with most values between –3 and 3 and a mean close to 0). Here, use standardized residuals.

Checking the Normality Assumption

To check the normality assumption, have Minitab construct a normal probability plot (NPP) of the standardized residuals.

- Choose **Stat > Basic Statistics > Normality Test**
- Double-click **C20 SRES2** as the "Variable"
- Verify that the default test for normality is Anderson-Darling
- Click **OK**

The NPP, with Minitab's default title, appears as shown in Figure 11-11. Many of the points are far from the line, although this may be associated with extremely low and high values of standardized residuals at each end of the line. The mean, standard deviation, and sample size appear in the upper right-hand corner. The results of the default Anderson-Darling normality test are also in the upper-right corner. The test's p-value is less than 0.005, so you reject the null hypothesis that the standardized residuals are normally distributed at any reasonable level of significance level such as .05. In order to perform inference on this data set you may have to eliminate some of these unusual observations.

FIGURE 11-11

NPP of standardized residuals with the Anderson-Darling normality test

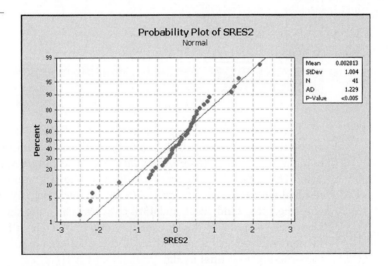

Using Scatterplots to Verify the Homoscedasity Assumption

The best single check for constant variance is a scatterplot with the standardized residuals on the vertical axis and the fitted values on the horizontal axis. In addition, you should construct separate scatterplots with the standardized residuals on the vertical axis and each predictor variable on the horizontal axis. There should be approximately the same amount of variation in the standardized residuals over the range of the horizontal variable in each plot; otherwise, the assumption is violated. Begin by plotting SRES2 against FITS2:

- Choose **Graph > Scatterplot** and double-click *Simple*
- Press ⌷F3⌷ to clear the previous selections
- Double-click **C20 SRES2** as the "Y variables" column for Graph 1
- Double-click **C23 FITS2** as the "X variables" column for Graph 1
- Click **OK**

The scatterplot, with Minitab's default title, appears as shown in Figure 11-12. There appears to be less variation for the fitted values between $20,000 and $24,000 than for the other fitted values. This may be a concern if you decide to use this as your final model.

FIGURE 11-12

Scatterplot of standardized residuals against fits

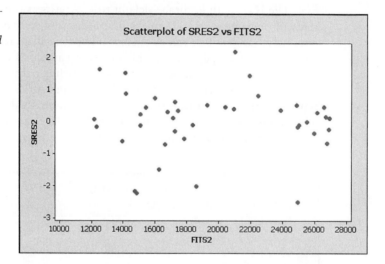

Note from the scatterplot that there are five outlying standardized residuals: one with a value greater than 2 and four with values less than –2. Recall that by holding your mouse cursor over a data point you can obtain a data tip about that point. Using this procedure you can identify the largest positive standardized residual with the 24th college and the four largest negative standardized residuals with the 12th, 22nd, 29th, and 45th colleges. These are the same five colleges identified (with Rs) in the Unusual Observations table of Figure 11-10. In the next section, you will examine all three of these colleges in more detail before constructing your final model.

From an examination of the scatterplots of SRES2 against %Enroll and SRES2 against %Top20%, you do not observe any extreme patterns of variation.

Although you construct scatterplots such as these to check the constant variation assumption, they are also useful in detecting other problems with the model. You should study any unusual pattern in a residual analysis because it may indicate an underlying weakness in the model.

Checking the Independence Assumption with a Time Series Plot

Usually, you check the independence assumption only when you know the order in which the data were collected. The appropriate technique is a time series plot of the data similar to the one you constructed in Tutorial 5 for the Dow Jones Composite data. Even though you are not aware of the order in which the colleges were selected, you can practice the technique by constructing a time series plot of the standardized residuals in C20 SRES2.

- Choose **Graph > Time Series Plot** and double-click *Simple*
- Double-click **C20 SRES2** as the "Series"
- Click **OK**

The time series plot, with Minitab's default title, appears as shown in Figure 11-13.

FIGURE 11-13

Time series plot of standardized residuals

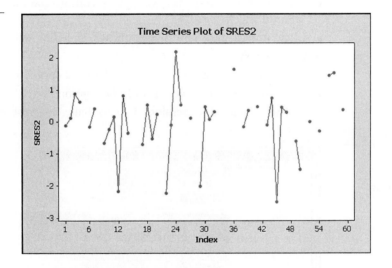

Time series plots show numerical data (in this case SRES2) on the y-axis against the passage of time or the sequence order on the x-axis. By default, points on the graph are displayed as symbols connected by lines, unless there are missing data present. (From Figure 11-10 you note that there are 19 cases containing missing values.)

There are no unusual patterns in this plot (for example, long runs of values with the same sign). However, again notice that there are five colleges that have large, positive or negative, standardized residuals. You will want to examine the data for these five colleges in more detail before you submit your final model.

▶ **Note** An additional way to check the independence assumption is by using the *Durbin-Watson statistic*, which is available in the Regression - Options dialog box. You will find an explanation of this statistic from Help in that dialog box. ▪

A convenient way of getting plots suitable for checking the various assumptions in regression is to obtain a Four in one graph from the Regression - Graphs dialog box. With this option you obtain:

1. An NPP of the residuals to check the normality assumption
2. A histogram of the residuals to check the normality assumption
3. A scatterplot of the residuals against the fitted values to check the constant variation assumption
4. A time series plot of the residuals to check the independence assumption

This option produces, on one display, three of the graphs that you had previously constructed. It does not contain the two graphs of the residuals against the predictor variables. To obtain this display for your current model:

- Choose **Stat > Regression > Regression**
- Verify that the "Response" variable is Tuition, and the "Predictors" are %Enroll and %Top20%
- Click **Storage**, clear all the check boxes, and click **OK**
- Click **Graphs**
- Click the "Residuals for Plots" **Standardized** option button
- Click the "Residual plots" **Four in one** option button

The completed Regression - Graphs dialog box is shown in Figure 11-14.

FIGURE 11-14

*The completed
Regression - Graphs
dialog box*

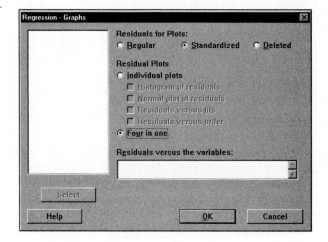

- Click **OK** twice

The Four in one graph appears as shown in Figure 11-15.

FIGURE 11-15

Four in one graph of standardized residuals

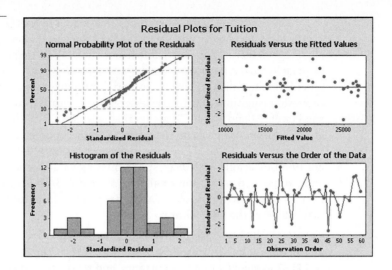

There are only two major differences between these displays and the displays that you had previously constructed. First, the NPP does not include any information in the upper right-hand corner. Second, the time series plot connects all of the non-missing standardized residuals.

11.4   *Examining Unusual Observations*

Now turn your attention to the six unusual observations identified in the table at the bottom of the output in Figure 11-10 and shown again in Figure 11-16.

FIGURE 11-16

Unusual observations table from Figure 11-10

```
Unusual Observations

Obs  %Enroll  Tuition    Fit  SE Fit  Residual  St Resid
 12     32.9     8963  14754     616     -5791    -2.18R
 22     29.3     8936  14862     645     -5926    -2.23R
 24     30.7    26871  21021     491      5850     2.18R
 25     77.6    26019  24891    1714      1128     0.53 X
 29     44.9    13198  18584     563     -5386    -2.01R
 45     27.7    18360  24967     723     -6607    -2.51R

R denotes an observation with a large standardized residual.
X denotes an observation whose X value gives it large influence.
```

The five unusual observations flagged with an R identify Massachusetts colleges whose standardized residuals have absolute values that exceed 2. The tuition at Bridgewater State College (Obs 12) is $8,963 but the regression equation predicts a value of $14,754, leaving a residual of negative $5,791. Unusual negative residuals are also associated with Framingham State College (Obs 22), Massachusetts College of Art (Obs 29), and Stonehill College (Obs 45). An unusual positive residual is associated with Hampshire College (Obs 24). These are the same five colleges identified through the data tips in the scatterplot of standardized residuals against fits shown in Figure 11-12.

The other unusual observation is Harvard University (Obs 25). It enrolls 77.6% of its accepted students and 98% of the class was in the top 20% of their High School class. (This last value is not shown.) These predictor values are associated with a large *HI (High Influence)*, or *leverage, value*. The HI value for an observation is a measure of the potential influence of that observation on the regression equation. Only those observations with HI values that exceed the smaller of .99 and $3(p + 1)/n$ are marked with an X and should receive special attention from the analyst. Here, p is the number of predictor variables, and n is the sample size. From the results in Figure 11-10, p = 2 and n = 41 because there are 19 observations with at least one missing value among the response variable and two predictors. So, the smaller of .99 and $3*3/41 = .2195$ is .2195.

To check why this observation has a substantial influence, you decide to take a look at the HI2 column.

■ Choose **Data > Display Data**

■ Double-click **C12 Tuition**, **C6 %Enroll**, **C11 %Top20%**, **C21 HI2**, and **C23 FITS2** as the "Columns, constants, and matrices to display" and click **OK**

The printed data appear in the Session window and in Figure 11-17 (you may have to scroll to see all of this output).

FIGURE 11-17

Data display containing HI values and four other variables

MTB > Print 'Tuition' '%Enroll' '%Top20%' 'HI2' 'FITS2'.

Data Display

Row	Tuition	%Enroll	%Top20%	HI2	FITS2
1	14800	32.7848	20	0.047490	15104.3
2	27240	44.1932	95	0.091245	26928.4
3	16500	31.5909	14	0.058588	14179.8
4	18945	28.6796	31	0.039932	17278.9
5	26858	65.0224	*	*	*
6	24544	36.8178	80	0.054030	24951.8
7	22050	38.0000	57	0.026990	20969.0
8	17715	48.9055	*	*	*
9	25026	32.9324	89	0.080870	26771.2
10	26228	27.1364	87	0.098326	26874.4
11	27076	27.1787	86	0.095477	26701.9
12	8963	32.9412	18	0.050788	14753.9
13	24620	20.3076	58	0.084186	22486.7
14	25020	34.3394	85	0.067842	25986.9
15	24830	33.1797	*	*	*
16	14458	54.3662	*	*	*
17	14744	30.0885	28	0.039716	16663.8
18	20718	33.3856	45	0.025736	19289.7
19	16412	28.0911	34	0.039411	17831.5
20	15704	32.9953	20	0.047439	15088.3
21	9068	33.8570	*	*	*
22	8936	29.3073	17	0.055770	14861.6
23	18134	51.8824	48	0.078937	18387.7
24	26871	30.6563	54	0.032250	21021.0
25	26019	77.5829	98	0.393660	24891.4
26	16100	25.3685	*	*	*
27	17425	37.0370	34	0.031449	17149.7
28	11857	*	*	*	*
29	13198	44.8669	46	0.042445	18583.9
30	27728	57.6385	99	0.163060	26580.7
31	12346	60.0000	15	0.210506	12183.9
32	17645	25.6497	27	0.052441	16832.8
33	26408	35.5337	*	*	*
34	16425	20.0741	*	*	*
35	21800	33.2268	*	*	*
36	16800	37.7837	7	0.079651	12523.1
37	20733	28.8823	*	*	*
38	11894	44.8052	9	0.100750	12326.4
39	18400	30.7047	33	0.034281	17463.1
40	9175	42.3449	*	*	*

Largest HI value

41	21668	25.1477	48	0.048718	20425.3
42	25610	69.2308	*	*	*
43	24742	42.3348	83	0.061715	25039.1
44	18000	33.9318	26	0.038664	16032.4
45	18360	27.7358	76	0.069933	24967.0
46	16616	26.3331	19	0.059599	15426.8
47	26892	36.5953	87	0.068981	26153.5
48	14433	31.0919	*	*	*
49	12357	45.5491	19	0.080465	13962.2
50	12283	35.2059	28	0.036451	16273.8
51	11892	42.8634	*	*	*
52	25504	44.8037	87	0.074299	25527.9
53	13650	39.9529	*	*	*
54	16494	21.8956	28	0.068090	17288.2
55	9636	41.1215	*	*	*
56	25790	25.0253	57	0.055117	21957.9
57	18195	38.5042	17	0.058475	14160.7
58	25540	44.8893	*	*	*
59	24890	28.4553	70	0.056227	23896.7
60	8653	36.1569	*	*	*

This display verifies that there is one potentially "influential" college (Obs 25) because its HI is 0.393660. All of the other HI values are smaller. If you remove an observation with a very large HI value from an analysis, you might obtain a strikingly different equation.

Cook's Distance

Cook's distance is a function of the standardized residuals and the HI values. Check Minitab's Help for its equation. It is an overall measure of the combined impact of each observation on the estimated regression coefficients. Large values signify unusually influential observations. It is customary to flag observations with Cook's distances that are greater than the 50th percentile of an F distribution. These observations usually have either large standardized residuals or a large HI value (or both), and hence a substantial impact on the regression equation. You can check the Data window to determine if there are any Cook's distance values greater than 0.8034 (the 50th percentile of the F distribution with 3 numerator DF (degrees of freedom) and 37 denominator DF). There are no such observations.

11.5   *Incorporating an Indicator (Dummy) Variable into a Model*

The results of both your residual analysis and your examination of unusual observations are cause for concern. You need to construct a better model. Thus far, you have considered only quantitative variables in your modeling. But there is also one qualitative variable, C2-T PubPriv, that you could include as a predictor in your regression equation. Recall that C2-T PubPriv is a text variable with the text values of Private and Public. Before you can include this variable, however, you must convert it to an *indicator (or dummy) variable*. Such a variable takes two values—typically 0 and 1.

The Calc > Make Indicator Variable command can be used to transform C2-T PubPriv into two indicator variables:

- Choose **Calc > Make Indicator Variables**
- Double-click **C2 PubPriv** as the "Indicator variables for" column

- Type **PP1 PP2** as the "Store results in" columns

The completed Make Indicator Variables dialog box should look as shown in Figure 11-18.

FIGURE 11-18

The completed Make Indicator Variables dialog box

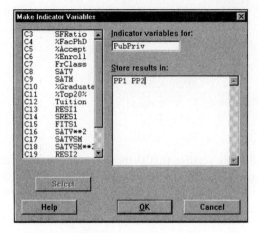

- Click **OK**
- Click the **Current Data Window** toolbar button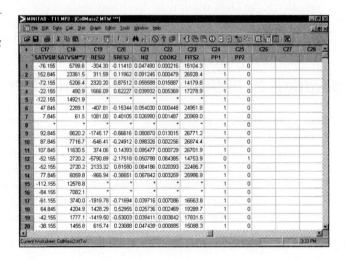

The Data window should resemble Figure 11-19 (you may have to scroll to see all of the contents of this Figure). Notice that the private Massachusetts colleges (with Private in C2-T PubPriv) have a 1 in C24 PP1 and a 0 in C25 PP2. Public colleges are identified by a 0 in C24 PP1 and a 1 in C25 PP2.

FIGURE 11-19

The Data window showing the two indicator (dummy) variables representing the qualitative variable PubPriv

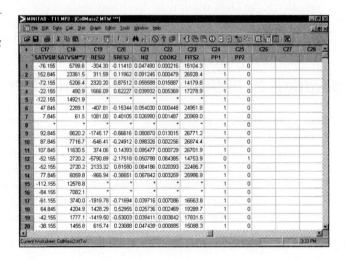

You can represent the qualitative variable C2-T PubPriv in your regression analysis by including one of the two indicator variables. You would have a serious collinearity problem if you included both variables because the two are exactly linearly related—they add up to 1. With the inclusion of this potentially important predictor, it is wise to consider all the other predictors in the initial model.

To fit a multiple linear regression:

- Choose **Stat > Regression > Regression**
- Verify that the "Response" variable is Tuition
- Clear the two current "Predictors"
- Highlight **C3 SFRatio**, **C4 %FacPhD**, **C5 %Accept**, **C6 %Enroll**, **C7 FrClass**, **C8 SATV**, **C9 SATM**, **C10 %Graduate**, **C11 %Top20%**, and click **Select** to select these variables as "Predictors"
- Double-click **C24 PP1** as another "Predictors" variable

♦ **Note** When choosing an extensive and contiguous list of variables, it is easier to highlight the list and click Select than it is to double-click each variable. ■

- Click **Graphs** and click the "Residual Plots" **Individual plots** option button to clear the Four in one option button
- Click **OK** twice

The Session window shows the regression output as shown in Figure 11-20.

FIGURE 11-20

Multiple linear regression with the nine quantitative and one indicator variables

```
MTB > Regress 'Tuition' 10  'SFRatio'-'%Top20%' 'PP1';
SUBC>    Constant;
SUBC>    VIF;
SUBC>    Brief 2.
```

Regression Analysis: Tuition versus SFRatio, %FacPhD, ...

```
The regression equation is
Tuition = - 2089 - 235 SFRatio + 69.1 %FacPhD + 26.2 %Accept - 61.3 %Enroll
          + 0.350 FrClass + 11.2 SATV + 12.9 SATM + 19.3 %Graduate
          + 39.6 %Top20% + 3534 PP1


25 cases used, 35 cases contain missing values
```

Predictor	Coef	SE Coef	T	P	VIF
Constant	-2089	10030	-0.21	0.838	
SFRatio	-234.5	137.9	-1.70	0.111	2.1
%FacPhD	69.13	41.76	1.66	0.120	2.7
%Accept	26.16	46.84	0.56	0.585	9.2
%Enroll	-61.26	49.76	-1.23	0.239	3.3
FrClass	0.3499	0.6223	0.56	0.583	1.8
SATV	11.22	17.20	0.65	0.525	16.3
SATM	12.88	15.97	0.81	0.433	15.5
%Graduate	19.27	52.42	0.37	0.719	4.3
%Top20%	39.60	54.82	0.72	0.482	22.6
PP1	3534	1711	2.07	0.058	1.8

```
S = 1747.75   R-Sq = 92.5%   R-Sq(adj) = 87.1%


Analysis of Variance
```

Source	DF	SS	MS	F	P
Regression	10	526413302	52641330	17.23	0.000
Residual Error	14	42764915	3054637		
Total	24	569178217			

```
Source      DF     Seq SS
SFRatio      1   223611501
%FacPhD      1   187678564
%Accept      1    12426244
%Enroll      1    44681589
FrClass      1    13307275
SATV         1    19693432
SATM         1     4473493
%Graduate    1     5421234
%Top20%      1     2087080
PP1          1    13032890


Unusual Observations

Obs  SFRatio  Tuition    Fit  SE Fit  Residual  St Resid
 36     20.0    16800  14238    1228      2562      2.06R
 45     17.0    18360  21023    1132     -2663     -2.00R

R denotes an observation with a large standardized residual.
```

You can be encouraged by the introduction of the indicator variable. There is a smaller S value and a higher R-Sq value than with your previous model. Still there remains a collinearity problem. Three of the predictor variables have a VIF over ten. You might simplify your model without decreasing its predictive ability by omitting these three variables. If you do this on your own, the resulting equation is:

$$\text{Tuition} = 6748 - 310\ \text{SFRatio} + 113\ \%\text{FacPhD} - 28.3\ \%\text{Accept} - 76.1\ \%\text{Enroll} + 0.976\ \text{FrClass} + 118\ \%\text{Graduate} + 4248\ \text{PP1}$$

And, while this model does not have a problem with collinearity, two population coefficients are individually insignificantly different from 0, as indicated by p-values greater than .05. You will obtain a regression equation without the predictors associated with these insignificant coefficients. The resulting equation is:

$$\text{Tuition} = 2583 - 329\ \text{SFRatio} + 140\ \%\text{FacPhD} - 64.2\ \%\text{Enroll} + 137\ \%\text{Graduate} + 3551\ \text{PP1}$$

You are more satisfied with this model, but there is still one insignificant population coefficient (for %Enroll) that is insignificantly different from 0 based upon a p-value of 0.086. Thus you may omit the %Enroll variable from the model. The new equation of this multiple linear regression is

$$\text{Tuition} = -468 - 284\ \text{SFRatio} + 153\ \%\text{FacPhD} + 109\ \%\text{Graduate} + 4722\ \text{PP1}$$

The Session window shows the regression results for this reduced model. These results are also shown in Figure 11-21.

FIGURE 11-21

*Multiple linear regression
with SFRatio, %FacPhD,
%Graduate, and PP1 as
the predictor variables*

```
MTB > Regress 'Tuition' 4 'SFRatio' '%FacPhD' '%Graduate' 'PP1';
SUBC>    Constant;
SUBC>    VIF;
SUBC>    Brief 2.
```

Regression Analysis: Tuition versus SFRatio, %FacPhD, %Graduate, PP1

```
The regression equation is
Tuition = - 468 - 284 SFRatio + 153 %FacPhD + 109 %Graduate + 4722 PP1


31 cases used, 29 cases contain missing values


Predictor    Coef  SE Coef      T      P   VIF
Constant     -468     3509  -0.13  0.895
SFRatio     -284.3    126.8  -2.24  0.034   1.6
%FacPhD     153.12    32.04   4.78  0.000   1.6
%Graduate   108.75    33.37   3.26  0.003   2.8
PP1           4722     1269   3.72  0.001   1.7


S = 1995.46   R-Sq = 90.0%   R-Sq(adj) = 88.5%


Analysis of Variance

Source          DF         SS        MS      F      P
Regression       4  930775805 232693951  58.44  0.000
Residual Error  26  103528336   3981859
Total           30 1034304140


Source       DF     Seq SS
SFRatio       1  419928257
%FacPhD       1  304384042
%Graduate     1  151329899
PP1           1   55133607


Unusual Observations

Obs  SFRatio  Tuition    Fit  SE Fit  Residual  St Resid
 10     12.0    26228  20813     433      5415     2.78R

R denotes an observation with a large standardized residual.
```

Interpreting the Regression Coefficient for an Indicator Variable

The output indicates a big improvement in S and R-Sq compared to your original
model with two predictors. There are statistically significant results for the F-test
and all four t-tests. In addition, there is now only one unusual observation and no
problem with collinearity. If you perform a residual analysis on this model, you
will find no obvious problems, except for the one outlying college.

Now take a look at the least squares/regression equation:

$$\text{Tuition} = -468 - 284 \text{ SFRatio} + 153 \text{ \%FacPhD} + 109 \text{ \%Graduate} + 4722 \text{ PP1}$$

The sample coefficient associated with the indicator variable PP1 has an
interesting interpretation. The predicted base tuition for a private Massachusetts
college (PP1 = 1) is \$4,722 more than that for a public college in the same state
(PP1 = 0), assuming the same student-faculty ratio, percentage of faculty with
PhDs, and percentage of students graduating.

▶ **Note** In general, if the original qualitative variable takes k different text or numeric values, the Calc > Make Indicator Variables command will create k indicator variables. For each value associated with the original variable, one of the indicator variables will take the value 1 when that value occurs and 0 otherwise. However, because these k new variables will add to 1, only k − 1 of them should be included in the regression model. ■

11.6 Performing Best Subsets Regression

You obtained the last regression model by trial and error. You can take a more systematic approach by first constructing a matrix plot for the ten variables under consideration:

- Choose **Graph** > **Matrix Plot** and double-click *Matrix of Plots, Simple*
- Highlight **C3 SFRatio**, **C4 %FacPhD**, **C5 %Accept**, **C6 %Enroll**, **C7 FrClass**, **C8 SATV**, **C9 SATM**, **C10 %Graduate**, and **C11 %Top20%**, and click **Select** to select these as "Graph variables"
- Double-click **C24 PP1** and **C12 Tuition** as additional "Graph variables"

▶ **Note** Selecting C12 Tuition as the last variable makes it easier for you to see the relationship of this response variable to the ten possible predictor variables. ■

- Click **Matrix Options** and click the "Matrix Display" **Lower left** option button

   Note that you can choose between a "Full", a "Lower left" or an "Upper right" matrix display.
- Click **OK** twice

   The matrix plot, with Minitab's default title, appears as shown in Figure 11-22.

FIGURE 11-22

Matrix plot with ten possible predictor variables and the response variable

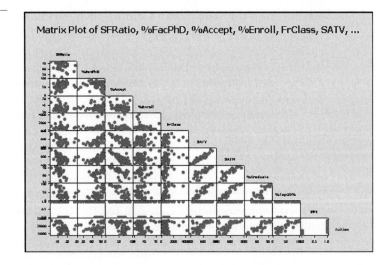

The graph contains a scatterplot for each pair of variables listed in the "Graph variables" text box. You are most interested in the lower row of plots; in each, Tuition is the variable on the vertical axis. These plots indicate that many of the predictors are linearly related to Tuition. Further, in the lower right-hand plot you can see that tuition tends to be much lower at public colleges (where PP1 = 0) than at private colleges (where PP1 = 1). This lower row of plots gives you some insight into which possible predictor variables you should include in the regression model. However, the relationship between the response variable and possible predictor variables can change depending upon the variables already in the model. A good plan is to begin your analysis with a popular variable-selection procedure called best subsets regression.

Best subsets regression generates regression models using the maximum R-squared criterion by first examining all one-predictor regression models and selecting the two models giving the largest R-Sq. Then Minitab displays information on these models. It next examines all two-predictor models, selects the two models with the largest R-Sq, and displays information on these two models. This process continues until the model contains all predictors. The technique includes only colleges that have no missing values for the response and predictor variables.

Best subsets regression is an efficient way to select a group of "best subsets" for further analysis by selecting the smallest subset that fulfills certain statistical criteria. You can use this output to select a reasonably small set of predictors that is associated with a reasonably large value of R-Sq.

Now, use best subsets regression to identify predictors for Tuition from the ten predictors considered previously:

- Choose **Stat > Regression > Best Subsets**
- Double-click **C12 Tuition** as the "Response" variable
- Highlight **C3 SFRatio, C4 %FacPhD, C5 %Accept, C6 %Enroll, C7 FrClass, C8 SATV, C9 SATM, C10 %Graduate**, and **C11 %Top20%** and click **Select** to select these variables as "Free predictors"
- Double-click **C24 PP1** as another "Free predictors" variable

The completed Best Subsets Regression dialog box is shown in Figure 11-23.

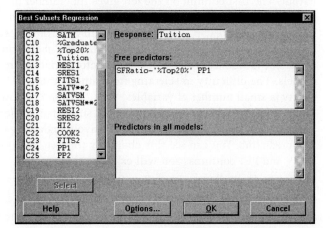

- Click **OK**

The output is shown in Figure 11-24.

FIGURE 11-24

Best subsets regression for Tuition

```
MTB > BReg 'Tuition' 'SFRatio'-'%Top20%' 'PP1' ;
SUBC>   NVars 1 10;
SUBC>   Best 2;
SUBC>   Constant.
```

Best Subsets Regression: Tuition versus SFRatio, %FacPhD, ...

Response is Tuition

25 cases used, 35 cases contain missing values

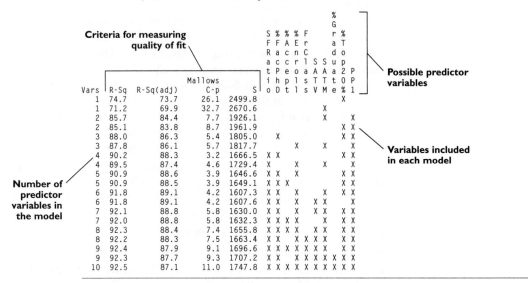

The output includes a note that the analysis was based upon only 25 of the 60 colleges that had complete data for Tuition and the ten possible predictor variables. These ten variables are listed vertically on the top right-hand portion of the output. The first column on the left gives the number of predictor variables in each model listed. For each model, the values for four criteria are given: R-Sq, R-Sq(adj), Mallows C-p, and S, the standard error of the estimate. Check Minitab Help for more information about the *Mallows C-p* criterion. The Xs to the right of the output indicate which of the predictor variables are included in the "best" models. The difficulty in selecting from among these 20 models is to balance simplicity (a small number of variables) with explanatory power (a high R-Sq value).

This best subsets regression output suggests a tuition model with SATM and PP1 as the predictors based upon its relatively high R-Sq value of 85.7 and only two predictors. You can see this choice by the two Xs in the third row and the SATV and PP1 columns. You will examine this model:

- Choose **Stat > Regression > Regression**
- Verify that the "Response" variable is Tuition
- Double-click **C9 SATM** and **C24 PP1** to replace the previous "Predictors"

- Click **OK**

Figure 11-25 contains the regression output for the model suggested by the Best Subsets Regression.

```
MTB > Regress 'Tuition' 2 'SATM' 'PP1';
SUBC>    Constant;
SUBC>    VIF;
SUBC>    Brief 2.
```

Regression Analysis: Tuition versus SATM, PP1

```
The regression equation is
Tuition = - 8256 + 37.5 SATM + 8198 PP1


58 cases used, 2 cases contain missing values


Predictor    Coef  SE Coef      T      P  VIF
Constant     -8256     2631  -3.14  0.003
SATM        37.471    4.860   7.71  0.000  1.1
PP1         8197.8    979.0   8.37  0.000  1.1


S = 3009.96   R-Sq = 75.9%   R-Sq(adj) = 75.0%

Analysis of Variance

Source           DF          SS         MS      F      P
Regression        2  1565281525  782640763  86.39  0.000
Residual Error   55   498291569    9059847
Total            57  2063573094


Source  DF      Seq SS
SATM     1   929992861
PP1      1   635288664


Unusual Observations

Obs  SATM  Tuition    Fit  SE Fit  Residual  St Resid
  5   410    26858  15305     872     11553     4.01R
 15   420    24830  15680     831      9150     3.16R

R denotes an observation with a large standardized residual.
```

Initially you may be surprised that the R-Sq is only 75.9%. It is lower than the 85.7% shown in the best subsets regression output (Figure 11-24). The difference is due to the fact that 58 colleges are now being considered in the model instead of 25. Still this model has many strengths. First, it has only two predictor variables. Second, there is no problem with collinearity and there are only two unusual observations. Both coefficients of the predictor variables have p-values of 0.000. This suggests that at any reasonable level of significance, both variables belong in the model.

Before settling on this model you should check whether the assumptions of linear regression are satisfied. You can repeat your multiple linear regression analysis with a Four in one graph, a scatterplot of the standardized residual against SATM, and a scatterplot of the standardized residual against PP1. From the Four in one display, you will see that two colleges are likely responsible for the violation of both the normality and the constant variation assumption. These two colleges are Atlantic Union College and Curry College, both of which have tuitions substantially higher than would be predicted from their average math SAT scores and their ownership status. You decide that these two colleges are sufficiently unusual in this regard that they should be excluded from the analysis. Here is one way to do this:

- Choose **Data > Subset Worksheet**
- Verify that the "Name of the New Worksheet" will be Subset of CollMass2.mtw
- Click the "Include or Exclude" **Specify which rows to exclude** option button
- Click the "Specify Which Rows to Exclude" **Row numbers** option button and type **5 15**

The completed Subset Worksheet dialog box is shown in Figure 11-26.

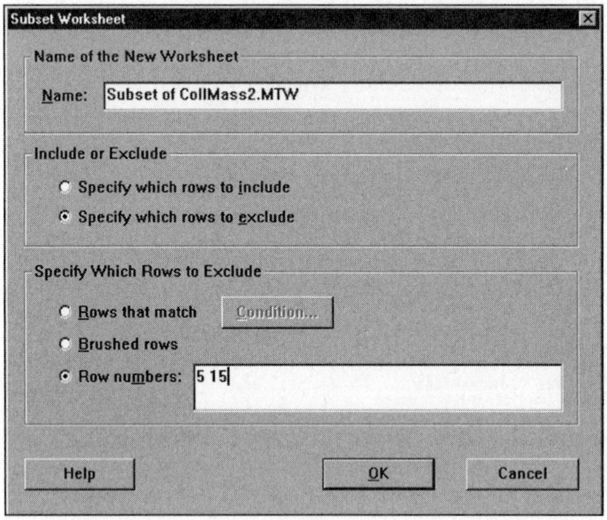

FIGURE 11-26

The completed Subset Worksheet dialog box

- Click **OK**

The new worksheet is now the current worksheet. It does not include the data from Atlantic Union and Curry Colleges. Now, repeat your multiple linear regression analysis (with the four in one graph and two requested residual analysis displays). The resulting output is present in the Session window and Figure 11-27.

FIGURE 11-27

Regression suggested by best subsets regression with two unusual colleges deleted

```
MTB > Regress 'Tuition' 2 'SATM' 'PP1';
SUBC>   GFourpack;
SUBC>   GVars 'SATM' 'PP1';
SUBC>   RType 2;
SUBC>   Constant;
SUBC>   VIF;
SUBC>   Brief 2.
```

Regression Analysis: Tuition versus SATM, PP1

```
The regression equation is
Tuition = - 13188 + 47.1 SATM + 7165 PP1


56 cases used, 2 cases contain missing values


Predictor    Coef  SE Coef      T      P  VIF
Constant   -13188     1972  -6.69  0.000
SATM       47.078    3.665  12.84  0.000  1.1
PP1        7165.0    705.7  10.15  0.000  1.1


S = 2128.10   R-Sq = 87.7%   R-Sq(adj) = 87.3%


Analysis of Variance

Source         DF          SS         MS       F      P
Regression      2  1719062231  859531115  189.79  0.000
Residual Error 53   240027866    4528828
Total          55  1959090097


Source  DF      Seq SS
SATM     1  1252158398
PP1      1   466903833


Unusual Observations

Obs  SATM  Tuition    Fit  SE Fit  Residual  St Resid
 22   600    26871  22223     342      4648      2.21R
 40   580    25610  21282     326      4328      2.06R
 51   519    13650  18410     376     -4760     -2.27R

R denotes an observation with a large standardized residual.
```

The Four in one graph is shown in Figure 11-28.

FIGURE 11-28

Four in one graph suggested by the best subsets regression with two unusual colleges deleted

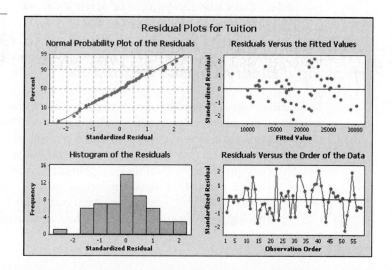

The output is encouraging. Based upon the analysis of the remaining 56 colleges, the coefficient of determination is 87.7% and the standard error of the estimate is $2128.10. The values for VIF (1.1 and 1.1) indicate no problem with collinearity. Moreover, all of the assumptions for multiple linear regression inference appear to be satisfied. (The standardized residuals are now normal based upon an Anderson-Darling test p-value of 0.916.) Hence the p-values (all are 0.000) associated with each predictor suggest that SATM and PP1 belong in the model. And, while there are three unusual observations the absolute values of their standardized residuals are all close to 2.

To conclude your regression analysis you can use this model to obtain tuition predictions for your school. It is a private school with an average SATM score of 539. Multiple linear regression yields a 95% two-sided prediction interval estimate of ($15,028, $23,676). You are pleased because your current tuition is within this interval.

You have completed your regression analysis on the tuition at Massachusetts colleges and are ready to submit your recommendations. Best subsets regression helped you to determine the best variables to use as predictors. You will recommend a model that uses average math SAT scores and that indicates whether or not a college is private to predict the tuition structure for the colleges. Your results will be valuable to the administration as it develops a strategic plan for deciding tuition rates and attracting students.

▶ **Note** Best subsets regression is one of two variable-selection procedures in Minitab. The other procedure is *stepwise regression* (Stat > Regression > Stepwise). It allows the user to perform three stepwise regression methods: Stepwise (forward and backward), Forward selection, and Backward elimination. Currently, most statisticians prefer to use best subsets regression instead of stepwise regression except when the number of predictors is too large to be accommodated by best subsets regression. Check Minitab's Help for further information about stepwise regression. ■

Before moving on to another type of regression, save your work:
- Click the **Save Project** toolbar button 🖫

11.7 Performing a Binary Logistic Regression

The administration is so pleased with your multiple linear regression model for tuition you are asked you to obtain a preliminary model for determining what variables explain the type of college (public or private). You accept the assignment but realize that you cannot use multiple linear regression because the response variable is not quantitative. PubPriv is a qualitative variable with two possible values. To obtain a model for this dichotomous variable, you can employ Minitab's Stat > Regression > Binary Logistic Regression command. *Binary logistic regression* performs logistic regression on a binary response variable. A *binary variable* has only two possible values, in this case, Public or Private.

Binary logistic regression is used to identify which variables best explain the distribution of two categories. It is also used to classify observations into one of the two categories. It does not use the method of least squares. More information about binary logistic regression is available in Minitab's Help. The first part of the overview Help screen is shown in Figure 11-29.

FIGURE 11-29

The Binary Logistic Regression Help window

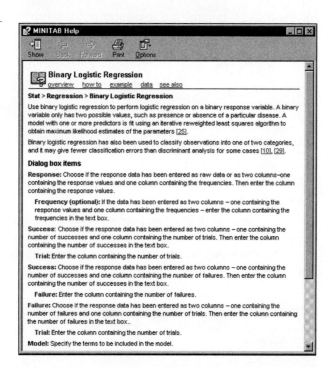

> ▶ **Note** As you saw in Section 11.5, qualitative variables can be included as predictor variables when they are transformed into indicator variables. However, a number of the assumptions associated with the standard regression model will be violated if such a variable is used as the response. For example, the constant variation assumption may be violated. In addition, there will be problems with the fitted values if you perform linear regression to predict the value of an indicator variable. ■

For this assignment, you focus on which of the three quantitative variables (C3 SFRatio, C7 FrClass, or C12 Tuition) are best at distinguishing between the two types of colleges. You will use most of Minitab's binary logistic regression defaults.

- Click the **Current Data Window** toolbar button 📖 until you return to *CollMass12.mtw*
- Choose **Stat > Regression > Binary Logistic Regression**
- Double-click **C2 PubPriv** as the "Response" variable and **C3 SFRatio**, **C7 FrClass**, and **C12 Tuition** as the "Model" variables

  The completed dialog box should look as shown in Figure 11-30.

FIGURE 11-30

*The completed Binary
Logistic Regression
dialog box*

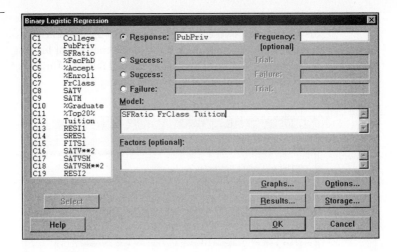

To focus on the essential aspects, print only part of the possible output:

- Click **Results**
- Click **Response information, regression table, log-likelihood, and test that all slopes equal 0** option button

This completed Binary Logistic Regression - Results dialog box is shown in Figure 11-31.

FIGURE 11-31

*The completed
Binary Logistic
Regression - Results
dialog box*

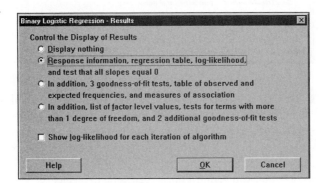

- Click **OK** twice

The results of this Brief 1 output are shown in Figure 11-32.

FIGURE 11-32

Binary logistic regression

```
MTB > BLogistic 'PubPriv' = SFRatio FrClass Tuition;
SUBC>    Logit;
SUBC>    Brief 1.
```

Binary Logistic Regression: PubPriv versus SFRatio, FrClass, Tuition

```
Link Function: Logit


Response Information

Variable  Value    Count
PubPriv   Public     12   (Event)
          Private    47
          Total      59

* NOTE * 59 cases were used
* NOTE * 1 cases contained missing values

Logistic Regression Table
                                                Odds      95% CI
Predictor       Coef     SE Coef      Z      P  Ratio  Lower  Upper
Constant     21.4695     10.9758   1.96  0.050
SFRatio    -0.0806637   0.303276  -0.27  0.790   0.92   0.51   1.67
FrClass     0.0019217  0.0020086   0.96  0.339   1.00   1.00   1.01
Tuition    -0.0016271  0.0008771  -1.86  0.064   1.00   1.00   1.00


Log-Likelihood = -4.150
Test that all slopes are zero: G = 51.297, DF = 3, P-Value = 0.000
```

The G-statistic of 51.297 is used to test the null hypothesis that all three population slopes are zero. It is similar to the F-statistic used in multiple linear regression. Its associated p-value of 0.000 suggests that at least one of the three quantitative variables belongs in the model.

In binary logistic regression, the Z-test plays the same role as the t-test for determining the significance of individual variable coefficients in multiple linear regression. For commonly used levels of significance (those less than .10), only the coefficient with a Z-statistic of –1.86 and associated p-value of 0.064 suggests that its predictor (Tuition) is a statistically significant predictor of whether a college is public or private. If you repeat the analysis with only tuition as a predictor, the p-value for Tuition is only 0.008.

You report to the administration that the tuition distinguishes whether a school is public or private. Also, you mention that, if the administration wishes, you may perform a more complete analysis of this model.

To finish this tutorial:

■ Click the **Save Project** toolbar button 🖫 and exit Minitab

Congratulations on finishing Tutorial 11! In the next tutorial you will look at ways of measuring the relationship between qualitative, rather than quantitative, variables.

Minitab Command Summary

This section describes the commands introduced in, or related to, this tutorial. Use Minitab's online Help for a complete explanation of all commands.

Minitab Menu Commands

Menu	Command	Description
Calc ➤	Standardize	Centers and/or scales columns of data.
	Make Indicator Variables	Creates indicator (dummy) variables that you can use in a regression analysis.
Stat ➤	Regression	
	➤ Regression	Performs simple or multiple linear regression, allowing for both descriptive and inferential analyses.
	➤ Stepwise	Performs stepwise regression using stepwise (forward and backward), forward selection, or backward elimination methods.
	➤ Best Subsets	Performs an all-possible subsets regression using the R-Squared criterion.
	➤ Binary Logistic Regression	Performs logistic regression on a two-possible-value (dichotomous) response variable.
Graph ➤	Matrix Plot	
	Matrix of Plots, Simple	Produces a scatterplot matrix of up to 20 variables at once.

Minitab Session Commands

Session Command	Description
brief k	Controls the amount of output from Regression and other commands displayed in the Session window. For regression, k = 0 produces no output and k = 3 displays the most. The default is k = 2.

Review and Practice

Matching Problems

Match the following terms to their definitions by placing the correct letter next to the term it describes.

_____ Time Series Plot

_____ SRES against FITS plot

_____ VIF

_____ Best Subsets

_____ Binary Logistic Regression

_____ Four in one graph

_____ Make Indicator Variable

_____ brief

_____ 3

_____ 2

a. The number of major linear regression assumptions

b. A display you can use to check the independence assumption

c. The number of indicator variables needed as predictors in order to represent a qualitative variable with three text values

d. A display you can use to check the major assumptions of linear regression

e. The Minitab Session command that you use to control the amount of output produced by the Regression command

f. Minitab notation for a measure of how much the variance of an estimated regression coefficient increases if collinearity is present

g. The Minitab command that examines all possible combinations of predictor variables

h. The Minitab command that allows one to model a dichotomous response variable

i. The Minitab command for creating dummy variables

j. A display you can use to check constant variation assumption

True/False Problems

Mark the following statements with a *T* or an *F*.

a. ____ High VIF values indicate the presence of collinearity.

b. ____ A Matrix Plot can produce an upper-right display.

c. ____ The Normality Test command provides a choice of two tests: the Anderson-Darling normality test and the Ryan-Joiner W test.

d. ____ It is possible for different sets of data to produce the same regression equation, standard error of estimate, and coefficient of determination.

e. ____ To perform a best subsets regression in Minitab, you should choose Stat > Regression > Regression.

f. ____ You create indicator variables so that you can include a qualitative variable as a response variable in a linear regression.

g. ____ The default output for a best subsets regression includes three models for each number of predictors.

h. ____ By default, Minitab identifies an observation that has an unusually large positive or negative standardized residual with an R and prints out the value of its standardized residual.

i. ____ By default, Minitab identifies an observation that has high leverage with an X and prints out its HI value.

j. ____ A normal probability plot is an excellent way to check one of the major linear regression assumptions.

Practice Problems

The practice problems instruct you to save your worksheets with a filename prefixed by P, followed by the tutorial number. In this tutorial, for example, use *P11* as the prefix. If necessary, use Help or Appendix A Data Sets, to review the contents of the data sets referred to in the following problems. Interpretations should use the language of the subject matter of the question. If you are using the Student version, be sure to close worksheets when you have completed a problem.

1. Open *DJC20012002.mtw*. (If you are using the Student version of Minitab, you will be asked to erase some columns because the size of this data file exceeds the limit of 10,000 elements (cells). For this problem you should erase all the columns except for Day, KO, HD, PG, and SBC.)

   a. Use the Graph > Scatterplot > *With Regression* command to obtain scatterplots for the four stock variables KO, HD, PG, and SBC against Day. In each case click Data View and click the Connect line option button. Why is it beneficial to use this option?

   b. Write a brief paragraph explaining how useful the simple linear regression model is for these four data sets based upon these displays.

2. Open *MnWage2.mtw*.

   a. Obtain a scatterplot of minimum wage against year. How would you characterize the pattern in the plot?

   b. Perform a simple linear regression to explain the minimum wage based upon year. What is the standard error of the coefficient of year? What is the value of R-Sq?

   c. Create a new variable, Year**2 and perform a quadratic regression to explain the minimum wage based upon Year and Year**2. There is a problem with this analysis due to extreme collinearity. Which statistics in the output suggest this problem? Verify the problem by computing the regression's VIFs and the correlation coefficient between these two predictor variables.

   d. Perform a quadratic regression to explain the minimum wage based upon a centered-year variable and a centered-year squared variable. Is there any problem with collinearity in this model? Explain briefly.

   e. Is the quadratic term worth adding? Explain briefly.

3. Open *Homes.mtw*.

   a. Perform a simple linear regression with Price as the response variable and Area as the predictor variable. What is the regression equation?

   b. Perform a complete residual analysis using the non-regression command techniques introduced in this tutorial. Are you comfortable with performing hypotheses and constructing interval estimates based upon this model? Explain briefly.

   c. Perform a residual analysis using the Stat > Regression > Regression command Graph option displays introduced in this tutorial. Explain differences, if any, between this output and that in part b.

   d. Discuss any unusual observations. Specifically, hold your mouse cursor over the point(s) in a plot of Price against Area to identify the unusual observation(s). Then obtain data tip information. How many have large absolute valued standardized residuals? How many have a large influence upon the regression equation?

e. Is there evidence of collinearity present in this model?

4. Open *TwoTowns.mtw*. Perform a multiple linear regression to explain Price based upon Town, Baths, HalfBaths, and Garages. You will have to transform the Town variable. What is the regression equation? Interpret the coefficients of each predictor.

5. Open *Murders.mtw*.

   a. Create four indicator variables for the qualitative variable Region. Name your indicator variables Region1, Region2, Region3, and Region4. What does each indicator variable represent?

   b. Perform a multiple linear regression to explain MurderRate based upon Region1, Region2, and Region3. What is the regression equation?

   c. What are this model's F-statistic value and corresponding p-value?

   d. Perform a one-way ANOVA with MurderRate as the response variable and Region as the factor (choose the Stat > ANOVA > One-Way command). What are this model's F-statistic value and corresponding p-value? They should equal the values you reported in part c.

   e. What can you conclude from these equalities? Explain briefly.

6. Open *Homes.mtw*.

   a. Perform a best subsets regression to obtain a model to estimate Price based upon the other variables. There is no correct answer! The trick is to balance simplicity (a small number of variables) with explanatory power (a large value of R-Sq). What predictor variables did you select?

   b. Perform a complete regression analysis on this model. Would you use the best subsets regression model based upon your examination of assumptions, unusual observations, and other criteria? Briefly explain your answer.

7. Open *Temco.mtw*. Use best subsets regression and linear regression to determine which of the variables in the data set affect Salary. Remember to transform each text variable into indicator variables. Be sure to consider the unusual observations and to perform a complete residual analysis.

8. Open *OpenHouse.mtw*.

   a. How many text values are in the Style variable? How many text values are in the Location variable?

   b. Construct indicator variables for each of the qualitative variables. What names did you choose for these indicator variables?

   c. Perform a multiple linear regression with Price as a response variable and the other two quantitative variables and the appropriate number of indicator variables as predictors. What is the resulting regression equation? Are you pleased with this multiple linear regression's standard error of the estimate, coefficient of determination, and p-values? Would you use the multiple linear regression model based upon your examination of assumptions, unusual observations, and other criteria? Explain briefly.

   d. Perform a best subsets regression to determine a model to estimate Price based upon the other variables. Does this model support your discussion in part c?

e. If the best subsets regression model differs from that in part c, perform a linear regression analysis on this model. What is the resulting regression equation? Are you pleased with this multiple linear regression's standard error of the estimate, coefficient of determination, and p-values? Would you use the best subsets regression model based upon your examination of assumptions, unusual observations, and other criteria? Briefly explain your answer.

f. Based upon your answer to part e, remove outlying houses from the data set and repeat the complete linear regression analysis using the same model. Would you use the best subsets regression model based upon your examination of assumptions, unusual observations, and other criteria? Briefly explain your answer.

g. Regardless of your answer to part f, for the median number of bedrooms and baths, modal location, and modal style, what is the 95% confidence interval for an average price and what is the 95% prediction interval for an individual price?

9. Open *MLBGameCost.mtw*. A game cost index is a linear combination of expenses a fan incurs at a typical sporting event. The game cost index for Major League Baseball (MLB) may be based upon one or more of the following costs: an average ticket, an average child's ticket, a beer, a soda, a hot dog, parking an automobile, a game program, and a souvenir cap. Use best subsets regression and multiple linear regression to determine the equation of this fan cost index. Interpret the coefficients in this equation.

10. Open *GasData.mtw*.

a. Use binary logistic regression to obtain a model for explaining Locale based upon Gallons, Cost, and MPG. What are the p-values associated with these predictors? Which of these three predictors would you remove from the model? Explain briefly.

b. Use binary logistic regression to obtain a model for explaining Locale based upon the remaining two predictors. What are the p-values associated with these predictors? Which of these two predictors would you remove from the model? Explain briefly.

c. Use binary logistic regression to obtain a model for explaining Locale based upon the remaining predictor. What's the p-value associated with this predictor?

d. Which of the three models do you prefer? Briefly explain your answer.

11. Open *Donner.mtw*. Consider Survive as the response variable. Use binary logistic regression to obtain three models to explain Survive. The first with Age as the predictor, the second with Gender as the predictor (you will have to create indicator variables for this qualitative variable), and the third with Age and Gender as the predictors. Which of these three models best explains whether or not an individual survived? Explain briefly.

On Your Own

Open *CollMass2.mtw*. Choose Stat > Regression > Stepwise to model Tuition based upon the nine quantitative variables and PP1, the indicator variable created in this tutorial. Use the command's three methods: Stepwise (forward and backward), Forward selection, and Backward elimination. (Please note that you need to type No in the Session window after each method.) What are the final regression equations for each method? Which of these equations do you prefer based upon their standard error of the estimates? Which of these equations do you prefer based upon their coefficients of determination? Which of these equations do you prefer based upon the number of predictors in the model? Taking all three criteria into account, which of these equations do you prefer? Provide a brief explanation for each of your choices.

Answers to Matching Problems

(b) Time Series Plot
(j) SRES against FITS plot
(f) VIF
(g) Best Subsets
(h) Binary Logistic Regression
(d) Four in one graph
(i) Make Indicator Variable
(e) brief
(a) 3
(c) 2

Answers to True/False Problems

(a) T, (b) T, (c) F, (d) T, (e) F, (f) F, (g) F, (h) T, (i) F, (j) T

Tutorial

Analyzing Qualitative Data

Most of the data you worked with in previous tutorials were quantitative. In this tutorial, you will use Minitab to make statistical decisions about qualitative data. First, you will perform a goodness-of-fit test by issuing a series of commands and executing a Minitab macro. Then you will test whether two qualitative variables are independent. In both cases, you use the chi-square distribution.

OBJECTIVES

In this tutorial, you learn how to:

- Compare an observed distribution of counts to a hypothesized distribution by performing a chi-square goodness-of-fit test
- Construct a Minitab Exec macro to perform a chi-square goodness-of-fit test
- Perform a chi-square test for the independence of two qualitative variables when the worksheet contains the contingency table counts
- Perform a chi-square test for the independence of two qualitative variables when the worksheet contains the raw data for the two variables
- Combine or omit categories so that the distributional assumptions underlying the chi-square test for independence are valid

Comparing an Observed Distribution of Counts to a Hypothesized Distribution

| CASE STUDY | MANAGEMENT—ENTREPRENEURIAL STUDIES (CONTINUED) |

Your management professor, Dr. Michaels, has assigned a term paper in which you are to report on the pattern of sales for small-scale retail outlets for an industry of your choice. You select the music industry. As part of your research, you discover the fact that your neighborhood music outlet sells approximately 1,000 items of recorded music per week.

One of the references used in your Sales Management course, *The New York Times 2003 Almanac*, presents the percentages of recorded music sales for 2003 associated with different music genres. (The percentages are adjusted slightly from the original so that they add to 100%.) In Tutorial 6 you simulated a week of music sales at a music outlet based upon the 2003 percentages. Here are the corresponding proportions:

Genre		Proportion of Sales
1.	Rock	0.271
2.	Pop	0.134
3.	Rap/Hip-Hop	0.126
4.	R&B/Urban	0.118
5.	Country	0.116
6.	Gospel	0.074
7.	Jazz	0.038
8.	Classical	0.035
9.	Other	0.088
	All	1.000

Anticipating the possibility that your instructor will ask for some kind of verification that your sample of 1,000 simulated values came from the population of music sales, you decide to perform a test. In this context a test to determine whether or not a sample comes from a specified population is the *chi-square goodness-of-fit test*. Here are its hypotheses:

Null Hypothesis, H_0: The sample came from the population specified in the above table.

Alternative Hypothesis, H_1: The sample came from a population other than the one specified.

Begin by opening Minitab:

- Open **Minitab** and maximize the Session window
- If necessary, **Enable Commands** and clear Output Editable from the Editor menu
- Click the **Save Project** toolbar button ▣ and, in the location where you are saving your work, save this project as **T12.mpj**

In Tutorial 6, you saved the project, *T6.mpj*, with a worksheet containing your simulated class data. The file *MusicData.mtw* contains data based upon the same procedures you performed in Tutorial 6. In this tutorial, you should work with this data file to ensure consistency.

- Open the worksheet **MusicData.mtw**
- If necessary, click the **Current Data Window** toolbar button ▦

The first three columns of the worksheet display the nine numeric values and the corresponding text values of the music genre (C1 GenreN and C2-T Genre, respectively) and the associated probabilities (C3 Probability). C4 Sample and C5-T TxSample list the numeric and text values for your random sample of 1,000 genres.

Minitab does not have a single command to perform the chi-square goodness-of-fit test, but you can combine a number of commands—each of which you have used before—to perform the test. You can use Minitab to perform the chi-square goodness-of-fit calculations using the formula

$$\chi^2_{k-1} = \sum \frac{(\text{Obs} - \text{Exp})^2}{\text{Exp}}$$

where Obs and Exp are the *observed* and *expected counts* of sales for each category (genre), respectively. You can then calculate the p-value by computing the area equal to or to the right of the observed chi-square value under a *chi-square distribution* with k – 1 = 9 – 1 = 8 DF (degrees of freedom), where k is the number of categories (genres in this case).

Begin by using the Tally Individual Variables command to compute the count of each genre in your sample:

- Choose **Stat > Tables > Tally Individual Variables**
- Double-click **C4 Sample** as the "Variables"
- Click **OK**

The Session window displays the counts, as shown in Figure 12-1.

FIGURE 12-1

Tally for the simulated music sample

```
MTB > Tally 'Sample';
SUBC>    Counts.
```

Tally for Discrete Variables: Sample

```
Sample   Count
    1      295
    2      137
    3      110
    4      111
    5      125
    6       80
    7       41
    8       23
    9       78
   N=     1000
```

Now store these nine sample values and their corresponding counts into the next two columns of your worksheet, C6 and C7, named Genre2 and Observed:

- Highlight **the nine rows of these data** (not the name or total rows), click your **right-hand mouse button**, and choose **Copy**
- Click the **Current Data Window** toolbar button 🔳
- Type **Genre2** and **Observed** as the names for columns C6 and C7
- Click in **C6, row 1**
- Click your **right-hand mouse button**, choose **Paste Cells**, and click **OK** to use spaces as delimiters in order to paste the data into the worksheet

Columns C6 Genre2 and C7 Observed now contain the genre numbers and the corresponding sample counts.

Next, obtain the expected counts (numbers of sales under H_0 for each genre) by multiplying the genre probabilities (stored in C2 Probability) by the sample size, 1,000.

- Choose **Calc > Calculator**
- Type **Expected** as the "Store result in variable"
- Enter, by clicking or typing, **Probability * 1000** as the "Expression" and click **OK**

The column C8 Expected now lists the expected counts for each genre: 271, 134, 126, 118, 116, 74, 38, 35, and 88. The Data window should resemble Figure 12-2.

FIGURE 12-2

*The Data window showing
the observed and expected
counts*

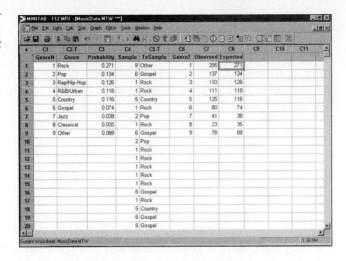

To obtain the value of the chi-square statistic, compute the sum of
(Obs – Exp)^2/Exp values for each genre:

- Click the **Edit Last Dialog** toolbar button 🖬
- Type **ChiSq** as the "Store result in variable"
- Enter, by clicking or typing, **SUM((Observed – Expected)**2/Expected)** as the
  "Expression" and click **OK**

The value of the chi-square statistic is 11.3119, as shown in the first cell of
C9 ChiSq.

Next, compute the DF associated with this statistic and store it as a constant (not as a variable):

- Click the **Edit Last Dialog** toolbar button 🖬
- Type **k1** as the "Store result in variable"
- Enter, by clicking or typing, **COUNT(Observed) – 1** as the "Expression" and
  click **OK**
- Click the **Session Window** toolbar button 🖬
- Type **name k1 'DF'** after the Minitab prompt ("MTB >") to name the DF con-
  stant, and press ⏎

The expression COUNT(Observed) – 1 will give you one less than the number of categories, which is the number of degrees of freedom for this test of fit.
Confirm that you have created this new constant:

- Type **info** and press ⏎

After the listing of the columns, the one constant K1 is listed as having the
value 8 and with the name DF.

To compute the p-value associated with the chi-square statistic (11.3119) in
C9 ChiSq:

- Choose **Calc > Probability Distributions > Chi-Square**
- Verify that the "Cumulative probability" option button is clicked
- Double-click **K1 DF** as the "Degrees of freedom"

- Double-click **C9 ChiSq** as the "Input column" and type **CumProb** as the "Optional storage" column
- Click **OK**
- Click the **Current Data Window** toolbar button 🗗

The cumulative probability in the first cell of C10 CumProb is equal to 0.815354. This is the probability of a value less than or equal to the chi-square statistic (11.3119). The p-value, however, is the probability of a value greater equal to or greater than 11.3119, so you must subtract 0.815354 from 1:

- Choose **Calc > Calculator**
- Type **PValue** as the "Store result in variable"
- Enter, by either clicking or typing, **1 – CumProb** as the "Expression" and click **OK**

Note that the p-value is equal to 0.184646 and is stored in C11 PValue. Now, obtain a summary of your chi-square goodness-of-fit test:

- Choose **Data > Display Data**
- Double-click **K1 DF** and **C7 Observed** through **C11 PValue** as the "Columns, constants, and matrices to display", and click **OK**

A well-designed display of your test results appears in the Session window, as shown in Figure 12-3.

<table>
<tr><td>FIGURE 12-3</td><td colspan="6">MTB > Print 'DF' 'Observed' - 'PValue'.</td></tr>
<tr><td>Chi-square goodness-of-fit test</td><td colspan="6">Data Display</td></tr>
<tr><td></td><td colspan="6">DF   8.00000</td></tr>
</table>

Row	Observed	Expected	ChiSq	CumProb	PValue
1	295	271	11.3119	0.815354	0.184646
2	137	134			
3	110	126			
4	111	118			
5	125	116			
6	80	74			
7	41	38			
8	23	35			
9	78	88			

Because the p-value for the chi-square statistic is 0.184646, you should not reject the null hypothesis at any reasonable significance level and should conclude that the results are consistent with the sample being drawn from the population of genres taken from *The New York Times 2001 Almanac*.

Above this output are the Session commands that generated this output, as shown in Figure 12-4. (The info output has been omitted.) These commands are also present in the History folder of the Project Manager.

```
MTB > Name C9 'ChiSq'
MTB > Let 'ChiSq' = SUM((Observed - Expected)**2/Expected)
MTB > Let k1 = COUNT(Observed) -1
MTB > name k1 'DF'
MTB > info
MTB > Name c10 "CumProb"
MTB > CDF 'ChiSq' 'CumProb';
SUBC>   ChiSquare 'DF'.
MTB > Name C11 'PValue'
MTB > Let 'PValue' = 1 - CumProb
MTB > Print 'DF' 'Observed' 'Expected' 'ChiSq' 'CumProb' 'PValue'.
```

In the next section, you will save and modify this sequence of commands in a special file. This will allow you to perform future chi-square goodness-of-fit tests using this same sequence of commands. If you do not wish to save and modify these commands, go to Section 12.3.

■ Click the **Save Project** toolbar button 📁

12.2 A Minitab Exec Macro for a Chi-Square Goodness-of-Fit Test

In addition to using Minitab's menu and Session commands, you can also write and store your own sets of commands in files called macros. In this section, you work with *Exec macros*, the simplest type of Minitab macros. The other types of macros available in Minitab are Global and Local macros. These are more powerful and available only in the Professional version. You can use Exec macros to perform repetitive analyses and statistical analyses that are not available as commands on Minitab's menus. This macro feature of Minitab essentially gives you the power to "program" in Minitab.

An Exec macro is a series of Minitab commands stored in a file. Exec macro files must be in text format. Creating Execs require you to know the Session command (and, where appropriate, subcommand) equivalent for menu commands. To find the Session command equivalent for any menu command, you can consult the Help for that command. To find the menu command equivalent for any Session command, you can also consult Minitab's Help (Help > Help > Reference > Session Commands > Session Commands by Function).

▶ **Note** Remember that the Editor > Command Line Editor command provides a quick way to execute commands used earlier in your session or pasted in from another file. Consult Help for further information. ■

Recall that the History folder records the Session command equivalent to every menu command you have issued. To see this:

■ Click the **Show History** toolbar button ▣

The History folder contains the Session commands shown in Figure 12-4, but without the Minitab prompts.

In this section you will use these commands in the History folder to create an Exec macro to perform a chi-square goodness-of-fit test. You will create the

Exec in the Notepad text editor that comes with Microsoft Windows and Minitab. Notepad was introduced in Tutorial 2. You can find this application on the Tools menu.

- Choose **Tools > Notepad**

  Notepad opens and displays a blank page.

▶ **Note** You can use any word processor to create an Exec macro file and then save it as a text file. However, Notepad is ideal for this purpose because it is a basic text editor that you can employ to create simple documents. It can also be opened without leaving Minitab. ▪

You can type the relevant Session commands into Notepad. However, an easier approach is to copy and paste them from Minitab's History folder:

- Click the **MINITAB** button at the bottom of your screen to return to Minitab
- Highlight the **Session commands in the History folder,** shown in Figure 12-4, and click the **Copy** toolbar button 🖻
- Click the **Untitled - Notepad** button at the bottom of your screen to return to Notepad
- Choose **Edit > Paste**

  A copy of the Minitab Session commands is now in the Notepad. (If your commands don't match those in Figure 12-4, you should edit them in Notepad to match Figure 12-4.)

  You will create the macro in three steps. First, change the commands so that the ChiSq, CumProb, and PValue variables will be placed in C3, C4, and C5, respectively, by the Exec Macro:

- Highlight the **C9** in the first line and type **C3**
- Highlight the **c10** in the sixth line and type **C4**
- Highlight the **C11** in the ninth line and type **C5**
- Delete the fifth line containing the word info

  Next, annotate the Exec macro so that the user knows its restrictions and exactly what the Exec macro expects to find in various columns. To add an annotation to an Exec macro, use the note command. Minitab ignores any information following the word "note", except to print the information when you run the macro.

- Press Ctrl+Home to place the cursor in the upper-left corner of the Notepad file
- Type **note Chi-Square Goodness-of-Fit Procedure** and press ↵
- Type **note Procedure assumes observed and expected counts are in** and press ↵
- Type **note columns 1 and 2 and are named Observed and Expected** and press ↵

  Finally, specify the end of your Exec macro instructions:

- Press Ctrl+End to place the cursor at the beginning of the line following the last line of text of the Notepad file
- Type **end**

  The Notepad window is shown in Figure 12-5.

FIGURE 12-5

Notepad with chi-square goodness-of-fit test Exec macro

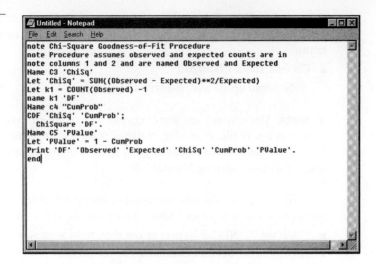

```
Untitled - Notepad
File  Edit  Search  Help
note Chi-Square Goodness-of-Fit Procedure
note Procedure assumes observed and expected counts are in
note columns 1 and 2 and are named Observed and Expected
Name C3 'ChiSq'
Let 'ChiSq' = SUM((Observed - Expected)**2/Expected)
Let k1 = COUNT(Observed) -1
name k1 'DF'
Name c4 "CumProb"
CDF 'ChiSq' 'CumProb';
  ChiSquare 'DF'.
Name C5 'PValue'
Let 'PValue' = 1 - CumProb
Print 'DF' 'Observed' 'Expected' 'ChiSq' 'CumProb' 'PValue'.
end
```

▶ **Note** You can also add an annotation to an Exec macro by replacing note with #. Information after the # symbol is not printed when you run an Exec macro. This is particularly helpful if you want to remind yourself of the function of commands in the macro, but you don't want these reminders printed when you run the macro. For instance, the notation "# This adds the squared (O – E) differences" is of no interest to a user of your Exec, but is a useful reminder when the creator of the Exec prints out the contents of the file. ■

Next, verify that each line is correct, then save the Exec macro in a file with the name *Chisqgof.mtb*. Minitab recognizes the *mtb* file extension as a text file containing Minitab commands.

- Choose **File > Save As**
- Click the "Save as type" **drop-down list arrow**, click on **All Files**, and, in the location where you are saving your work, save the Exec macro as a file named *Chisqgof.mtb*
- Choose **File > Exit**

  Now return to the Current Data window of Minitab:

- Click the **MINITAB** button at the bottom of your screen to return to Minitab
- Click the **Current Data Window** toolbar button ▦

  Now, try running the Exec macro using the observed and expected music genre counts. The macro is expecting these to be in C1 and C2 and will place the results in C3, C4, and C5. So, to avoid disturbing the current worksheet, place the contents of C7 Observed and C8 Expected, into a new worksheet:

- Highlight **the contents of C7 Observed and C8 Expected, including the column names**
- Click the **Copy** toolbar button ▣
- Choose **File > New**
- Click **OK**

- Verify that the C1 column name cell is highlighted and click the **Paste** toolbar button 📋

The new worksheet should look like as shown in Figure 12-6 with the observed counts in C1 Observed and the expected counts in C2 Expected.

FIGURE 12-6

The new worksheet with the observed and expected counts

You are now ready to run the Exec macro.

- Choose **File > Other Files > Run an Exec**

The first Run an Exec dialog box is as shown in Figure 12-7.

FIGURE 12-7

The first Run an Exec dialog box

The dialog box has a default of 1 time to execute the macro. Because this is the number of times you want to run your Exec macro, move to select the file containing your Exec macro:

- Click **Select File**
- In the second Run an Exec dialog box, navigate to the location where you are saving your work to see the file *Chisqgof.mtb* (if it is not in the window already)
- Double-click on the file ***Chisqgof.mtb***

Opening the file causes the macro to run. The Session window contains the results shown in Figure 12-8.

```
MTB > Execute "A:\Chisqgof.mtb" 1.
Executing from file: A:\Chisqgof.mtb
Chi-Square Goodness-of-Fit Procedure
Procedure assumes observed and expected counts are in
columns 1 and 2 and are named Observed and Expected
```

Data Display

```
DF   8.00000
```

Row	Observed	Expected	ChiSq	CumProb	PValue
1	295	271	11.3119	0.815354	0.184646
2	137	134			
3	110	126			
4	111	118			
5	125	116			
6	80	74			
7	41	38			
8	23	35			
9	78	88			

The output is identical to the output produced in Figure 12-3, except for two lines explaining that this is a macro and the three lines of notes that were included in the macro. Minitab has executed each command you entered in your Exec macro. For example, it reports the same chi-square statistic and p-value as before.

Your results should be the same. If they aren't, re-enter the commands in Notepad, verifying each line as you type it, and then save the file again. When Notepad asks whether you want to replace the existing *Chisqgof.mtb*, click Yes. Notepad erases your previous version and saves the new version using this name. Run the Exec macro and make sure that it works properly before continuing to the next section.

You now have a chi-square goodness-of-fit test Exec macro; it is essentially a new Minitab command. Anytime you need to perform a chi-square goodness-of-fit test you can use this Exec macro—as long as columns 1 and 2 contain the observed and expected values and are named Observed and Expected.

At this point, save your project.

■ Click the **Save Project** toolbar button

12.3 The Chi-Square Test for Independence for Two Qualitative Variables in a Contingency Table

CASE STUDY | **HUMAN RESOURCES— EMPLOYMENT STATISTICS**

Employed in a school's human resource department, you are presented with a five-by-four contingency table by your supervisor. It contains a summary of the school's 788 employees classified by job status (Administration, Faculty, Security, Service, and Staff) and ethnicity (Asian, Black, Hispanic, and White). She asks you to perform a test to determine if these two classifications are independent of each other. If independence were rejected, it would suggest that the distribution of ethnicity differed by position. You may assume that these summarized data come from a random sample over time.

You'll be working with the following contingency table:

Job Status	Ethnicity				Total
	Asian	Black	Hispanic	White	
Admin	8	12	3	280	303
Faculty	17	4	1	230	252
Security	0	2	0	16	18
Service	0	9	31	54	94
Staff	7	5	1	108	121
Total	32	32	36	688	788

Before analyzing these data, you have to place the values in the 20 contingency table cells in a new worksheet. You do not enter the Ethnicity row, the Job Status column, the Total column, and the four column totals present in the last row. The row position corresponds to the five job positions. Minitab will compute the totals based upon the 20 contingency cell counts that you entered.

First, create a new worksheet:

- Choose **File > New**
- Click **OK**
- Name C1 **Asian** and enter **8, 17, 0, 0, 7** in this column's first five cells
- Name C2 **Black** and enter **12, 4, 2, 9, 5** in this column's first five cells
- Name C3 **Hispanic** and enter **3, 1, 0, 31, 1** in this column's first five cells
- Name C4 **White** and enter **280, 230, 16, 54, 108** in this column's first five cells

The Data window contains the four columns that you entered as shown in Figure 12-9.

FIGURE 12-9

The contingency table counts entered into C1–C4

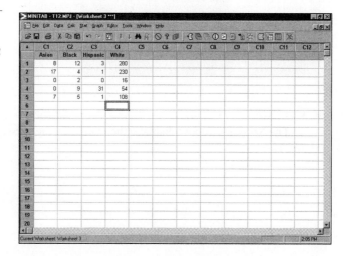

To test for the independence of the job status and ethnicity variables, based upon these summarized data, first formulate the hypotheses:

Null Hypothesis, H_0: The qualitative variables, job status and ethnicity, are independent.

Alternative Hypothesis, H_1: The qualitative variables, job status and ethnicity, are dependent.

As with the chi-square goodness-of-fit test, the *chi-square test for independence* involves comparing the observed counts (in C1–C4) with the expected counts computed assuming the null hypothesis is true. For each cell the expected count is simply the product of the row total count and the column total count divided by the grand total count. The test statistic is obtained by using the formula given in Section 12.1 with the sum being over all cells of the contingency table (20 in this example).

To run the test:

- Choose **Stat > Tables > Chi-Square Test (Table in Worksheet)**
- Double-click **C1 Asian, C2 Black, C3 Hispanic,** and **C4 Whit**e as the "Columns containing the table"

The completed Chi-Square Test (Table in Worksheet) dialog box will look as in Figure 12-10.

FIGURE 12-10

The completed Chi-Square Test (Table in Worksheet) dialog box

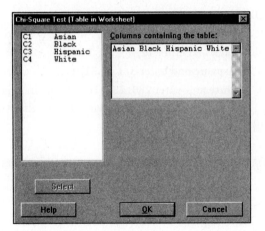

- Click **OK**

The Session window displays the results of the chi-square test, as shown in Figure 12-11.

```
MTB > ChiSquare 'Asian' 'Black' 'Hispanic' 'White'.
```

Chi-Square Test: Asian, Black, Hispanic, White

```
Expected counts are printed below observed counts
Chi-Square contributions are printed below expected counts

      Asian  Black  Hispanic   White  Total
  1       8     12         3     280    303
      12.30  12.30     13.84  264.55
      1.506  0.008     8.493   0.903

  2      17      4         1     230    252
      10.23  10.23     11.51  220.02
      4.474  3.797     9.600   0.453

  3       0      2         0      16     18
       0.73   0.73      0.82   15.72
      0.731  2.203     0.822   0.005

  4       0      9        31      54     94
       3.82   3.82      4.29   82.07
      3.817  7.037   166.073   9.601

  5       7      5         1     108    121
       4.91   4.91      5.53  105.64
      0.886  0.002     3.709   0.053

Total    32     32        36     688    788

Chi-Sq = 224.171, DF = 12
WARNING: 3 cells with expected counts less than 1. Chi-Square approximation
         probably invalid.
8 cells with expected counts less than 5.
```

Minitab prints the chi-square statistic (Chi-Sq = 224.171), but it also prints two warning messages at the bottom of the output. First, it warns you that the chi-square approximation is probably invalid because 3 cells have expected cells less than 1 (the values 0.73, 0.73, and 0.82). Thus, it does not print the p-value corresponding to the test statistic. It also reports that 8 cells have expected counts less than 5. You can obtain further information about *small expected cell counts* by examining Minitab's Help.

- Click the **Edit Last Dialog** toolbar button
- Click **Help**
- Click on **see also** and on "Related Topics" **Cells with small expected frequencies**
  The Help window appears as shown in Figure 12-12.

FIGURE 12-12

The Cells with Small
Expected Frequencies Help
window

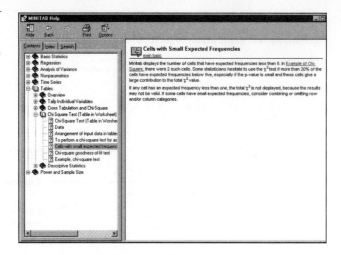

Minitab's Help tells you that you should not use the chi-square test for independence if any expected frequency (count) is less than 1. It also points out the common advice against using this test if more than 20% of the cells have expected frequencies below 5. (These are the only distributional assumptions underlying the chi-square test for independence required for the test to be valid.) The last sentence in Help suggests combining or omitting row or column categories as a way of dealing with small expected frequencies.

- Click the **Close** button ![X]
- Click **Cancel**

A possible solution to this problem is to combine the Asian, Black, and Hispanic columns into a new NonWhite column.

- Click the **Current Data Window** toolbar button ![icon]
- Choose **Calc > Row Statistics**
- Verify that the default "Statistic" is Sum
- Double-click **C1 Asian**, **C2 Black**, and **C3 Hispanic** as the "Input variables"
- Type **NonWhite** as the "Store result in" column
- Click **OK**

Now C5 NonWhite contains the sum of the non-white employees for each job status category (23, 22, 2, 40, and 13).

Ethnicity is now defined as either White or Non-White. Now, run the chi-square test with C4 and C5.

- Choose **Stat > Tables > Chi-Square Test (Table in Worksheet)**
- Double-click **C5 NonWhite** and **C4 White** as the "Columns containing the table"
- Click **OK**

The Session window displays the results of the chi-square test, as shown in Figure 12-13.

FIGURE 12-13

Chi-square test without expected cell count problem

```
MTB > ChiSquare  'NonWhite' 'White'.

Chi-Square Test: NonWhite, White

Expected counts are printed below observed counts
Chi-Square contributions are printed below expected counts

          NonWhite   White  Total
    1           23     280    303
             38.45  264.55
             6.209   0.903

    2           22     230    252
             31.98  220.02
             3.114   0.453

    3            2      16     18
              2.28   15.72
             0.035   0.005

    4           40      54     94
             11.93   82.07
            66.057   9.601

    5           13     108    121
             15.36  105.64
    Total      100     688    788

Chi-Sq = 86.791, DF = 4, P-Value = 0.000
1 cells with expected counts less than 5.
```

Minitab does not print a warning for these data even though it does indicate that there is one cell with an expected count less than 5 (2.28 for security employees who are non-white). But because 1 cell out of 10 (or 10% of the cells) is not greater than 20% of the cells, you can feel comfortable in reporting these chi-square analysis results.

Minitab prints the expected counts and individual chi-square contributions below the observed counts for each cell. These individual chi-square contributions are summed to obtain the chi-square statistic, 86.791.

From these statistics, you can see which cells contribute the most to the chi-square total. Note that fewer than expected non-whites are employed as administrators and faculty, while more than expected non-whites are employed as service personnel. In addition, fewer than expected whites are employed as service personnel.

The degrees of freedom of the chi-square statistic (DF = 4) and the test's p-value (P-Value = 0.000) appear after the test statistic. DF is equal to the product of the table's number of rows minus 1, multiplied by the table's number of columns minus 1, in this case, $(4-1)*(5-1) = 12$. Based upon this small p-value (less than .0005), you would reject H_0 for any reasonable level of significance. You can inform your supervisor that the data suggest that job status and ethnicity are not independent. You also mention that a major contribution of the chi-square statistic comes from those employees in the service job status category.

At this point, save your project.

■ Click the **Save Project** toolbar button 🖫

The Chi-Square Test for Independence for Two Qualitative Variables Using Raw Data

You ask your supervisor if you can examine the actual raw data from which the table of summarized data was constructed. Using the raw data, you can perform additional analyses, such as easily calculating contingency table percentages with the Stat > Tables > Cross Tabulation and Chi-Square command. You can also compute the chi-square statistic from these data with this command.

◆ **Note** Even though you can combine a chi-square test of independence with combinations of row, column, and total percentage distributions, for clarity and ease of interpretation, it is usually better to separate the two analyses. ■

Your supervisor tells you that the data are stored in the file *EmployeeInfo.mtw*.

- Open the worksheet *EmployeeInfo.mtw*

There are two text variables in the Data window, C1-T JobStatus and C2-T Ethnicity. Before performing the chi-square test, you have to combine the Asian, Black, and Hispanic categories into a non-white category to avoid the problem of cells with small expected counts that you encountered in the previous section. To make this change:

- Choose **Data > Code > Text to Text**
- Double-click **C2 Ethnicity** as the "Code data from columns" variable
- Type **Ethnicity2** as the "Into columns" variable
- Type **Asian** as the "Original values (eg, red "light blue")" text value, press ⒯ₐᵦ, type **NonWhite** as the "New" text value (be sure to match the case of the values), and press ⒯ₐᵦ
- Type **Black** as the "Original values (eg, red "light blue")" text value, press ⒯ₐᵦ, type **NonWhite** as the "New" text value, and press ⒯ₐᵦ
- Type **Hispanic** as the "Original values (eg, red "light blue")" text value, press ⒯ₐᵦ, and type **NonWhite** as the "New" text value

The completed Code - Text to Text dialog box should resemble Figure 12-14.

FIGURE 12-14

*The completed
Code - Text to Text
dialog box to create
the Ethnicity2 variable*

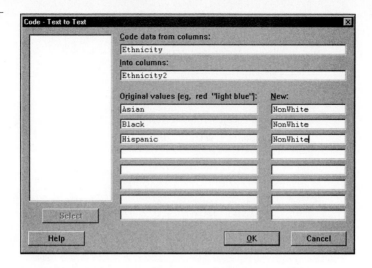

- Click **OK**

    You are ready to create the contingency table from the raw data in C1-T Job Status and C3-T Ethnicity2 and to test whether these variables are independent:

- Choose **Stat > Tables > Cross Tabulation and Chi-Square**
- Double-click **C1 JobStatus** as the "Categorical variables for rows" column and **C3 Ethnicity2** as the "Categorical variables for columns" column
- Click **Chi-Square**
- Click the **Chi-Square analysis**, **Expected cell counts**, and **Each cell's contribution to the Chi-Square statistic** check boxes

    The completed Cross Tabulation - Chi-Square dialog box should resemble that in Figure 12-15.

FIGURE 12-15

*The completed Cross
Tabulation - Chi-Square
dialog box*

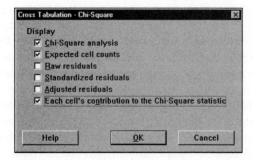

- Click **OK** twice

    Minitab displays the contingency table and chi-square test results in the Session window, as shown in Figure 12-16.

FIGURE 12-16

*Cross-tabulation
contingency table
and chi-square test*

```
MTB > XTABS 'JobStatus' 'Ethnicity2';
SUBC>    Layout 1 1;
SUBC>    Counts;
SUBC>    DMissing 'JobStatus' 'Ethnicity2';
SUBC>    ChiSquare;
SUBC>    Expected;
SUBC>    XResiduals.
```

Tabulated statistics: JobStatus, Ethnicity2

```
Rows: JobStatus   Columns: Ethnicity2

           NonWhite    White     All

Admin            23      280     303
              38.45   264.55  303.00
             6.2093   0.9025       *

Faculty          22      230     252
              31.98   220.02  252.00
             3.1143   0.4527       *

Security          2       16      18
               2.28    15.72   18.00
             0.0354   0.0051       *

Service          40       54      94
              11.93    82.07   94.00
            66.0566   9.6012       *

Staff            13      108     121
              15.36   105.64  121.00
             0.3613   0.0525       *

All             100      688     788
             100.00   688.00  788.00
                  *        *       *

Cell Contents:       Count
                     Expected count
                     Contribution to Chi-square


Pearson Chi-Square = 86.791, DF = 4, P-Value = 0.000
Likelihood Ratio Chi-Square = 64.155, DF = 4, P-Value = 0.000

* NOTE * 1 cells with expected counts less than 5
```

Minitab now displays the count, expected count, and contribution to chi-square for each cell of the contingency table, similar to the output in Figure 12-13. The Pearson chi-square statistic (86.791), the DF for this test (4), and the p-value (0.000) appear after the contingency table. As before, on the basis of the p-value, you reject H_0 because it is less than any reasonable level of significance. Your data suggest that job status and non-white/white ethnicity are not independent. (The Cross Tabulation and Chi-Square Test also contains the likelihood ratio chi-square statistic, along with its DF and p-value. It leads to a similar statistical decision. More information about this statistic is available in Minitab's Help.)

By looking at the third entry in each cell in Figure 12-16, you can see the cells' contribution to the Chi-Square value. Notice that the contribution (66.0566) made by NonWhite Service employees dwarfs the other contributions and accounts for the large Chi-Square value. The contribution is as large as it is because there are many more NonWhite Service employees (40) than would have been expected (11.93).

▶ **Note** If you want your contingency table to be formatted with ethnicity as the column variable and job status as the row variable, then you need only to switch the C3-T Ethnicity2 and C1-T Job Status entries in the first two "Categorical variables" text boxes of the Cross Tabulation and Chi-Square dialog box. ■

Another method for investigating the reasons for the significant test result is to obtain the row percentages of the contingency table.

- Click the **Edit Last Dialog** toolbar button 🔳
- Click **Chi-Square**
- Clear the Chi-Square analysis, Expected cell counts, and Each cell's contribution to the Chi-Square statistic check boxes and click **OK**
- Click the "Display" **Row percents** check box and click **OK**

Minitab displays the contingency table with row percents in the Session window, as shown in Figure 12-17.

FIGURE 12-17

Cross-tabulation contingency table with row percents

```
MTB > XTABS 'JobStatus' 'Ethnicity2';
SUBC>   Layout 1 1;
SUBC>   Counts;
SUBC>   RowPercents;
SUBC>   DMissing 'JobStatus' 'Ethnicity2'.
```

Tabulated statistics: JobStatus, Ethnicity2

```
Rows: JobStatus   Columns: Ethnicity2

           NonWhite   White     All

Admin            23     280     303
               7.59   92.41  100.00

Faculty          22     230     252
               8.73   91.27  100.00

Security          2      16      18
              11.11   88.89  100.00

Service          40      54      94
              42.55   57.45  100.00

Staff            13     108     121
              10.74   89.26  100.00

All             100     688     788
              12.69   87.31  100.00

Cell Contents:      Count
                    % of Row
```

From this output you note that for all job categories except Service, the percentage of employees who are non-white is small, varying from 7.59% for Admin to 11.11% for Security. However, for the Service category 42.55% are non-white. This substantial difference explains much of the significantly large value for the chi-square statistic.

▶ **Note** As mentioned in Tutorial 4 you can obtain percentages from a contingency table by using the Stat > Tables > Cross Tabulation and Chi-Square command. First, you have to construct for each cell of the contingency table three columns: a categorical variable to designate the row of the cell, a categorical variable to designate the column of the cell, and a numeric variable with the corresponding counts of each cell. Then, after choosing the first two variables, you choose the third variable for the Frequencies column and check the counts and percents you wish to display. ■

- Click the **Save Project** toolbar button 🖫 and exit Minitab.

Congratulations on finishing this tutorial! You performed chi-square goodness-of-fit tests and chi-square tests of independence. In addition, you created an Exec macro to perform a statistical analysis that was not available on Minitab's menu. This tutorial has looked at the relationship between qualitative variables and so there were no assumptions about the variables' parameters. In the next tutorial, on nonparametric methods, you will study other techniques that require limited or no assumptions about the distribution of the variable or variables under investigation.

Minitab Command Summary

This section describes the commands introduced in, or related to, this tutorial. Use Minitab's online Help for a complete explanation of all commands.

Minitab Menu Commands

Menu	Command	Description
File ➤	Other Files	
	➤ Run an Exec	Executes the commands in an Exec file a user-specified number of times.
Edit ➤	Command Line Editor	Provides a quick way to execute commands used earlier in your session copied and pasted from another file.
Data ➤	Code	
	➤ Text to Text	Creates a new column whose text values are based on the text values in another column.
Stat ➤	Tables	
	➤ Cross Tabulation and Chi-Square	Constructs contingency tables using raw data consisting of numbers or text. These tables may contain counts, row percentages, column percentages, and total percentages. It also performs chi-square test analysis from raw data.
	➤ Chi-Square Test (Table in Worksheet)	Performs chi-square test analysis when the worksheet contains the contingency table cell counts (summarized data).

Minitab Session Commands

Session Command	Description
end	Ends the storage of Minitab commands in an Exec macro.
name	Provides a name for a constant or column.
note	Displays messages to the user on the screen during the execution of an Exec macro.
#	Allows for internal documentation to an Exec macro.

Review and Practice

Matching Problems

Match the following terms to their definitions by placing the correct letter next to the term it describes.

_____ DF

_____ Expected count

_____ Observed count

_____ Chi-Square Test (Table in Worksheet)

_____ Run an Exec

_____ Less than 1

_____ Cross Tabulation and Chi-Square

_____ Less than 5

_____ Number of categories minus 1

_____ (number of rows minus 1)*(number of columns – 1)

a. A command on the Tables menu that generates the Chi-Square Test of independence for two qualitative variables

b. The number of sample items with a given characteristic

c. A statistic presented by default in the Chi-Square Test (Table in Worksheet) output

d. Expected cell counts for which you should not perform a chi-square test

e. A command on the Tables menu that generates the chi-square test of independence from summarized data

f. Minitab notation for the degrees of freedom of a chi-square statistic

g. Minitab command to invoke a macro

h. Expected cell counts for which you may not perform a chi-square test

i. Degrees of freedom for a chi-square test for independence

j. Degrees of freedom for a chi-square goodness-of-fit test

True/False Problems

Mark the following statements with a *T* or an *F*.

a. ____ You use the Fit command to perform chi-square goodness-of-fit tests in Minitab.

b. ____ When computing p-values for the chi-square goodness-of-fit test Exec macro, you must always store the degrees of freedom as a constant.

c. ____ To combine columns prior to using the Chi-Square Test (Table in Worksheet) command, use the Minitab Code command.

d. ____ Row percents, Column percents, and Total percents are Minitab check boxes available on the Cross Tabulation and Chi-Square dialog box that produce percentage summaries.

e. ____ You can use text data columns with the Cross Tabulation and Chi-Square command but not the Tally Individual Variables command.

f. ____ When you are using the Tally Individual Variables command, you can obtain each cell's chi-square value and each cell's standardized value by clicking the appropriate options.

g. ____ Minitab always provides the chi-square's p-value when you perform a chi-square test of independence.

h. ____ When you use Notepad to create files, it saves them in text format.

i. ____ In addition to the chi-square statistic, Minitab also prints the contribution of each cell when you choose the Chi-Square Test (Table in Worksheet) command.

j. ____ The Minitab History folder is useful in constructing an Exec macro.

Practice Problems

The practice problems instruct you to save your worksheets with a filename prefixed by P, followed by the tutorial number. In this tutorial, for example, use *P12* as the prefix. If necessary, use Help or Appendix A Data Sets, to review the contents of the data sets referred to in the following problems. Interpretations should use the language of the subject matter of the question. If you are using the Student version, be sure to close worksheets when you have completed a problem.

1. Open *Lotto.mtw*. Many states have lotto games. In a certain state, you pay $1 and choose six numbers between 1 and 47. If your numbers match those selected, you win. Each Friday, the state publishes a summary of how many times each number has been drawn in all of the drawings. Assuming the drawings are random, you would expect to see each number the same number of times. *Lotto.mtw* contains the state lotto information. Rows 1 through 47 contain the frequency with which that number has been drawn over the past three years of a particular state's biweekly lotto drawing. Determine if there is any indication that the drawings are not random.

2. Create an Exec macro that prints out the mean, standard deviation, variance, and mean absolute deviation (MAD) for sample data in C1. Document the macro by using the note Session command and by naming the variable and constants.

3. Create an Exec macro that will compute and print out the sample size necessary to estimate an unknown population mean, μ, given the desired level of confidence, the desired margin of error, and the value for the standard deviation. Assume that these three values are stored in the first three rows of C1. You will find a discussion of using Minitab to make this computation in Section 7.3. Document the macro by using the note Session command and by naming the variable and constants.

4. Open *Prof.mtw*. Assume that these data can be considered a random sample of students at this school.

   a. Code the instructor's rating variable with these intervals: 1 to 2, from 2 to 2.5, from 2.5 to 3, from 3 to 3.5, and from 3.5 to 4. You might code the ratings in the following order: (3.5:4) code to 5, (3:3.5) code to 4, (2.5:3) code to 3, (2:2.5) code to 2, and (1:2) code to 1. The order is important because Minitab codes sequentially. This ensures that the boundary points are coded to the higher value. What are the counts of each interval?

   b. Analyze the data to determine whether the level of the course and the instructor's rating are independent.

5. Open *Carphone.mtw*. Determine whether the presence of a car phone and the occurrence of an accident are independent. Carefully explain your conclusion by referencing appropriate row and/or column percents. What assumption is implicit in this analysis?

6. Open *Stores2.mtw*. Assume that these summarized data were obtained from a random sample of shoppers.

   a. For each store, compute the percentage distribution of the five income categories.

   b. Determine whether store and salary are independent when there are five income categories.

   c. For each store, compute the percentage distribution when there are three income categories: Under35K, 35KTo49.9K, and 50KPlus. How are they related to the percentages for each of these income categories for each store? In part a?

d. Determine whether store and salary are independent when there are three income categories: Under35K, 35KTo50K, and 50K&Over.

e. Based upon the p-values from parts b and d, which contingency table produced the most significant result? Explain briefly.

7. Open *EMail.mtw*. This data set begins and ends on a Saturday.

   a. Construct a new variable that represents the day of the week (Sunday, Monday, Tuesday, Wednesday, Thursday, Friday, and Saturday). For each day of the week how many days were recorded?

   b. Determine if the incoming e-mails were uniformly distributed over the seven days. Does the result surprise you?

   c. Determine if the outgoing e-mails were uniformly distributed over the seven days. Does the result surprise you?

   d. Determine if the incoming e-mails were uniformly distributed over the five work days (Monday through Friday). Does the result surprise you?

   e. Determine if the outgoing e-mails were uniformly distributed over the five work days (Monday through Friday). Does the result surprise you?

   f. What assumption is implicit in this analysis?

8. Open *PulseA.mtw*. Assume that these 92 students can be considered a random sample of 18- to 21-year-olds in college.

   a. Construct a contingency table with Gender as the column variable and Smokes as the row variable. Include column percents. What percentage of each gender smokes?

   b. Perform a chi-square test of independence for these two variables. Comment on your answer.

   c. Type the four cell counts into two new columns and use the Stat > Tables > Chi-Square Test (Table in Worksheet) command to check if you get the same result as in part b. In what ways does the output in parts b and c differ?

9. Open *Infants.mtw*. Analyze the data to determine whether smoking status and ethnicity are independent. Note that you will have to combine categories to perform this analysis. Provide an interpretation of your analysis. You may assume that these data can be considered a random sample of mothers.

10. Open *MLBGameCost.mtw*. You may assume this is random sample over time.

   a. Create a new text variable that indicates whether a team is a member of the American or National League. (You may need to obtain this classification from the Internet or your local library.)

   b. Create a new variable that indicates whether a team's Fan Cost Index (FCI) is above or below the median FCI.

   c. Perform a chi-square test of independence for these two variables. Comment on your answer.

On Your Own

For a class assignment three teams randomly sample undergraduates to determine if they favor an increased activities fee. Each decides to sample a different number of undergraduates. Here are the results of their random sampling:

Team 1

Favor?	Gender Male	Female
Yes	60	90
No	75	75

Team 2

Favor?	Gender Male	Female
Yes	80	120
No	100	100

Team 3

Favor?	Gender Male	Female
Yes	160	240
No	200	200

For each team determine if favoring the increased activities fee is independent of gender at a significance level of .10. For each team determine if favoring the increased activities fee is independent of gender at a significance level of .05. For each team determine if favoring the increased activities fee is independent of gender at a significance level of .01. For each team calculate the row percentages of its contingency table. Describe the similarities and/or differences between the three sets of row percentages. Are you concerned with these row percentages and your hypothesis tests? Explain briefly.

Answers to Matching Problems

(f) DF
(c) Expected count
(b) Observed count
(e) Chi-Square Test (Table in Worksheet)
(g) Run an Exec
(d) Less than 1
(a) Cross Tabulation and Chi-Square
(h) Less than 5
(j) Number of categories minus 1
(i) (number of rows minus 1)*(number of columns minus 1)

Answers to True/False Problems

(a) F, (b) T, (c) F, (d) T, (e) F, (f) F, (g) F, (h) T, (i) T, (j) T

13

Analyzing Data with Nonparametric Methods

Until now, the techniques you have used to make inferences about parameters have assumed random samples drawn from populations that were approximately normally distributed. In this tutorial, you check data for randomness and analyze data from populations that may not be normally distributed. Techniques that require less stringent assumptions about the nature of the population probability distributions are called *nonparametric* statistical methods.

OBJECTIVES

In this tutorial, you learn how to:

- Check a sampling process for randomness
- Compare the median of a population to a hypothesized value
- Estimate the median of a population
- Compare the medians of two populations
- Compare the medians of more than two populations

13.1 The Runs Test for Randomness

CASE STUDY	METEOROLOGY— SNOWFALL

Since the late 1880s, the U.S. government has collected snowfall measurements. Professor Mandel asks your meteorology class to determine whether the yearly amount of snowfall between 1892 and 2000 for Boston followed a random pattern. The presence of a non-random pattern would be of interest to many people, such as municipal authorities who allocate money for snow removal. The data are stored in *YearlySnow.mtw*.

Begin by opening Minitab:

- Open **Minitab** and maximize the Session window
- If necessary, **Enable Commands** and clear Output Editable from the Editor menu
- Click the **Save Project** toolbar button 🖫 and, in the location where you are saving your work, save this project as *T13.mpj*
- Open the worksheet *YearlySnow.mtw*

C1 Year lists the year of each measurement, and C2 Snowfall shows the amount of snowfall in inches. (C3 Rain is the amount of water, in inches, that would be equivalent to the actual amount of snowfall.)

It is difficult to determine by scrolling down to look at the data, whether the sequence of 109 snowfall measurements in C1 is random. Because these data vary over time, construct a time series plot to get an idea of how snowfall varies:

- Choose **Graph > Time Series Plot** and double-click *Simple*
- Double-click **C2 Snowfall** as the "Series" column
- Click **Time/Scale** and type **1892** as the "Index" for "All Start Values"

This last step will ensure that the graph's horizontal axis will begin with 1892, the first year of the series. Without this step, the horizontal axis would begin with the default, 1. The completed Time Series Plot - Time/Scale dialog box should look as shown in Figure 13-1.

FIGURE 13-1

The completed Time Series Plot - Time/Scale dialog box

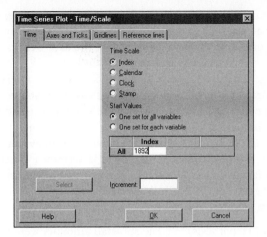

■ Click **OK** twice

The plot, with Minitab's default title, is shown in Figure 13-2.

FIGURE 13-2

Time series plot of Snowfall

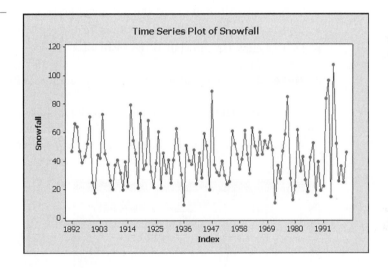

By default, each snowfall value is connected to the next one with a straight line. The plot indicates a lot of variation but no apparent trend in the amount of snowfall with a few possible outlying (that is, unusual) values. Even with the plot, however, it is difficult to judge whether the pattern of snowfall over the 109 years is a random pattern (a pattern in which what happens in one year does not affect what happens in subsequent years).

▶ **Note** An alternative to a Time Series Plot of a variable is a scatterplot of the variable against a time variable. ■

A *runs test* can be used to decide whether the order of a sequence of numbers is random, that is, determined by chance. In some of the tests considered in these tutorials, you have to assume that the data were randomly ordered. The runs test can be used to verify that assumption.

In general, a *run* is a sequence of consecutive observations that share the same characteristic. In this context that characteristic is either falling above a specified constant value or falling on or below the specified constant. The default constant in Minitab is the sample mean. Thus, a sequence of four consecutive snowfall amounts above the mean followed by an amount below the mean would constitute a run of size 4. The runs test determines whether the snowfall data have too few or too many runs compared to what you would expect to get if snowfall was "generated" by a truly random process.

First set up the hypotheses:

Null Hypothesis, H_0: The snowfall amounts were "generated" by a random process.

Alternative Hypothesis, H_1: The snowfall amounts were "generated" by a process that was not random.

Now, use Minitab to test the null hypothesis.

- Choose **Stat > Nonparametrics > Runs Test**
- Double-click **C2 Snowfall** as the "Variables"

▶ **Note** Minitab only allows numeric variables for its Runs Tests. ■

The completed Runs Test dialog box should look as shown in Figure 13-3. Note that Minitab gives you the option of specifying some constant other than the mean of the observations as the reference value for determining runs.

FIGURE 13-3

The completed Runs Test dialog box

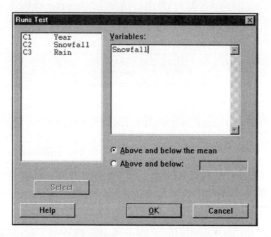

- Click **OK**

The output appears in the Session window, as shown in Figure 13-4.

FIGURE 13-4

Runs test for Snowfall

```
MTB > Runs 'Snowfall'.
```

Runs Test: Snowfall

```
Runs test for Snowfall

Runs above and below K = 42.2229

The observed number of runs = 49
The expected number of runs = 55.2752
51 observations above K, 58 below
P-value = 0.225
```

Minitab used the sample mean, $K = 44.2229$, as the constant. In all, there were 49 observed runs. (It is time consuming but you can verify this from the raw data.) When there are too many or too few runs, the sample is considered to be in non-random order.

Minitab calculates that the expected number of runs is 55.2752, assuming a random process, and displays the number of observations above the constant (51) and below the constant (58). The observed number of runs is close enough to the expected number of runs to produce a p-value for this test of 0.225, so you cannot reject H_0 at the 0.05 level of significance.

You can report to your meteorology class that you cannot reject the hypothesis that the pattern of the amount of yearly snowfall is random. Hence, city planners, for example, have little evidence upon which to base a request for more or less money for snow removal in future years.

▶ **Note** Tutorial 14 discusses additional ways to analyze a time series. ∎

■ Click the **Save Project** toolbar button 🖫

The runs test is a nonparametric test in that it makes no assumptions about the distribution of the data. There is no widely-used comparable parametric procedure. Many nonparametric procedures, however, were developed precisely for use in situations where a comparable parametric procedure might be invalid. The one-, two-, and multi-sample procedures covered in Tutorials 7, 8, and 9, respectively, are all parametric procedures. The t-tests in tutorials 7 and 8, for example, assumed that the data follow a normal distribution, as did the F-test in a one-way ANOVA in Tutorial 9. In the next six sections you will examine nonparametric procedures that may be used if these normality assumptions are violated. For each nonparametric procedure you will use the same case study used to illustrate the comparable parametric procedure.

13.2 | Testing a Hypothesis About the Population Median Using the Sign Test

CASE STUDY	CASE STUDY: HEALTH CARE— CEREAL AND CHOLESTEROL (CONTINUED)

Recently 14 male patients suffering from high levels of cholesterol took part in an experiment to examine whether a diet that included oat bran would reduce cholesterol levels. Each was randomly assigned to a diet that included either corn flakes or oat bran. After two weeks, their low-density lipoprotein (LDL) cholesterol levels were recorded. Each man then repeated this process with the other cereal. The 14 pairs of LDL results are recorded in a data file called *Chol.mtw*. As a rookie analyst at the clinic where the research was conducted, you have been asked to analyze these data. (You may assume these men are a random sample of all men suffering from high levels of cholesterol.)

In Tutorial 8 you performed a one-tailed paired t-test on the 14 pairs of LDL results that suggested the use of oat bran significantly reduces the mean LDL cholesterol level. However, because you are considering writing up the results of the experiment for publication, you decide to bolster your analysis with two nonparametric tests (the sign test and the Wilcoxon test).

The *sign test* is the simplest nonparametric analog to the t-test, because it depends solely on the sign of the difference between an observation, the hypothesized median, and the assumption of a random sample. You'll be working with the 14 pairs of LDL results that are in the data file *Chol.mtw*.

- Open the worksheet **Chol.mtw**

    In this data set, C1 CornFlke and C2 OatBran contain the LDL level for each of the 14 patients in the study for each cereal, respectively.

    First, compute the 14 (CornFlke − OatBran) differences:

- Choose **Calc > Calculator**
- Type **Diff** as the "Store result in variable"
- Enter, by clicking or typing, **CornFlke − OatBran**, as the "Expression"
- Click **OK**

    Minitab stores the 14 differences in C3 Diff.

    Now, form the hypotheses:

    Null Hypothesis, H_0: Population median difference = 0.

    Alternative Hypothesis, H_1: Population median difference > 0.

    To perform the sign test:

- Choose **Stat > Nonparametrics > 1-Sample Sign**
- Double-click **C3 Diff** as the "Variables"
- Click the **Test median** option button

    Because you are testing a median difference of 0, use the default of 0.0.

- Click the "Alternative" **drop-down list arrow** and click on **greater than**

The completed 1-Sample Sign dialog box should look as shown in Figure 13-5.

FIGURE 13-5

The completed 1-Sample Sign dialog box

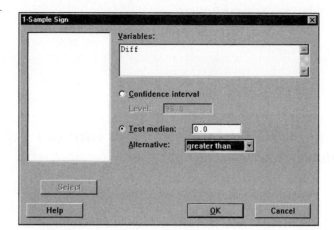

- Click **OK**

The output in the Session window is shown in Figure 13-6.

FIGURE 13-6

Sign test on the 14 differences

```
MTB > Name C3 'Diff'
MTB > Let 'Diff' = 'CornFlke' - 'OatBran'
MTB > STest 0.0 'Diff';
SUBC>    Alternative 1.
```

Sign Test for Median: Diff

Sign test of median = 0.00000 versus > 0.00000

```
        N  Below  Equal  Above       P  Median
Diff   14      2      0     12  0.0065  0.3600
```

Of the 14 differences, 2 were below 0 and 12 were above. The p-value for this test is 0.0065, so you can reject the null hypothesis at the .05 or .01 level of significance. The sample median of 0.3600 suggests that the median LDL cholesterol level is higher with a diet of corn flakes than with a diet of oat bran.

The sign test uses the binomial distribution with n = 14 (in this case) and p = 0.5 to compute the p-value, 0.0065. (Binomial probabilities were discussed in Section 6.1.) You can calculate the p-value by computing the (cumulative) probability of two or fewer successes in a binomial experiment, with n = 14 and p = .5:

- Choose **Calc > Probability Distributions > Binomial**
- Verify that the Cumulative probability option button is clicked
- Type **14** as the "Number of trials" and **.5** as the "Probability of success"
- Click the **Input constant** option button
- Type **2** as the "Input constant" and click **OK**

The output, shown in Figure 13-7, confirms that the probability is indeed approximately 0.0065. If H_0 was true, the probability of getting two or fewer differences less than 0 is 0.0064697.

FIGURE 13-7

Obtaining the p-value as a binomial probability

```
MTB > CDF 2;
SUBC>   Binomial 14 .5.
```

Cumulative Distribution Function

```
Binomial with n = 14 and p = 0.5

x   P( X <= x )
2    0.0064697
```

13.3 Estimating the Population Median with the 1-Sample Sign Confidence Interval Estimate

You can use Minitab's 1-Sample Sign procedure to find the default 95% two-sided confidence interval for the unknown population median difference in the LDL cholesterol level.

- Choose **Stat > Nonparametrics > 1-Sample Sign**
- Click the **Confidence interval** option button
- Click **OK**

    The resulting confidence interval(s) are shown in Figure 13-8.

FIGURE 13-8

The 1-Sample Sign confidence intervals

```
MTB > SInterval 95.0 'Diff'.
```

Sign CI: Diff

```
Sign confidence interval for median

                             Confidence
                  Achieved    Interval
       N  Median  Confidence  Lower   Upper   Position
Diff  14  0.3600  0.9426      0.1000  0.7200  4
                  0.9500      0.0995  0.7226  NLI
                  0.9871      0.0900  0.7700  3
```

The output provides three confidence intervals. The top interval has achieved a confidence level of 0.9426 (below 0.95) while the bottom one has achieved a confidence level of 0.9871 (above 0.95). These intervals are based upon the 14 differences. For instance, 0.1000 is the fourth smallest difference (in position 4) and 0.7200 is the fourth largest difference—or the eleventh largest (in position 4, counting down). The achieved confidence level for the top interval is $P(4 \le X \le 11) = 1 - 2*P(X \le 3) = 1 - 2*0.0287 = 0.9426$. Here, X is the binomial random variable with n = 14 and p = .5. Similarly, 0.0900 and 0.7700 are, respectively, the third smallest and the third largest differences, and the achieved level of confidence for the bottom interval is $P(3 \le X \le 12) = 0.9871$. A process called *nonlinear interpolation (NLI)* is used to produce the middle confidence interval (0.0995, 0.7226) with the desired level of confidence.

> ◆ **Note** You can use the Data > Sort command to sort the 14 differences. The binomial probability $P(X \le 3) = 0.0287$ can be obtained by using choosing Calc > Probability Distributions > Binomial. ■

These results suggest that you can be 95% confident that the population median difference in the median LDL cholesterol level for the two cereals lies between 0.0995 and 0.7226.

13.4 Testing Hypotheses About the Population Median Using the Wilcoxon Test

Next, you move on to analyze these same cholesterol data using the Wilcoxon test and the confidence interval. The *Wilcoxon test* (which is sometimes called the *Wilcoxon Signed Ranks test*) is based upon the ranks of the differences from the hypothesized median and is more powerful than the sign test. It only assumes a random sample of data. The hypotheses are the same as for the sign test:

Null Hypothesis, H_0: Population median difference = 0.

Alternative Hypothesis, H_1: Population median difference > 0.

To perform the Wilcoxon test:

- Choose **Stat > Nonparametrics > 1-Sample Wilcoxon**
- Double-click **C3 Diff** as the "Variables"
- Click the **Test median** option button
  Because you are testing a median difference of 0, use the default of 0.0.
- Click the "Alternative" **drop-down list arrow** and click on **greater than**
  The completed 1-Sample Wilcoxon dialog box—which looks remarkably similar to the 1-Sample Sign dialog box—should look as shown in Figure 13-9.

FIGURE 13-9

The completed 1-Sample Wilcoxon dialog box

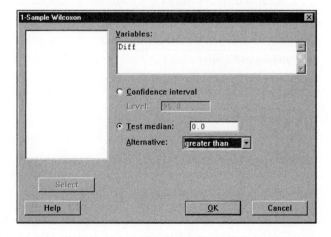

- Click **OK**

The output is shown in Figure 13-10.

FIGURE 13-10

Wilcoxon test for the differences in cholesterol levels

```
MTB > WTest 0.0 'Diff';
SUBC>   Alternative 1.
```

Wilcoxon Signed Rank Test: Diff

```
Test of median = 0.000000 versus median > 0.000000
```

	N	N for Test	Wilcoxon Statistic	P	Estimated Median
Diff	14	14	93.0	0.006	0.3900

The p-value for this test is 0.006, so you would again reject the null hypothesis at the .05 or the .01 level of significance. The estimated median of 0.3900 suggests that the median LDL cholesterol level is higher with a diet of corn flakes than with a diet of oat bran. Recall, however, that the sign test provided the sample median difference as 0.3600. The value 0.3900 is the median of the pairwise averages. (*Pairwise averages*—sometimes called *Walsh averages*—are the averages for each possible pair of values taken from the 14 differences, including the averages of each value with itself.)

13.5 Estimating the Population Median with the Wilcoxon Confidence Interval Estimate

The 1-Sample Wilcoxon procedure can be used to obtain a 95% two-sided confidence interval for the population median difference.

- Click the **Edit Last Dialog** toolbar button 🔲
- Click the **Confidence interval** option button and click **OK**

Minitab displays the 94.8% confidence interval for the population median as (0.130, 0.610), as shown in Figure 13-11.

FIGURE 13-11

The 1-Sample Wilcoxon confidence intervals

```
MTB > WInterval 95.0 'Diff'.
```

Wilcoxon Signed Rank CI: Diff

	N	Estimated Median	Achieved Confidence	Confidence Interval Lower	Upper
Diff	14	0.390	94.8	0.130	0.610

For the Wilcoxon procedure, the level of confidence that Minitab generates won't exactly match your request. The interval is based upon the distribution of signed ranks that, like the binomial distribution, is discrete. In this case, the closest you can get to 95% is 94.8% confidence. You can be 94.8% confident that the population median difference in the median LDL cholesterol level for the two cereals lies between 0.130 and 0.610.

You can feel reassured that all three procedures (the paired-t, sign, and Wilcoxon) suggest the conclusion that the use of oat bran appears to reduce the LDL cholesterol level.

If you worked through the one- and two-sample t-tests in Tutorials 7 and 8, you will recall that for those procedures confidence intervals were provided automatically and the type of interval matched the form of the alternative hypothesis—one-sided intervals with one-tailed tests and two-sided intervals with a two-tailed test. By contrast, in the corresponding nonparametric tests you had the option of selecting either a test or a confidence interval but not both. Moreover, Minitab provides only two-sided intervals. But, suppose you wanted a one-sided confidence interval? Suppose, for instance, in the example above you wanted a 95% upper one-sided confidence interval for the median difference in cholesterol levels. A simple way to compute such an interval is to obtain a 90% two-sided confidence interval and take the lower bound of this interval as the lower bound for a 95% upper one-sided confidence interval. Here is the needed sequence.

- Click the **Edit Last Dialog** toolbar button 🔲
- Click the **Confidence interval** option button
- Type **90** as the "Level" to replace the default value of 95.0 and click **OK**

The output is shown in Figure 13-12.

FIGURE 13-12

90% two-sided Wilcoxon confidence interval to assist in computation of 95% upper one-sided confidence interval

```
MTB > WInterval 90 'Diff'.
```

Wilcoxon Signed Rank CI: Diff

		Estimated	Achieved	Confidence Interval	
	N	Median	Confidence	Lower	Upper
Diff	14	0.390	89.7	0.180	0.580

The achieved confidence level is 89.7% rather than 90%. So, an approximate 95% upper one-sided confidence interval for the median difference in cholesterol level is 0.180 to ∞.

Before going on to the next case study, save your work.

- Click the **Save Project** toolbar button 💾

13.6 | Comparing the Medians of Two Independent Populations

| CASE STUDY | SOCIOLOGY—AGE AT DEATH (CONTINUED) |

A sociology professor, Dr. Ford, is interested in examining the extent to which women living in affluent areas of the United States live longer than women in the country as a whole. Dr. Ford has obtained the age at death for all of the residents of an affluent suburb of Boston, who died in the year 2001. These data are stored in a file called *AgeDeath.mtw*.

In Tutorial 8, you investigated the related question, whether there is a difference in the population mean ages at death between females and males, using a two-independent-sample t-test and concluded that though there was a statistically significant difference based upon the sample, the necessary normality assumption was not satisfied. Now you investigate whether there is a difference in the population median ages at death between women and men, based upon the same data, using a test that does not assume normality.

In Tutorial 8, you performed t-tests for the equality of two population means, assuming near normality. The *Mann-Whitney test* is a nonparametric counterpart to this t-test. It compares two population medians using sample ranks based upon two independent samples. The procedure also computes the corresponding point and confidence interval estimates of the difference between these medians.

The age at death data is in the data file *AgeDeath.mtw*.

- Open the worksheet ***AgeDeath.mtw***

C3-T Gender contains the gender of 150 of the 151 people (recall that the gender of one of the 151 was unknown); C5 Age contains the age at death of the 86 females and 64 males.

For the Mann-Whitney nonparametric procedure Minitab requires that the two samples be in different columns. Hence, you must unstack the age data into two columns: the ages at death for the females and for the males.

- Choose **Data > Unstack Columns**
- Double-click **Age** as the "Unstack the data in" column and **Gender** as the "Using subscripts in" column
- Click the "Store unstacked data" **After last column in use** option button and click **OK**

There are now two new variables in your worksheet: C6 Age_Female and C7 Age_Male.

The appropriate hypotheses for the Mann-Whitney test are:

Null Hypothesis, H_0: The population median age at death for females living in affluent areas and the population median age at death for males living in affluent areas are equal.

Alternative Hypothesis, H_1: The population median age at death for females living in affluent areas and the population median age at death for males living in affluent areas are not equal.

Now, use Minitab to test the null hypothesis.

- Choose **Stat > Nonparametrics > Mann-Whitney**
- Double-click **C6 Age_Female** as the "First Sample" and **C7 Age_Male** as the "Second Sample"

By default the command also produces a 95% confidence interval for the difference between the two population median ages at death. The default alternative hypothesis is "not equal"—this is what you want. The completed Mann-Whitney dialog box looks as shown in Figure 13-13.

FIGURE 13-13

*The completed
Mann-Whitney dialog box*

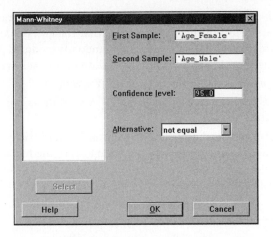

■ Click **OK**

The results of the test and confidence interval appear in the Session window and are shown in Figure 13-14.

FIGURE 13-14

*Mann-Whitney test and
confidence interval for the
age at death data*

```
MTB > Mann-Whitney 95.0 'Age_Female' 'Age_Male';
SUBC>   Alternative 0.
```

Mann-Whitney Test and CI: Age_Female, Age_Male

```
              N   Median
Age_Female   86   87.000
Age_Male     64   83.000


Point estimate for ETA1-ETA2 is 5.000
95.0 Percent CI for ETA1-ETA2 is (3.000,8.001)
W = 7424.0
Test of ETA1 = ETA2 vs ETA1 not = ETA2 is significant at 0.0004
The test is significant at 0.0004 (adjusted for ties)
```

ETA which appears frequently in the output is the seventh letter of the Greek alphabet. The symbol for ETA is sometimes used to represent the population median.

Minitab first lists the sample median for each gender and then displays the point estimate (5 years) for the difference (ETA1 – ETA2) in the population medians. Note that this point estimate is not equal to the difference of the two medians. More information about how Minitab computes this estimate is available in Minitab's Help. The 95% confidence interval for ETA1 – ETA2 ranges from 3.000 years to 8.001 years. Minitab reports a value for the Mann-Whitney test statistic of 7424.0 and a p-value of 0.0004, so you can reject H_0 at the .05 level. The data suggest that there is a significant difference between the median ages at death for affluent men and affluent women, and that the median age at death for women exceeds that for men.

Note that this p-value differs from the one produced by the two-independent-sample t-test; this isn't surprising, because the two tests are not equivalent.

Although the p-values differ, the conclusions are the same.

The Mann-Whitney test assumes that the two populations have the same shape (which need not be normal and randomly sampled data). You can examine the first assumption by constructing a dotplot for each set of ages:

- Choose **Graph > Dotplot > *Multiple Y's, Simple***
- Double-click **C6 Age_Female** and **C7 Age_Male** as the "Graph variables" and click **OK**

The resulting dotplots, with Minitab's default title, are shown in Figure 13-15. Both are skewed to the left, with the ages of the females greater than the ages of the males. This graph is consistent with the assumption that the two populations have the same shape.

FIGURE 13-15

Dotplots for the ages at death for affluent women and affluent men

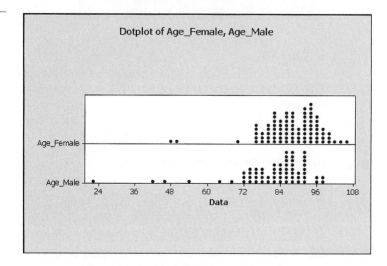

▶ **Note** This is further evidence that multiple dotplots are an effective graphical tool for comparing groups. Another equally effective graph is a multiple boxplot display. ■

- Click the **Save Project** toolbar button 🖫

13.7 Comparing the Medians of K Independent Populations Using the Kruskal-Wallis Test

In Tutorial 9, you used analysis of variance (ANOVA) to compare the means of more than two populations. To perform ANOVA, you had to make the parametric assumption of normality of each population. The *Kruskal-Wallis H-test* is a non-parametric counterpart of the one-way ANOVA and is a generalization of the Mann-Whitney test.

CASE STUDY	CHILD DEVELOPMENT— INFANT ATTENTION SPANS (CONTINUED)

You are a psychology student taking a course in child development. As part of your project for the course, you have been examining how different designs vary in their abilities to capture an infant's attention. You devised an experiment to see whether an infant's attention span varies with the type of design on a mobile. You randomly divided a group of 30 three-month-old infants into five groups of six each. Each group was shown a mobile with one of five multicolored designs: A, B, C, D, or E. Because of time constraints, each infant could be shown only one design. You recorded the median time (in seconds) that the infant spent looking at the design. In a Minitab worksheet, you recorded the times in C1 Times and the designs that each infant saw in C2-T Design. The worksheet is saved in the file *Baby.mtw*.

Here, you will run a Kruskal-Wallis test on these data to test whether there is a difference among the population median attention times for the five designs. You will decide to use this test because the ANOVA is particularly sensitive to departures from normality.

- Open the worksheet **Baby.mtw**

The 30 attention times are in C1 Time. The designs are listed in C2-T Design.

One of the times, 5.6 seconds for mobile E, must be discarded because the observer was not paying attention.

- Type * in row 27 of C1 Time as the value to replace 5.6.

The Kruskal-Wallis procedure requires that the samples be random and independent, with five or more measurements in each (or at least six measurements in one). You must list the data in one "Response" column, and the subscripts identifying the sample in which an observation belongs in a "Factor" column. For your data, the times in C1 are the response values and the designs in C2-T are the factors.

First, formulate the two hypotheses:

Null Hypothesis, H_0: The population median attention times for the five designs are equal.

Alternative Hypothesis, H_1: The population median attention times for the five designs are not equal.

To perform the Kruskal-Wallis test:

- Choose **Stat > Nonparametrics > Kruskal-Wallis**
- Double-click **C1 Time** as the "Response" and **C2 Design** as the "Factor"

The completed Kruskal-Wallis dialog box should look as shown in Figure 13-16.

FIGURE 13-16

The completed Kruskal-Wallis dialog box

- Click **OK**

The output in the Session window is shown in Figure 13-17.

FIGURE 13-17

Kruskal-Wallis test for the attention times

```
MTB > Kruskal-Wallis 'Time' 'Design'.

Kruskal-Wallis Test: Time versus Design

29 cases were used
1 cases contained missing values


Kruskal-Wallis Test on Time

Design    N  Median  Ave Rank      Z
A         6  10.400      19.8   1.56
B         6  11.000      23.8   2.83
C         6   9.000      10.0  -1.62
D         6   7.750       6.3  -2.80
E         5  10.100      15.1   0.03
Overall  29              15.0

H = 16.56   DF = 4   P = 0.002
H = 16.59   DF = 4   P = 0.002  (adjusted for ties)
```

Minitab first reports summarized statistics for each design. Design B has the largest median time (11.000 seconds); Design D has the smallest (7.750 seconds). The value of the Kruskal-Wallis H-statistic is 16.56. The p-value corresponding to 16.56 is 0.002. Because there are ties—two or more observations with the same value—Minitab provides another Kruskal-Wallis H-statistic value adjusted for ties, as well as the corresponding p-value. Based upon this p-value, you reject the null hypothesis that the population median attention times are the same for the five designs.

The Kruskal-Wallis test, similar to the Mann-Whitney, assumes that the populations all have the same shape and randomly sampled data. To check the first assumption, obtain boxplots of the times for the five designs:

■ Choose **Graph > Boxplot > *One Y, With Groups***

■ Double-click **C1 Time** as the "Graphs variables" column and **C2 Design** as the "Categorical variables for grouping (1-4, outermost first)" column

■ Click **OK**

The graph, with Minitab's default title, is shown in Figure 13-18.

FIGURE 13-18

Boxplots for attention times by design

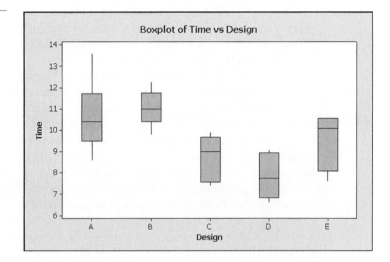

The shapes of the five boxplots are nearly all different, but it is important not to read too much into this. Remember that each boxplot is based upon only five or six values, so the shape can be significantly affected by a single value. You should be concerned, but the evidence against the assumption of equal-shaped populations is not conclusive.

You could use Minitab to perform *Mood's Median test* (Stat > Nonparametric > Mood's Median Test), which tests the same hypotheses as the Kruskal-Wallis test. The former is less sensitive to outliers than the Kruskal-Wallis test, but it is less powerful (less likely to detect important differences) for data from many distributions, including the normal. The Kruskal-Wallis test is more widely used than Mood's Median test.

Minitab can perform nonparametric tests other than those described here. For example, the *Friedman test* (Stat > Nonparametric > Friedman), is a nonparametric alternative to the two-way ANOVA with one observation per cell. This test is appropriate in the context of a randomized block experiment. You will find references to this type of experimental design in most introductory applied statistics textbooks and in Minitab's Help nonparametric references. Use Help to find out about several other nonparametric measures that Minitab will compute—Pairwise Averages, Pairwise Differences, and Pairwise Slopes.

Congratulations on completing this tutorial! You have made use of many of the most popular nonparametric techniques. The next tutorial on time series data builds on the simple example in Section 13.1.

- Click the **Save Project** toolbar button 🖫 and exit Minitab

---

MINITAB *at work*

MEDICAL DIAGNOSTICS

Although most of us associate heart failure with older people, it can affect children and infants when the main artery of the heart, the aorta, becomes obstructed and blood cannot flow freely. Doctors treat these blockages by inserting small balloons into the constricted aorta and inflating them to clear the obstruction.

A pediatrician at a midwestern university hospital used the balloon treatment on 30 children, aged 14 days to 13 years, with obstructed aortas. He monitored their recovery carefully to see what factors would predict whether the operation was a success. One factor the pediatrician used was the measurement of aorta diameter at five points. In a normal aortic valve, the diameters at all five points would be approximately the same and hence the variance in diameter measurements would be close to zero.

The pediatrician used Minitab's Mann-Whitney test to compare diameter variances before and after treatment of two groups of patients: those who eventually required a second balloon treatment and those who did not. The results indicated that the children who would completely recover showed a greater decline in diameter variance from pretreatment levels than those who would require more surgery. This result gave the pediatrician another diagnostic tool to predict whether more treatment would be needed at a later date for a particular patient. Being able to predict patients' future needs enables doctors to better treat them in the present.

Minitab Command Summary

This section describes the commands introduced in, or related to, this tutorial. Use Minitab's online Help for a complete explanation of all commands.

Minitab Menu Commands

Menu	Command	Description
Stat ➤	Nonparametrics	
	➤ 1-Sample Sign	Performs a nonparametric 1-sample sign inference (test or confidence interval) on a median.
	➤ 1-Sample Wilcoxon	Performs a nonparametric Wilcoxon 1-sample inference (test or confidence interval) on a median.
	➤ Mann-Whitney	Performs 2-sample nonparametric inference (test and confidence interval) on two medians.
	➤ Kruskal-Wallis	Performs a nonparametric comparison of two or more medians.
	➤ Mood's Median Test	Performs a nonparametric comparison of two or more medians.
	➤ Friedman	Performs a nonparametric analysis of a randomized block experiment.
	➤ Runs Test	Performs a test to evaluate random order of data.
Graph ➤	Dotplot	
	Multiple Y's, Simple	Produces multiple dotplots on the same graph based upon data in different columns.
	Boxplot	
	One Y, With Groups	Produces multiple boxplots on the same graph based upon data in one column and grouping variable(s) in one or more columns.

Review and Practice

Problems

: following terms to their definitions by placing the correct letter next to the term it describes.

_____ 1	a. A series of Minitab statistical techniques that does not assume that the population(s) possesses any particular distribution
_____ ETA1 – ETA2	b. Minitab notation for the difference between two population medians
_____ Nonparametric	c. A sequence of consecutive observations that share the same characteristic
_____ K	d. Minitab notation for the column in which the data are stored for the Kruskal-Wallis command
_____ Estimated median	e. Minitab notation for the test statistic from the Kruskal-Wallis test
_____ Run	f. Minitab notation for the constant used in the runs test to identify a run
_____ Factor	g. Minitab name for the statistic equal to the middle value of all pairwise averages of data values
_____ Response	h. Minitab dialog box notation for the subscript column used in the Kruskal-Wallis command
_____ Friedman	i. Minitab command that performs a nonparametric analysis of a randomized block experiment
_____ Mann-Whitney	j. Minitab command that is the nonparametric counterpart of the two-independent-sample t-test with independent samples

True/False Problems

Mark the following statements with a T or an F.

a. _____ The Runs Test command requires numeric data.

b. _____ You can always use Minitab's nonparametric commands to obtain confidence intervals that correspond to exact levels of confidence.

c. _____ The Wilcoxon command allows you to perform only two-sided tests.

d. _____ The Runs Test command allows you to obtain a confidence interval for the population median.

e. _____ By default, the Wilcoxon command provides confidence intervals as part of the output.

f. _____ The Kruskal-Wallis command requires the data to be in one column with a subscript identifier in another.

g. _____ The default null hypothesis value for the median in the Wilcoxon command is zero.

h. _____ The Runs Test command uses the sample median as the default value to identify runs.

i. _____ The Mann-Whitney command assumes that the two samples of data are stored in different columns.

j. _____ The Friedman command performs an analysis that is a nonparametric alternative to the two-way ANOVA with one observation per cell.

Practice Problems

The practice problems instruct you to save your worksheets with a filename prefixed by P, followed by the tutorial number. In this tutorial, for example, use *P13* as the prefix. If necessary, use Help or Appendix A Data Sets, to review the contents of the data sets referred to in the following problems. Interpretations should use the language of the subject matter of the question. If you are using the Student version, be sure to close worksheets when you have completed a problem.

1. Open *YearlySnow.mtw*. The time series plot of these data in Figure 13-2 suggest that there are some snowfall amounts that are outliers. If you are testing for randomness in this situation it may be better to use the median as the constant rather than the mean. Perform a runs test for the randomness of the snowfall data by specifying K equal to the median. Does your test result differ much from that in Section 13-1?

2. Open *Election2.mtw*.

   a. Use a runs test to determine if the Democratic percentage of the presidential vote can be viewed as a random process. Use the mean as the constant.

   b. Repeat part a for the Republican percentage of the presidential vote.

   c. Are your conclusions the same in each case? Write a brief account of your conclusions in parts a and b.

3. Open *Salary02.mtw*. The median salary for full professors at this type of institution in 2002 was approximately $76,400.

   a. Use a sign test to determine if this institution's median salary is comparable to other similar institutions with a significance level of .05.

   b. Use Minitab's cumulative binomial probability command to compute the p-value in part a. Briefly explain how you computed the p-value.

   c. Use a 95% sign confidence interval to determine if this institution's median salary is comparable to other similar institutions.

   d. Use a Wilcoxon test to determine if this institution's median salary is comparable to other similar institutions with a significance level of .05.

   e. Use a 95% Wilcoxon confidence interval to determine if this institution's median salary is comparable to other similar institutions.

   f. Comment briefly on the similarities and differences of the results from the sign and the Wilcoxon procedures.

   g. Is it reasonable to assume that these data can be considered a random sample? How does your answer affect your conclusion?

4. Open *AgeDeath.mtw*. Use the 1-Sample Wilcoxon procedure to test whether the median age at death for all women living in an affluent area is significantly less than 90 years.

5. Open *Force.mtw*. Assume that the undergraduate women were randomly selected.

   a. Compute the difference, Force5 – Force45, in C3 and name the difference Diff.

   b. The physical therapist who recorded these data knows that such differences will be considerably greater than zero but wonders if they will exceed 3 pounds. Use the 1-Sample sign test to suggest that the difference will be greater than 3 pounds.

c. Answer part b using the 1-Sample Wilcoxon test.

6. Open *Salary02.mtw*.

   a. For each of the three professorial ranks, perform a Mann-Whitney test for a significant difference between the median male and female salaries for 2002. In each case construct dotplots to examine whether the two distributions of salary have the same shape.

   b. For the college newspaper, write a paragraph summarizing your results in part a. If necessary, your report should indicate that a test was not possible because the assumptions were violated.

7. Open *Rivers.mtw*. Regard the observations as random samples over time.

   a. Determine if the median temperature of the river is different at site 2, which is directly up river from the power plant, than at site 3, which is directly down river from the power plant's discharge from its cooling towers. (You must unstack the Temp column to create new columns that contain site 2 and site 3.)

   b. Construct two dotplots to examine whether or not the two populations have the same shape. Do they? What is the implication if the dotplots have quite different shapes? Explain briefly.

8. Open *Textbooks.mtw*. Regard the 17 general education texts and the 19 business/economics texts as random samples of their respective types of texts.

   a. Obtain a 90% two-sided confidence interval for the difference between the population median number of pages for business/economics texts and the corresponding population median for general education texts.

   b. Interpret your confidence interval in part a for a group of publishing executives.

   c. Obtain boxplots of the number of pages by type of textbook. What do your plots suggest about the assumption that the two populations have the same shape?

   d. The output you obtained in part a included a p-value. What were the implicit hypotheses in this case? What conclusion does your p-value suggest?

   e. Obtain an interval plot for these data. Does your plot support your conclusion in part d? Explain briefly.

9. Open *Homes.mtw*. This problem considers both parametric and nonparametric tests.

   a. From these data on 150 randomly selected homes create a text variable corresponding to the three categories: homes with 1 bath, homes with 1.5 or 2 baths, and homes with more than 2 baths. Call the text values 1Bath, 1.5to2Baths, and More2Baths. How many homes are in each category?

   b. Test whether the mean acreage is the same for these three types of homes.

   c. Test whether the median acreage is the same for these three types of homes.

   d. Construct a display showing a boxplot of acreage for each category. What do your plots suggest about the appropriateness of your tests in parts b and c?

10. Open *Infants.mtw*. Assume that the infants were randomly selected in addition to being randomly assigned.

    a. Use the Kruskal-Wallis test to test whether there are significant differences among the median ages for the three ethnic groups. Use $\alpha = .05$. State your conclusion.

    b. Obtain boxplots of the ages of the three groups to check the assumption that the distribution of age is the same for each group. Comment on the appropriateness of your test in part a.

11. Open *Baby.mtw*. Use Mood's Median test to determine whether the population median attention times for the five designs are equal. Remember to delete the observation in row 27. How does your conclusion compare to the conclusion from the Kruskal-Wallis test considered in the tutorial? Explain briefly.

On Your Own

Minitab does not have a command that computes the Spearman's Rank Correlation Coefficient (rho). This is because the coefficient is fairly straightforward to compute by using existing commands. In fact, Spearman's rho is the correlation between the ranks associated with two variables. Write a short macro that computes and stores the ranks associated with two specified columns of data and then computes the correlation between the two sets of ranks. You may assume no ties.

Answers to Matching Problems

(e) H
(b) ETA1 – ETA2
(a) Nonparametric
(f) K
(g) Estimated median
(c) Run
(h) Factor
(d) Response
(i) Friedman
(j) Mann-Whitney

Answers to True/False Problems

(a) T, (b) F, (c) F, (d) F, (e) F, (f) T, (g) T, (h) F, (i) T, (j) T

14

Time Series Analysis

In Tutorials 10 and 11, you used regression analysis to develop models relating a response variable to one or more predictor variables. When your data consist of observations (responses) recorded over equally spaced intervals of time, regression methods often prove inadequate. With such data, error terms are frequently correlated rather than independent and the data itself is often cyclical in form rather than linear or quadratic. Minitab offers a range of specialized *time series tools* that help you select an appropriate model and predict or forecast future values for the series.

OBJECTIVES

In this tutorial, you learn how to:

- Perform a trend analysis to model and forecast a time series
- Perform a classical decomposition to model and forecast a time series
- Identify a model using autocorrelation and partial autocorrelation plots
- Transform a time series by lagging and by taking differences
- Use Box-Jenkins autoregressive integrated moving average (ARIMA) techniques to model and forecast a time series
- Compare the classical decomposition and ARIMA models

CASE STUDY **ENVIRONMENT—
TEMPERATURE VARIATIONS**

During the summer of 1988, which was one of the hottest on record in the Midwest, a graduate student in environmental science conducted a study for the state Environmental Protection Agency (EPA). She studied the impact of an electric generating plant along a river. The EPA was most concerned with the plant's use of river water for cooling. Scientists feared that the plant was raising the river's temperature and endangering its aquatic life. The EPA established a site directly downstream from the cooling discharge outlets of the plant at which they measured the water temperature hourly for 95 consecutive hours.

In this case study, you act as an environmental science graduate student working for the EPA on this project. You will use time series methods to analyze the problem by performing a time series analysis in the eight phases given in the following list:

Phase I   Perform a linear trend analysis on the time series.

Phase II   Perform a classical decomposition on the time series using a multiplicative model.

Phase III   Apply the model to forecast future values.

Phase IV   Investigate the autocorrelation and partial autocorrelation structure of the time series to determine an appropriate Box-Jenkins ARIMA model.

Phase V   Transform the data (if necessary) using lags and differences to obtain a stationary time series.

Phase VI   Fit an adequate Box-Jenkins ARIMA model.

Phase VII   Apply the Box-Jenkins ARIMA model to forecast future values.

Phase VIII   Examine the strengths and weaknesses of the Box-Jenkins ARIMA models by comparing forecasts based upon the model to the forecasts produced by the decomposition method.

Begin by opening Minitab:

- Open **Minitab** and maximize the Session window
- If necessary, **Enable Commands** and clear Output Editable from the Editor menu
- Click the **Save Project** toolbar button 🖫 and, in the location where you are saving your work, save this project as *T14.mpj*
- Open the worksheet *Riverc.mtw*
- If necessary, click the **Current Data Window** toolbar button 🖽

C1 Hour lists the time of day of each measurement, using a 24-hour clock. C2 Temp gives the temperature measurement in °C. The other columns in *Riverc.mtw* contain additional measurements taken at the same time as the temperature. For more information about these other measurements, refer to the description of this file in Help or in Appendix A Data Sets.

In Tutorials 10 and 11, you used linear regression to model and forecast a response variable based upon one or more predictor variables. Plots were used to provide initial insight into the relationship among variables. When your data consist of observations (responses) recorded over equally spaced intervals of time the observations form a *time series*. In this case, regression methods often prove inadequate. Minitab provides a number of techniques especially designed to accomplish the same goals. They appear on the Time Series menu of the Stat menu.

To obtain an initial picture of any trends in the temperature data over time, use the Stat > Time Series > Trend Analysis command. This command is similar to the Stat > Regression > Fitted Line Plot command for simple linear regression because it presents a useful combination of graphical and numerical information. It allows you to fit one of four models to a time series: a linear model, a quadratic model, an *exponential growth model*, or an *S-curve model*. You can store the fitted values and the residuals. In the context of *trend analysis*, the residuals are called the *detrended values*.

▶ **Note** In the context of time series, predicted (or fitted) values are the values obtained from the model corresponding to existing values of the series. By contrast, *forecasts* are obtained by inserting future values for the series in the model. ■

An additional option is to generate forecasts of future values of the series. For now, perform a trend analysis on the temperatures, using the Minitab default settings.

- Choose **Stat > Time Series > Trend Analysis**
- Double-click **C2 Temp** as the "Variable"

The completed Trend Analysis dialog box is as shown in Figure 14-1. Note that the default is a linear model.

FIGURE 14-1

The completed Trend Analysis dialog box

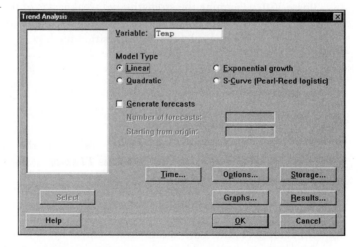

- Click **OK**

The Trend Analysis plot, with Minitab's default title, appears as shown in Figure 14-2.

FIGURE 14-2

Trend analysis plot

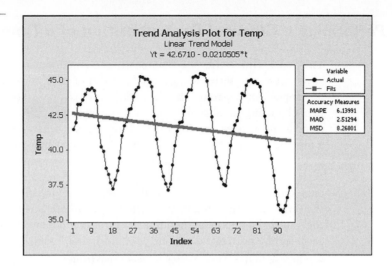

Minitab plots the temperature on the vertical axis and the hour on the horizontal axis—labeled "Index". Successive points are connected with a straight line. The plot indicates a cyclical pattern beginning with temperatures at 2 P.M. around 42 degrees, which rise for the next eight hours and then fall until they reach a low of about 37 degrees at 8 A.M. (Index equal to 18.) This cycle repeats through the next 24-hour period, but with some variation. The peaks and valleys occur roughly 24 hours apart. This suggests that your time series model should include a periodic component. In time series terminology, this is more commonly called a *seasonal component*, even though the cycle might be over hours or days.

The *trend line* fitted to these data is also printed. Its equation is below the title, Yt = 42.6710 – 0.0210505*t. The line suggests a slight negative trend in the data. In the upper right-hand corner of the plot are three measures of the adequacy of the fitted model:

1. *Mean Absolute Percentage Error (MAPE)*
2. *Mean Absolute Deviation (MAD)* (which you calculated by a series of Minitab commands in Tutorial 3)
3. *Mean Squared Deviation (MSD)*

The equation and these three measures are also reported in the Session window. (You can get the formulas for these measures from the Glossary in Minitab's Help.)

The strong seasonal component and the relatively small value of the slope coefficient (–0.02105) strongly suggest a more complex model be tried. A classical decomposition model will isolate the linear trend, the seasonal, and the error components of the series.

▶ **Note** Instead of the trend analysis that you just performed, you could have obtained a related graph by choosing the Stat > Time Series > Time Series Plot command or the Graph > Time Series Plot command, or even the Graph > Scatterplot command. Because the Trend Analysis provides so much more information than these other graphs, using this command is an example of using the Minitab tool best suited to your purpose. ■

14.2 Performing a Classical Decomposition of a Time Series

The Stat > Time Series > Decomposition command performs a *classical decomposition* on a time series using either a multiplicative or an additive model. Classical decomposition separates the time series into trend, seasonal, and error components by using trend analysis and moving averages. Check Help to verify that trend analysis is essentially the method of least-squares and to find a definition of a moving average. You can also generate forecasts based upon your model.

Based upon your examination of the time series plot, use the Stat > Time Series > Decomposition command to obtain the trend, seasonal, and error components of the default *multiplicative model*. In this model, an observation in the series is represented in the form $T*S*E$, where T is the trend component, S is the seasonal component, and E is the error component. You will also generate forecasts for the next two days (one per hour for 48 hours).

- Choose **Stat > Time Series > Decomposition**
- Double-click **C2 Temp** as the "Variable"
- Type **24** as the "Seasonal length"
- Click the **Generate forecasts** check box and type **48** as the "Number of forecasts"
- Click **Storage** and click the **Forecasts** check box

▶ **Note** The "Number of forecasts" text box in the Decomposition dialog box and the Forecasts check box in the Decomposition - Storage dialog box are available only when the Generate forecasts check box in the Decomposition dialog box is clicked. ∎

- Click **Help** in the Decomposition - Storage dialog box to obtain a brief explanation of the quantities that can be stored in this analysis
- Close the Help window to return to the dialog box and click **OK**

    The completed Decomposition dialog box should resemble that shown in Figure 14-3.

FIGURE 14-3

The completed Decomposition dialog box

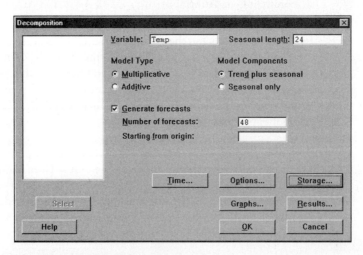

- Click **OK**

Minitab displays the results of the decomposition in the Session window and in three Graph windows. It also adds a column, C6 FORE1, which contains the forecasts.

- Click the **Session Window** toolbar button 🔲

- Scroll up in the Session window to view the initial results of the decomposition analysis (without the forecasts) that are shown in Figure 14-4

FIGURE 14-4

Initial time series decomposition for temperature

```
MTB > Name c6 "FORE1"
MTB > Decomp 'Temp' 24;
SUBC>    Forecasts 48;
SUBC>     Fstore 'FORE1';
SUBC>    First 1.
```

Time Series Decomposition for Temp

Multiplicative Model

```
Data     Temp
Length   95
NMissing 0
```

Fitted Trend Equation

Yt = 41.8801 - 0.00464900*t

Seasonal Indices

Period	Index
1	1.02184
2	1.03788
3	1.04638
4	1.05760
5	1.06893
6	1.07382
7	1.07150
8	1.07513
9	1.07606
10	1.07264
11	1.06164
12	1.02143
13	0.97266
14	0.96158
15	0.93218
16	0.91477
17	0.89585
18	0.88712
19	0.89433
20	0.92279
21	0.95440
22	0.98236
23	0.99757
24	0.99954

Accuracy Measures

```
MAPE  1.56767
MAD   0.63471
MSD   0.91941
```

Minitab provides the trend line equation Yt = 41.8801 − 0.00464900*t and the seasonal indices, which you can combine to obtain predicted (for given t) and forecasted (for future t) values. The indices have been normalized to average 1. Note that the first 12 indices are above 1 and the second 12 below 1. The output also indicates the three measures you can use to determine the accuracy of the fitted model: MAPE (1.56767), MAD (0. 63471), and MSD (0.91941). Minitab repeats these values in Figure 14-5, which displays a plot of the actual, predicted, and forecasted values along with the trend line, with Minitab's default title.

To access the three graphs generated with this command:

- Click the **Show Graphs Folder** toolbar button 🖻
- Double-click on **Time Series Decomposition Plot for Temp**

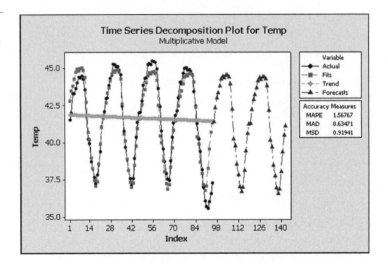

Notice the legend in the upper right-hand corner. It helps you to distinguish between the actual observations, the fitted values obtained from the multiplicative model, the values predicted from the trend line, and the forecasted values. The graph depicts a slight downward trend, with a 24-hour seasonal component. It also shows the divergence of the predicted values from the actual values at the top of each of the four peaks and at the fourth valley.

Open the graph that provides a component analysis for Temp (shown in Figure 14-6).

- Double-click on **Decomposition - Component Analysis for Temp** in the Project Manager window

FIGURE 14-6

Plots showing component analysis for Temp

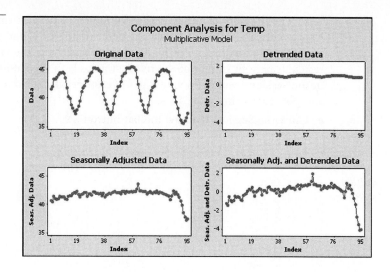

These graphs show plots of the original data, detrended data, seasonally adjusted data, and seasonally adjusted and detrended data. In the plot of the detrended data, the (slight) negative linear trend has been removed. The plot of the seasonally adjusted data shows what the series would look like if there was no seasonal effect. The fourth graph shows the residuals after the linear trend and the seasonal effects have been removed. The unusual pattern in the lower two plots dramatizes the fact that the last valley shown in the plot of the original data is deeper than all of the preceding valleys.

The final graph, shown in Figure 14-7, provides a seasonal analysis for Temp.

- Double-click on **Decomposition - Seasonal Analysis for Temp** in the Project Manager window

FIGURE 14-7

Plots showing seasonal analysis for Temp

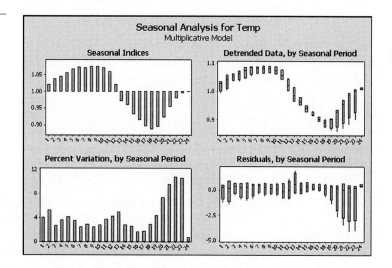

The first plot shows the seasonal indices. The next three contain plots of, respectively, the detrended data, the percent variation, and the residuals. In all four plots, the horizontal axis is the seasonal period. Note the large number of negative residuals present in four of the five last periods in the Residuals plot. Scroll down in the Session window to view the forecasts resulting from the decomposition analysis, shown in Figure 14-8.

■ Click the **Session Window** toolbar button 🖬

FIGURE 14-8

Time series decomposition forecasts for temperature

Forecasts

Period	Forecast
96	41.4146
97	42.3340
98	42.9937
99	43.3409
100	43.8007
101	44.2648
102	44.4623
103	44.3614
104	44.5066
105	44.5401
106	44.3935
107	43.9336
108	42.2649
109	40.2422
110	39.7792
111	38.5589
112	37.8341
113	37.0478
114	36.6824
115	36.9766
116	38.1488
117	39.4512
118	40.6022
119	41.2265
120	41.3030
121	42.2199
122	42.8779
123	43.2242
124	43.6827
125	44.1455
126	44.3425
127	44.2419
128	44.3867
129	44.4200
130	44.2738
131	43.8152
132	42.1509
133	40.1337
134	39.6719
135	38.4548
136	37.7321
137	36.9478
138	36.5834
139	36.8768
140	38.0458
141	39.3447
142	40.4926
143	41.1152

These forecasts, which were plotted in Figure 14-5, display the same pattern as your data. They peak between periods 102 and 105. Period 104 corresponds to 9:00 P.M. (the same time a peak occurred in your data).

After this decomposition analysis, you should still be concerned about the divergence of the predicted values from the actual values at the top of each peak and at the bottom of the fourth valley, all shown in Figure 14-5. Another approach to analyzing time series data is to fit ARIMA models. Such models are based to an extent upon the autocorrelational structure in the data. In the next section you will examine autocorrelation plots that will guide you in selecting the appropriate ARIMA model.

Before going on, save your work.

- Click the **Save Project** toolbar button 💾

14.3 | *Autocorrelation and Partial Autocorrelation Plots*

For time series data, autocorrelation and partial autocorrelation measure the degree of relationship between observations k time periods, or *lags*, apart. *Autocorrelation* is a correlation coefficient. However, instead of correlation between two different variables, the correlation is between two values of the same variable k time periods apart. The *partial autocorrelation* at a lag of k is the correlation between residuals at time t from an autoregressive model and observations at lag k with terms for all intervening lags present in the autoregressive model. So, partial autocorrelations measure the strength of a linear relationship after removing the effects of the earlier lagged terms.

Autocorrelation

Plots of both these quantities provide valuable information to help you identify an appropriate ARIMA model. Begin by obtaining an autocorrelation plot.

- Choose **Stat > Time Series > Autocorrelation**
- Double-click **C2 Temp** as the "Series"
- Click the **Number of lags** option button
- Type **24** as the "Number of lags"

You have told Minitab to compute and plot 24 autocorrelations (recall that 24 is the approximate cycle length).

The completed Autocorrelation Function dialog box should look as shown in Figure 14-9.

FIGURE 14-9

The completed Autocorrelation Function dialog box

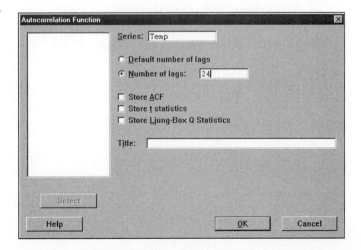

- Click **OK**

The autocorrelation function plot, with Minitab's default title, is shown in Figure 14-10.

FIGURE 14-10

Autocorrelation function plot for Temp

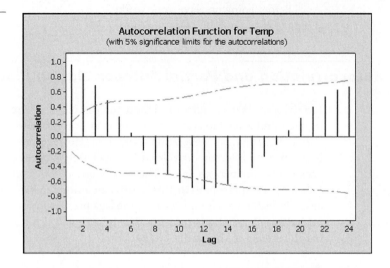

The autocorrelation function plot, indicated by vertical lines at each of the 24 lags, resembles a sine pattern. This suggests that temperatures close in time are strongly and positively correlated, while temperatures about 12 hours apart are highly negatively correlated. (The large positive autocorrelation at lag 24 is indicative of the 24-hour seasonal component.) The pattern in the autocorrelation plot suggests that your ARIMA model may have an *autoregressive (AR) component.* (An AR process of order k is one in which the current value of the series can be modeled as a linear combination of the k most recent past values for the series.) The band of 5% significance limits running along the top and bottom of the chart are, in fact, connected 95% confidence intervals for the corresponding population autocorrelations for each lag. All of the autocorrelations, except for the first three and several of the negative ones, are within the appropriate confidence interval.

Now, look in the Session window to see the actual correlations, shown in Figure 14-11.

- Click the **Session Window** toolbar button ▦

FIGURE 14-11

Autocorrelations, associated t-statistics, and associated Ljung-Box Q statistics for each lag

```
MTB > ACF 'Temp';
SUBC>    Lags 24.
```

Autocorrelation Function: Temp

Lag	ACF	T	LBQ
1	0.949284	9.25	88.34
2	0.835845	4.87	157.57
3	0.674533	3.21	203.14
4	0.477057	2.06	226.19
5	0.261378	1.08	233.18
6	0.038862	0.16	233.34
7	-0.172223	-0.70	236.44
8	-0.355473	-1.44	249.83
9	-0.504385	-2.00	277.09
10	-0.610853	-2.33	317.54
11	-0.673287	-2.43	367.27
12	-0.695611	-2.37	420.99
13	-0.678601	-2.19	472.74
14	-0.622735	-1.91	516.86
15	-0.531927	-1.57	549.45
16	-0.409360	-1.18	569.00
17	-0.262960	-0.75	577.17
18	-0.096963	-0.27	578.29
19	0.076070	0.21	578.99
20	0.242756	0.69	586.23
21	0.393810	1.11	605.55
22	0.524039	1.45	640.21
23	0.617273	1.68	688.98
24	0.661579	1.75	745.79

The output in the Session window lists the autocorrelations (ACF), associated t-statistics (T), and associated *Ljung-Box Q (LBQ)* statistics for each lag. The t-statistics can be used to test the null hypothesis that the autocorrelation at a specific lag equals zero. The LBQ statistics can be used to test the null hypothesis that the autocorrelations for all lags up to a specified value are equal to zero. The three positive, large autocorrelations at lags 1, 2, and 3 suggest an AR component of order 2 or 3 (that is, the AR component should include 2 or 3 of the most recent past values).

Partial Autocorrelations

Now construct a plot of the partial autocorrelation function to gather additional information about an appropriate ARIMA model.

- Choose **Stat > Time Series > Partial Autocorrelation**
- Double-click **C2 Temp** as the "Series"
- Click the **Number of lags** option button
- Type **24** as the "Number of lags"

The completed Partial Autocorrelation Function dialog box should look as shown in Figure 14-12.

FIGURE 14-12

The completed Partial Autocorrelation Function dialog box

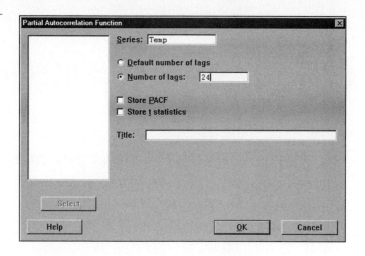

■ Click **OK**

The partial autocorrelation function plot, with Minitab's default title, is shown in Figure 14-13.

FIGURE 14-13

Partial autocorrelation function plot for Temp

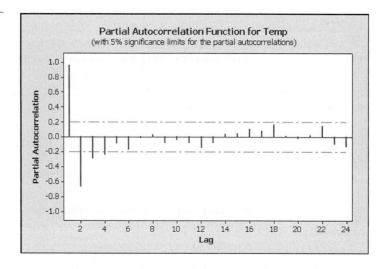

The partial autocorrelation plot shows two large partial autocorrelations (spikes) at lags 1 and 2. Both far exceed the bound of their corresponding 95% confidence intervals. This also suggests an AR component of degree order 2. Later in this tutorial, you will fit an ARIMA model based upon your findings thus far.

◆ **Note** A related Minitab command is Stat > Time Series > Cross Correlation. It computes and plots the correlation between one time series and another lagged time series. ■

Transforming a Time Series

The river data are fairly well-behaved, as evidenced by the trend analysis plot in Figure 14-2. If you were to draw a horizontal line at the mean of the series, you would see that the data oscillate with approximately the same cyclical pattern. A time series like this, whose mean and variance do not change with time, is called *stationary*.

Conversely, a series that grows or declines systematically over time is called *non-stationary*. However, by lagging and differencing a non-stationary time series, you can usually make it stationary. To practice using these components, you will obtain the river data's lags and differences, even though they appear almost stationary.

Lagging Data

You can think of lagging as *time-shifting*. Each element in a time series is shifted to a later point so that it lags behind by the number of time units you specify. When you lag a column in Minitab and store the lagged data, the new column is identical to the former one, except that the entries are shifted down, or lagged, by a specified number of rows. The empty rows at the beginning of the resulting column are filled with *s to indicate missing numeric values. Minitab uses a default lag of 1. To lag the temperature data:

- Choose **Stat > Time Series > Lag**
- Double-click **C2 Temp** as the "Series"
- Type **Lags** as the "Store lags in" column

    The completed Lag dialog box should look as shown in Figure 14-14.

FIGURE 14-14

The completed Lag dialog box

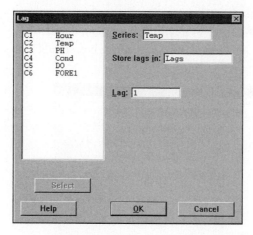

- Click **OK**
- Click the **Current Data Window** toolbar button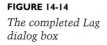

Compare the temperature data in C2 Temp with the data in C7 Lags in Figure 14-15. Note that a value in C7 Lags is one row lower than the corresponding value in C2 Temp. The first entry in C7 Lags is a missing numeric value.

FIGURE 14-15

The Data window showing lagged temperature data in C7 Lags

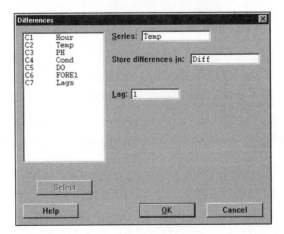

	C1 Hour	C2 Temp	C3 PH	C4 Cond	C5 DO	C6 FORE1	C7 Lags	C8	C9	C10	C11	C12
1	1400	41.52	8.67	0.92	2.92	41.4146	*					
2	1500	41.99	8.65	0.94	2.34	42.3340	41.52					
3	1600	43.29	8.64	0.95	2.78	42.9937	41.99					
4	1700	43.29	8.67	0.95	2.39	43.3409	43.29					
5	1800	43.63	8.74	0.96	2.54	43.8007	43.29					
6	1900	44.01	8.82	0.96	3.31	44.2648	43.63					
7	2000	44.39	8.85	0.96	3.43	44.4623	44.01					
8	2100	44.35	8.83	0.97	3.98	44.3614	44.39					
9	2200	44.48	8.88	0.96	4.85	44.5066	44.35					
10	2300	44.27	8.84	0.97	4.74	44.5401	44.48					
11	2400	43.55	8.84	0.95	5.09	44.3935	44.27					
12	100	41.77	8.81	0.93	5.36	43.9336	43.55					
13	200	40.25	8.80	0.97	5.57	42.2649	41.77					
14	300	39.92	8.76	1.11	5.12	40.2422	40.25					
15	400	38.73	8.73	0.99	5.07	39.7792	39.92					
16	500	38.27	8.69	0.99	4.64	38.5589	38.73					
17	600	37.68	8.66	1.01	4.55	37.8341	38.27					
18	700	37.25	8.62	1.01	4.15	37.0478	37.68					
19	800	37.85	8.61	1.00	3.89	36.6824	37.25					
20	900	38.52	8.58	0.99	3.71	36.9766	37.85					

Computing Differences

The Stat > Time Series > Differences command computes differences of observations that are a specified lag apart. For example, if you specify a lag of 1, Minitab computes differences between adjacent values in the time series and stores them in a column. To apply the Stat > Time Series > Differences command to the temperature data with differences of lag 1:

- Choose **Stat > Time Series > Differences**
- Double-click **C2 Temp** as the "Series"
- Type **Diff** as the "Store differences in" column

The completed Differences dialog box should look as shown in Figure 14-16.

FIGURE 14-16

The completed Differences dialog box

- Click **OK**

C8 Diff contains the differences between adjacent values in the time series for C2 Temp, as shown in Figure 14-17. Because you used the same lag to create C7 Lags, C8 Diff is equivalent to the difference between columns C2 Temp and C7 Lags. (For example, 0.47 = 41.99 – 41.52.)

FIGURE 14-17

The Data window showing differenced temperature data in C8 Diff

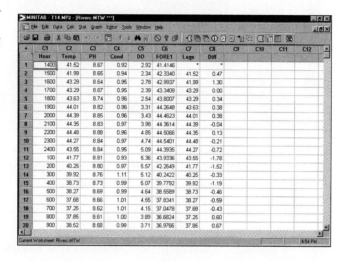

Recall that the shape of your time series plot indicates whether the time series is stationary. Consider taking differences when the time series plot indicates a non-stationary time series. You can then use the resulting differenced data as your time series, plotting the differenced time series to determine if it is stationary.

Before going on, save your work.

- Click the **Save Project** toolbar button 🖫

14.5 | *Performing a Box-Jenkins ARIMA Analysis of a Time Series*

Models that allow you to build in trend, seasonality, and autocorrelation are called *ARIMA (autoregressive integrated moving average)* or *Box-Jenkins models*. Your examination of the autocorrelation and partial autocorrelation plots suggested that you should consider an ARIMA model with an AR component of order 2, called AR 2, and a 24-hour seasonal AR component of order 1, called SAR 1. Hoping to keep the model as simple as possible, first try a model with AR 2 and no seasonality.

To do this, use the Stat > Time Series > ARIMA command to estimate the parameters for this model and investigate its fit.

- Choose **Stat > Time Series > ARIMA**
- Double-click **C2 Temp** as the "Series"
- Type **2** as the "Autoregressive: Nonseasonal" order

The completed ARIMA dialog box should resemble Figure 14-18.

FIGURE 14-18

The completed ARIMA dialog box

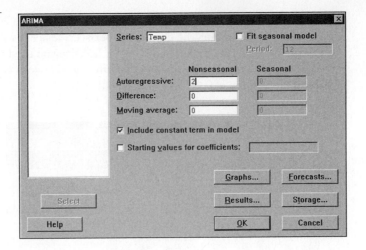

▶ **Note** The columns in the dialog box next to Autoregressive, Difference, and Moving average correspond to the nonseasonal and seasonal components of the ARIMA model, respectively. ■

■ Click **OK**

If necessary, scroll up in the Session window to view the results of the ARIMA analysis, the first half of which should resemble Figure 14-19.

FIGURE 14-19

Iteration estimates from the ARIMA analysis

```
MTB > ARIMA 2 0 0 'Temp';
SUBC>   Constant;
SUBC>   Brief 2.
```

ARIMA Model: Temp

Estimates at each iteration

Iteration	SSE		Parameters	
0	556.993	0.100	0.100	33.408
1	460.739	0.250	0.020	30.500
2	376.654	0.400	-0.064	27.719
3	301.680	0.550	-0.147	24.946
4	235.673	0.700	-0.231	22.176
5	178.602	0.850	-0.315	19.408
6	130.452	1.000	-0.399	16.643
7	91.073	1.150	-0.482	13.865
8	60.786	1.300	-0.566	11.110
9	39.320	1.450	-0.650	8.351
10	26.746	1.600	-0.734	5.611
11	22.933	1.733	-0.810	3.202
12	22.919	1.742	-0.815	3.069
13	22.919	1.742	-0.816	3.060

Relative change in each estimate less than 0.0010

Minitab fits the model you specify using an iterative process that involves using initial estimates of the parameters to obtain new estimates. The process stops when there are very small changes in the values for the estimates from one iteration to the next or when the estimates do not converge after 25 iterations. Minitab provides a history of this iterative process, with the sum of squared errors

(SSE) and the parameter estimates for each iteration. In this example, the estimates converge satisfactorily after 13 iterations. In Figure 14-19, the parameter estimates correspond to the two AR 2 components and the constant parameters, respectively.

▶ **Note** Minitab uses the common notation (pdq) × (PDQ) (S) to specify an ARIMA model. Here (pdq) stands for a nonseasonal model, (PDQ) for a seasonal model, and S for the seasonality. The value p is the order of the autoregressive component, d the number of differences, and q the order of the moving average component of the model. P, D, and Q are similarly defined for the seasonal model. S is the length of a "season". You have constructed a (2, 0, 0) (0, 0, 0) (0) ARIMA model. In the absence of a seasonal model, Minitab shortens the model to 2 0 0, which you can see as part of the Session command in Figure 14-19. ∎

The second half of the output displays the final parameter estimates in detail, as shown in Figure 14-20.

FIGURE 14-20

Final parameter estimates and modified Box-Pierce (Ljung-Box) chi-square statistics from the ARIMA analysis

```
Final Estimates of Parameters

Type          Coef  SE Coef       T      P
AR    1     1.7421   0.0629   27.70  0.000
AR    2    -0.8156   0.0639  -12.76  0.000
Constant  3.05969  0.05129   59.66  0.000
Mean      41.6201   0.6977


Number of observations:  95
Residuals:     SS =  22.8842 (backforecasts excluded)
               MS =   0.2487  DF = 92


Modified Box-Pierce (Ljung-Box) Chi-Square statistic

Lag              12      24      36      48
Chi-Square     25.2    48.7    58.4    76.7
DF                9      21      33      45
P-Value       0.003   0.001   0.004   0.002
```

Minitab estimates the first autoregressive coefficient (AR 1) as 1.7421, with a standard deviation of 0.0629 and a t-statistic of 27.70. The second autoregressive coefficient (AR 2) is estimated at –0.8156, with a standard deviation of 0.0639 and a t-statistic of –12.76. The estimated constant term is 3.05969.

You may conclude that both the AR 1 and AR 2 parameters are significantly different from zero because the t-statistics are so large (and the corresponding p-values so small). You could compute the estimated temperature at time T using the following model:

$$\text{Temp(at time T)} = 3.05969 + 1.7421 * \text{Temp(at time T} - 1)$$
$$- .8156 * \text{Temp(at time T} - 2)$$

The last table in Figure 14-20 shows modified Box-Pierce (Ljung-Box) chi-square statistics, which measure how well the model fits the data. For the following hypotheses, chi-square statistics, and corresponding p-values are computed at lags of 12, 24, 36, and 48:

Null Hypothesis, H_0: The specified ARIMA model fits the data.

Alternative Hypothesis, H_1: The specified ARIMA model does not fit the data.

The p-values for these tests are 0.003, 0.001, 0.004, and 0.002, respectively, indicating that each chi-square statistic is significant. In each case you should reject the null hypothesis. The output suggests that your specified ARIMA model does not fit the data at most standard levels of significance.

Constructing a Seasonal Model

Based upon this lack of fit and the pattern of the data in Figure 14-2, you should consider a new ARIMA model, (0, 0, 0) (1, 0, 0) (24), that includes a seasonal component.

- Click the **Edit Last Dialog** toolbar button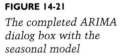
- Verify that the "Series" is Temp
- Click the **Fit seasonal model** check box and type **24** as the "Period"
- Type **0** as the "Autoregressive: Nonseasonal" order
- Type **1** as the "Autoregressive: Seasonal" order

The modified ARIMA dialog box should look as shown in Figure 14-21.

FIGURE 14-21

The completed ARIMA dialog box with the seasonal model

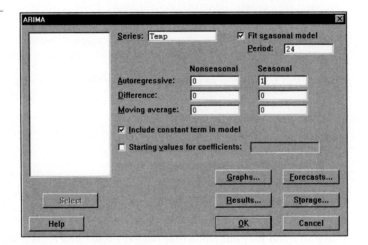

- Click **OK**

The output in the Session window, shown in Figure 14-22, alerts you that Minitab is unable to estimate this model.

FIGURE 14-22

*Session window showing
seasonal model attempt*

```
MTB > ARIMA 0 0 0 1 0 0 24 'Temp';
SUBC>   Constant;
SUBC>   Brief 2.
```

ARIMA Model: Temp

```
Estimates at each iteration

Iteration     SSE      Parameters
        0  715.280  0.100  37.584
        1  570.425  0.250  31.278
        2  443.907  0.400  25.001
        3  333.375  0.550  18.729
        4  237.159  0.700  12.468
        5  154.825  0.850   6.222
        6  106.324  1.000   0.000

Unable to reduce sum of squares any further

* ERROR * Model cannot be estimated with these data.
```

At this point you consult with an EPA statistician. After examining some additional plots, the statistician suggests that you try ARIMA (2, 0, 0) (1, 1, 0) (24). This is a model with a nonseasonal autoregressive component of order 2, a seasonal autoregressive component of order 1, one seasonal difference, and a seasonality component of 24.

- Click the **Edit Last Dialog** toolbar button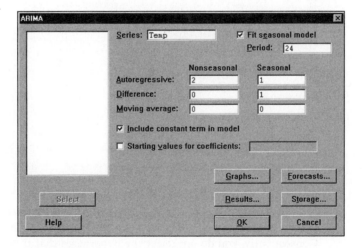
- Type **2** as the "Autoregressive: Nonseasonal" order
- Type **1** as the "Difference: Seasonal" order

The modified ARIMA dialog box should resemble Figure 14-23.

FIGURE 14-23

*The completed ARIMA
dialog box for the new
model*

- Click **OK**

The Session window contains the results of your new model. They are shown in Figure 14-24.

FIGURE 14-24

*Final parameter estimates
from the ARIMA results
for the new model*

```
MTB > ARIMA 2 0 0 1 1 0 24 'Temp';
SUBC>   Constant;
SUBC>   Brief 2.
```

ARIMA Model: Temp

```
Estimates at each iteration

Iteration     SSE          Parameters
        0  77.8840  0.100   0.100   0.100  -0.002
        1  57.9793  0.250   0.097   0.116  -0.014
        2  42.3752  0.400   0.086   0.126  -0.022
        3  30.5046  0.550   0.070   0.126  -0.027
        4  21.9906  0.700   0.050   0.108  -0.029
        5  16.5296  0.850   0.025   0.059  -0.026
        6  13.8547  1.000  -0.004  -0.058  -0.002
        7  13.7880  1.002  -0.002  -0.107   0.003
        8  13.6610  1.003  -0.002  -0.146  -0.002
        9  13.5726  1.003  -0.000  -0.178  -0.026
       10  13.4953  1.002   0.000  -0.224  -0.021
       11  13.4727  1.001   0.001  -0.255  -0.020
       12  13.4635  1.000   0.002  -0.276  -0.020
       13  13.4592  0.999   0.003  -0.291  -0.020
       14  13.4566  0.998   0.005  -0.300  -0.020
       15  13.4536  0.997   0.006  -0.307  -0.020
       16  13.4484  0.995   0.008  -0.311  -0.020
       17  13.4367  0.994   0.010  -0.314  -0.020
       18  13.4064  0.992   0.014  -0.316  -0.018
       19  13.3537  0.991   0.020  -0.317  -0.019
       20  13.3468  0.991   0.024  -0.325  -0.030
       21  13.2687  0.989   0.034  -0.321  -0.028
       22  13.2538  0.991   0.043  -0.327  -0.050
       23  13.1961  0.960   0.096  -0.369  -0.082
       24  13.1644  0.960   0.097  -0.370  -0.082
       25  12.9807  0.967   0.093  -0.369  -0.082

** Convergence criterion not met after 25 iterations **


Final Estimates of Parameters

Type           Coef     SE Coef      T      P
AR    1      0.9671      0.1245    7.77  0.000
AR    2      0.0927      0.1422    0.65  0.517
SAR  24     -0.3690      0.1526   -2.42  0.018
Constant -0.0820070  -0.0728330    1.13  0.264


Differencing: 0 regular, 1 seasonal of order 24
Number of observations:  Original series 95, after differencing 71
Residuals:    SS = 12.5825 (backforecasts excluded)
              MS =  0.1878  DF = 67


Modified Box-Pierce (Ljung-Box) Chi-Square statistic

Lag            12      24      36      48
Chi-Square   13.9    26.7    39.7    49.4
DF              8      20      32      44
P-Value     0.085   0.144   0.164   0.266
```

The modified Box-Pierce (Ljung-Box) chi-square statistics seem much better for this model. The p-values are now equal to 0.085, 0.144, 0.164, and 0.266. Consequently, you do not reject the hypotheses that this model fits the data.

◆ **Note** from Figure 14-24 that the iterative process used to obtain parameter estimates was stopped after 25 iterations. To continue the process to possibly obtain more accurate parameter estimates, you could return to the ARIMA dialog box and enter the final estimates (0.9671, 0.0927, –0.3690, and –0.0820070) as the starting values for coefficients and obtain a new round of iterations. ∎

14.6 Forecasting with ARIMA

Once you determine the model, the final step in the fitting process is to use your model to forecast future values. You will predict the temperatures for the next 48 hours and then store the forecasts and the corresponding prediction limits.

- Click the **Edit Last Dialog** toolbar button ▣
- Click **Forecasts**
- Type **48** as the "Lead" value
- Type **AForecast** as the "Storage: Forecasts"
- Type **LAForecast** as the "Storage: Lower limits"
- Type **UAForecast** as the "Storage: Upper limits"

    The completed ARIMA - Forecasts dialog box should resemble Figure 14-25.

FIGURE 14-25

*The completed
ARIMA - Forecasts
dialog box*

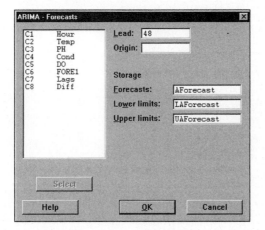

- Click **OK** twice

    For periods 96 to 143, Minitab displays forecasts and the corresponding 95% confidence limits. These forecasts and confidence limits, which are based upon your ARIMA model, are shown in Figure 14-26.

FIGURE 14-26

New ARIMA Model forecasts for temperature

```
Forecasts from period 95
                   95 Percent Limits
Period   Forecast     Lower     Upper   Actual
   96     37.1269   36.2773   37.9764
   97     37.6252   36.4434   38.8071
   98     38.3148   36.8453   39.7843
   99     38.0197   36.2856   39.7538
  100     38.0810   36.0934   40.0686
  101     38.1874   35.9512   40.4236
  102     37.7947   35.3109   40.2784
  103     37.1791   34.4461   39.9122
  104     36.7916   33.8055   39.7777
  105     36.1620   32.9174   39.4066
  106     35.4526   31.9427   38.9626
  107     33.9669   30.1834   37.7504
  108     32.2496   28.1834   36.3159
  109     30.3913   26.0318   34.7508
  110     28.4575   23.7933   33.1217
  111     26.7029   21.7215   31.6843
  112     24.8784   19.5662   30.1906
  113     23.0213   17.3636   28.6790
  114     21.4321   15.4132   27.4510
  115     20.2532   13.8563   26.6502
  116     19.7013   12.9083   26.4943
  117     19.5550   12.3468   26.7631
  118     19.1921   11.5483   26.8359
  119     18.7081   10.6071   26.8092
  120     17.5175    8.7447   26.2903
  121     16.9067    7.4597   26.3537
  122     16.4104    6.2642   26.5566
  123     14.7810    3.9096   25.6525
  124     13.6136    1.9880   25.2391
  125     12.2498   -0.1612   24.6608
  126     10.3824   -2.8480   23.6128
  127      8.2033   -5.8831   22.2897
  128      6.1506   -8.8311   21.1322
  129      3.7823  -12.1366   19.7011
  130      1.2285  -15.6724   18.1294
  131     -2.2698  -20.2005   15.6608
  132     -5.9691  -24.9803   13.0421
  133     -9.9589  -30.1046   10.1868
  134    -14.0681  -35.4056    7.2694
  135    -18.1506  -40.7407    4.4394
  136    -22.5373  -46.4442    1.3697
  137    -27.0944  -52.3864   -1.8024
  138    -31.5523  -58.3015   -4.8030
  139    -35.7192  -64.0021   -7.4364
  140    -39.5750  -69.4722   -9.6777
  141    -43.1511  -74.7481  -11.5540
  142    -47.0313  -80.4185  -13.6441
  143    -51.1369  -86.4097  -15.8642
```

Plotting the ARIMA Forecasts

To get a clearer picture of both the forecasts and the corresponding 95% confidence intervals, construct a multiple time series plot.

- Choose **Graph > Time Series Plot** and double-click *Multiple*
- Double-click **C9 AForecast, C10 LAForecast,** and **C11 UAForecast** as the "Series"
- Click **OK**

    To change the label on the vertical axis of the plot from the default "Data":

- Hold your cursor over the vertical axis label of the plot, click the **right-hand mouse button**, and choose **Edit Y Axis Label**
- Type **ARIMA Forecasts** as the "Text" to replace the default and click **OK**

    The final version of the graph, with Minitab's default title, is shown in Figure 14-27.

FIGURE 14-27

Multiple time series plot of new ARIMA model forecasts

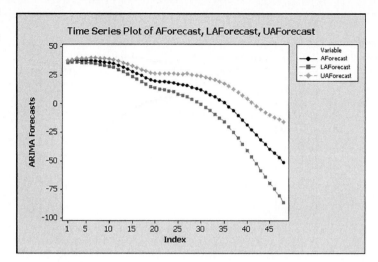

Your forecasts of river temperature indicate a limited amount of seasonality. Your initial 95% forecast intervals are fairly narrow, with widths of roughly 2 degrees. Later, the interval widths increase to approximately 70 degrees. This makes sense because confidence intervals increase in width as you forecast further into the future. Your forecasts indicate an unreasonable cooling of the river in later periods. The legend in the upper right-hand corner of the graph window allows you to easily identify the three variables.

14.7 | Comparing the Two Forecasting Models

The forecasts from the classical decomposition (in C6 FORE1) and from the ARIMA (2, 0, 0) (1, 1, 0) (24) models are different. First, observe that the initial values of the forecasts for the ARIMA model are much closer to what you would expect based upon the last few observations. Second, each set of forecasts produces a distinct pattern, which you can examine by constructing another multiple time series plot with both the FORE1 and AForecast forecast variables.

- Click the **Edit Last Dialog** toolbar button ▣
- Highlight **LAForecasts** and **UAForecasts** in the "Series" text box
- Double-click **C6 FORE1** to replace LAForecasts and UAForecasts as the "Series" and click **OK**

    Now change the label on the vertical axis of the plot.

- Hold your cursor over the vertical axis label of the plot, click the **right-hand mouse button**, and choose **Edit Y Axis Label**
- Type **Forecasts from Two Models** as the "Text" and click **OK**

    The graph, with Minitab's default title, is shown in Figure 14-28.

FIGURE 14-28

Time series plot comparing forecasts from two models

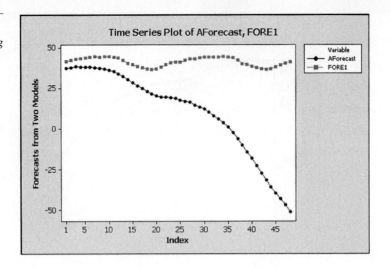

The legend identifies the two time series. Note that the classical decomposition preserves the seasonality observed in the original data over the 48 forecasts much better than the ARIMA model. It also produces forecast values similar in value and magnitude to the observed data.

You wonder if there might be an even better model for this time series. The EPA statistician suggests you might try an ARIMA (2, 1, 0) (1, 0, 0) (24) model. Other possibilities include models using moving averages, single exponential smoothing, double exponential smoothing, or seasonal exponential smoothing. You can implement these procedures by using the *Moving Average*, *Single Exp Smoothing*, *Double Exp Smoothing*, and *Winters' Method* commands, respectively, on the Time Series menu. See Minitab's Help for more information about these time series procedures.

▶ **Note** The assumptions for the confidence intervals and tests in this tutorial have not been stated in this tutorial. You should check with a statistician to learn about them. ∎

In this tutorial, you analyzed a time series with classical decomposition and Box-Jenkins ARIMA techniques. You used a variety of plots, together with lagging and differencing transformations, to help you identify possible models. The interactive nature of Minitab makes it possible to try as many preliminary models as your patience allows before you settle on a final model and begin to make forecasts. The techniques you have seen in this tutorial provide a powerful and flexible set of forecasting tools that can model virtually any set of time series data.

Congratulations! You have completed another tutorial. You can explore total quality management (TQM) tools in the last tutorial.

■ Click the **Save Project** toolbar button 🖫 and exit Minitab

Minitab Command Summary

This section describes the commands introduced in, or related to, this tutorial. Use Minitab's online Help for a complete explanation of all commands.

Minitab Menu Commands

Menu	Command	Description
Stat ➤	Time Series	
	➤ Time Series Plot	Produces a time series plot for one or more columns of data against time.
	➤ Trend Analysis	Uses trend analysis to fit a particular type of trend line to a time series and forecasts future values of the series.
	➤ Decomposition	Performs classical decomposition on a time series and forecasts future values of the series.
	➤ Moving Average	Uses moving averages to smooth out the noise in a time series and forecasts future values of the series.
	➤ Single Exp Smoothing	Uses single exponential smoothing to smooth out the noise in a time series and forecasts future values of the series.
	➤ Double Exp Smoothing	Performs Holt or Brown double exponential smoothing for a time series and forecasts future values of the series.
	➤ Winters' Method	Performs Holt-Winters seasonal exponential smoothing for a time series and forecasts future values of the series.
	➤ Differences	Computes the differences between values of a time series that are a specified number of rows apart and stores them in a new column.

Menu	Command	Description
Stat ➤	Time Series	
	➤ Lag	Shifts the data in a column down a specified number of rows and stores them in a new column.
	➤ Autocorrelation	Computes and produces a graph of the autocorrelations of a time series.
	➤ Partial Autocorrelation	Computes and produces a graph of the partial autocorrelations of a time series.
	➤ Cross Correlation	Computes and produces a graph of the cross correlations between two time series.
	➤ ARIMA	Fits a specified autoregressive integrated moving average model to time series data and forecasts future values of the series.
Graph ➤	Time Series Plot	
	Multiple	Produces a time series plot for one or more columns of data against time.

Review and Practice

Matching Problems

Match the following terms to their definitions by placing the correct letter next to the term it describes.

_____ Multiplicative

_____ Order 1 moving average component

_____ Time Series Plot

_____ Lag

_____ Number of lags

_____ Order 1 autoregressive component

_____ Decomposition

_____ Ljung-Box

_____ Stationary

_____ Trend Analysis

a. The option that allows you to specify the number of autocorrelations and partial autocorrelations to compute and display in the autocorrelation and partial autocorrelation plots

b. The default decomposition model

c. The command that shifts a particular column of data down a specified number of rows and saves the results in another column

d. (1, 0, 0) (0, 0, 0) (0)

e. A command found on both the Graph menu and the Stat > Time Series menu

f. (0, 0, 1) (0, 0, 0) (0)

g. The command that allows you to separate a time series into linear trend and error components, and provides forecasts

h. A time series whose mean and variance do not change with time

i. The name of the chi-square statistic you use to test the fit of an ARIMA model

j. The command that allows you to separate a time series into linear trend, seasonal, and error components, and provide forecasts

True/False Problems

Mark the following statements with a *T* or an *F*.

a. ____ When you use the Differences command, storing the results in a column is optional.

b. ____ You always produce forecasts of future events when you use the ARIMA command.

c. ____ A time series plot connects sequential points with lines.

d. ____ Before Minitab can compute differences, you must issue the Lag command.

e. ____ The ARIMA output contains p-values that determine the significance of the ARIMA model's fit.

f. ____ An additive model is the default model type with the Decomposition command.

g. ____ By default the Decomposition command produces nine plots in three different windows.

h. ____ Autocorrelation and partial autocorrelation plots always include a test of significance.

i. ____ MADE, MAP, and MSD are three measures you use to determine the accuracy of the fitted values in classical decomposition.

j. ____ Use the Trend Analysis command when there is no seasonal component to your time series.

Practice Problems

The practice problems instruct you to save your worksheets with a filename prefixed by P, followed by the tutorial number. In this tutorial, for example, use *P14* as the prefix. If necessary, use Help or Appendix A Data Sets, to review the contents of the data sets referred to in the following problems. Interpretations should use the language of the subject matter of the question. If you are using the Student version, be sure to close worksheets when you have completed a problem.

1. Open *YearlySnow.mtw*. This data set contains the amount of snowfall in Boston for each of the 109 years from 1892 to 2000. When it comes to snowfall, older Bostonians often suggest that "winters are not what they used to be". One implication of the remark is that the city used to get a lot more snow in the distant past. Perform a trend analysis to investigate whether there is any truth to this suggestion. Write a paragraph outlining your findings.

2. Open *SP5002.mtw*.

   a. Construct a time series plot of the Value variable. What does the plot suggest about the behavior of the series over time?

   b. Perform a trend analysis. Is there a trend component to the series? Is there a seasonal component? Briefly explain your answers.

   c. Perform a classical decomposition. Is there a trend component? Is there a seasonal component? Briefly explain your answers.

3. Open *Rivera.mtw*.

   a. Construct a simple time series plot of PH. Describe any unusual patterns in this time series.

   b. The values of pH should range between 0 and 14. A researcher should certainly exclude any negative pH values and any other values considered sufficiently unusual. In addition, a researcher should examine in detail any other usual observations to determine whether or not they belong in the analysis. Suppose it has been determined to exclude all of the PH values less than 7.5. Use the Data > Code > Numeric to Numeric command to change these unusual values to Minitab's missing value code. Name the resulting variable PH2. How many missing values are present in PH2?

   c. Construct a simple time series plot of PH2. Briefly explain how the missing data are represented.

d. Based upon the PH2 plot, should you perform one or two trend analyses on these data? Explain briefly.

e. Analyze the trend in these data based upon your answer to part d. Briefly explain your results.

> ▶ **Note** The data set *Rivers.mtw* also contains these same negative PH values. ■

4. Open *DJC20012002b.mtw*. For this problem you will perform trend analyses on a variable and eight quarterly means of the same variable.

a. Perform a linear trend analysis on the MMM variable. What is the resulting equation?

b. To obtain the quarterly means you must first transform the Day variable to a text variable that identifies each quarter. Use the Data > Code > Date/Time to Text command to create a new variable named Quarter by coding 1/1/2001 : 3/31/2001 to 2001Q1, 4/1/2001 : 6/30/2001 to 2001Q2, and so on. How many days are in each quarter?

c. To obtain the quarterly means use the Stat > Basic Statistics > Store Descriptive Statistics command. What is the default name of the variable that includes these means?

d. Perform a linear trend analysis on the quarterly means variable. What is the resulting equation?

e. Discuss any differences between the equations in parts a and d. Which linear trend analysis is most appropriate for a presentation? Explain briefly.

5. Open *Riverc.mtw*. In Section 14.4 you obtained the lagged column and the differences column for Temp with Stat > Time Series commands. You can obtain these same columns using the Calc > Calculator command.

a. Use the Calc > Calculator command to obtain a column identical to Lags in this tutorial.

b. Use the Calc > Calculator command to obtain a column identical to Diff in this tutorial.

6. Open *Sleep.mtw*.

a. Create a variable that is a lag 1 of Sleep1. Name this variable Sleep1Lag1.

b Regress Sleep1 against Sleep1Lag1. Are you surprised by the results of this regression. Explain briefly.

c. Create a variable that is a difference of lag 1 of Sleep1. Name this variable Sleep1Diff1.

d. Regress Sleep1 against Sleep1Diff1. Are you surprised by the results of this regression. Explain briefly.

7. Open *Riverc.mtw*. Reanalyze the time series in this tutorial using an ARIMA (2, 1, 0) (1, 0, 0) (24) model. How does it compare to the decomposition and new ARIMA models considered in the tutorial?

8. Open *Riverc.mtw*. In addition to the river's temperature, the state EPA measured three other river characteristics: pH, conductivity, and dissolved oxygen content. The pH variable (PH) indicates the acidity or alkalinity of the water. Perform a time series analysis of pH. Your analysis should follow the approach used in this tutorial. That is, for each characteristic, plot the series and examine the need for a classical decomposition model, fit and forecast a classical decomposition model, obtain the autocorrelation and partial autocorrelation plots, and, finally, fit and forecast an appropriate ARIMA model.

9. Open *Riverb.mtw*. The state EPA monitored several sites in the river study. Site B was directly upriver from the plant's intake pipes. Redo the analysis described in this tutorial using Site B's data.

10. Open *MonthlySnow.mtw*. As the name suggests, this data set contains monthly snowfall figures (for Boston). Amounts are available from 1890 to 2000.

    a. Use the Data > Copy > Columns to Columns command or the Data > Subset Worksheet command to isolate the snowfall amounts for the years 1980 to 2000. Briefly explain how you isolated these data.

    b. For these data, obtain autocorrelation and partial autocorrelation plots and fit an appropriate ARIMA model.

11. Open *EMail.mtw*. This data set contains the number of emails sent and received by a faculty member each day for more than a year.

    a. Use the Data > Copy > Columns to Columns command to isolate the number of emails sent (Out) and received (In) between August 18 and December 17 (including these two dates). This constitutes just over one college semester.

    b. For the emails sent, obtain autocorrelation and partial autocorrelation plots and fit an appropriate ARIMA model.

    c. For the emails received, obtain autocorrelation and partial autocorrelation plots and fit an appropriate ARIMA model.

    d. Write a brief report highlighting the differences (if any) between the models you derived in parts b and c.

12. Open *Riverc2.mtw*, which contains the results of the work you performed in this tutorial. Stack the actual values of the studied time series on top of the fitted values from the classical decomposition and the fitted values from the final ARIMA model. Produce a multiple time series plot of both variables. What does this plot show you?

On Your Own

The data set *DJC20012002a.mtw* contains the value of the Dow Jones Composite (DJC) price index for each of the 518 working days in the years 2001 and 2002. Perform a time series analysis of these data following the steps outlined in this tutorial. Write a brief report summarizing your results.

Answers to Matching Problems

(b) Multiplicative
(f) Order 1 autoregressive component
(e) Time Series Plot
(c) Lag
(a) Number of lags
(d) Order 1 autoregressive component
(j) Decomposition
(i) Ljung-Box
(h) Stationary
(g) Trend Analysis

Answers to True/False Problems

(a) F, (b) F, (c) T, (d) F, (e) T, (f) F, (g) T, (h) T, (i) F, (j) T

Total Quality Management Tools

Total quality management (TQM) is a philosophy used by many businesses and individuals to improve their processes. The foundation of TQM is making decisions based upon accurate and timely data. In this tutorial, you will examine some of Minitab's quality tools that help detect quality problems and improve processes. These tools include cause-and-effect diagrams, Pareto charts, and various kinds of control charts. Many of the tools are adaptations of techniques—bar charts and scatterplots, for example—that were examined in earlier tutorials.

OBJECTIVES

In this tutorial, you learn how to:

- Create a cause-and-effect diagram
- Create a Pareto chart
- Construct an $\overline{X}$ chart to monitor the mean value of a process
- Construct a range chart to monitor variability in a process
- Construct an individuals chart to monitor individual observations from a process
- Construct a moving range chart to monitor variability in a process from individual observations
- Construct a proportion chart to monitor the proportion from a process

15.1 Creating a Cause-and-Effect Diagram

CASE STUDY	EDUCATION— FACULTY SURVEY

You did a preliminary analysis of the results of a faculty survey taken by the student government. You have recorded a number of flaws in the survey process and plan an article for the student government newsletter that will summarize your results. You are looking for a structure for organizing the reasons for the flawed process. Minitab offers such a structure, called a *cause-and-effect* (sometimes referred to as a *fishbone* or *Ishikawa*) *diagram*, that helps you do this.

Begin by opening Minitab:

- Open **Minitab** and maximize the Session window
- If necessary, **Enable Commands** and clear Output Editable from the Editor menu
- Click the **Save Project** toolbar button 🖫 and, in the location where you are saving your work, save this project as *T15.mpj*

The cause-and-effect diagram lets you graphically represent the causes that influence a problem. The graph is arranged in a structure resembling the skeleton of a fish (hence the alternative name, fishbone) and allows for six main causes. By default, these causes are labeled Personnel, Machines, Materials, Methods, Measurements, and the Environment, but some or all can be changed. Experience suggests that these generic labels are suitable for a wide variety of situations. The graph is customized by adding to the diagram specific examples of each cause that may be contributing to the problem.

For the survey process, you use Minitab's default causes. The specific examples (reasons for the flawed survey) associated with each cause are stored in a file called *Process.mtw*:

- Open the worksheet *Process.mtw*
- If necessary, click the **Current Data Window** toolbar button 🖽

There are six text variables—C1-T Personnel, C2-T Machines, C3-T Material, C4-T Methods, C5-T Measurements, and C6-T Environment—each listing possible reasons for a flawed survey process.

To construct a cause-and-effect diagram:

- Choose **Stat > Quality Tools > Cause-and-Effect**
- Double-click **C1 Personnel** as the "Causes" for Branch 1
- Repeat this procedure for the next five "Causes" as shown in Figure 15-1

The labels don't need to be changed because you are using Minitab's defaults. If labels other than these defaults are needed, you can type them after entering the corresponding cause.

- Type **Flawed Process** as the "Effect"
- Type **Cause-and-Effect Diagram for Faculty Surveys** as the "Title"

The completed Cause-and-Effect dialog box should resemble Figure 15-1.

FIGURE 15-1

The completed Cause-and-Effect Diagram dialog box

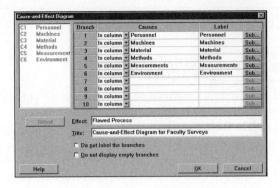

- Click **OK**

    Minitab produces a cause-and-effect diagram as shown in Figure 15-2.

FIGURE 15-2

Cause-and-effect diagram

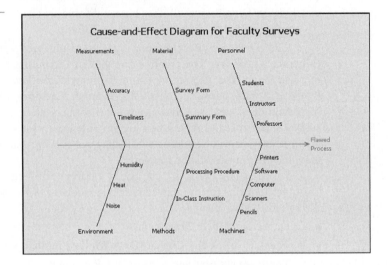

You are pleased with this graphical display. It presents specific examples of the numerous causes that can affect a flawed survey process. You can use this display as the starting point for a brainstorming session with colleagues to come up with solutions to these flaws (or even more areas of concern).

▶ **Note** For convenience, Minitab has two menus devoted to quality tools, Stat > Quality Tools and Stat > Control Charts. The former includes cause-and-effect diagrams and Pareto charts. The latter produces control charts, which are fundamental quality tools, discussed in Sections 15.3 to 15.7. ■

15.2 | Creating a Pareto Chart

The cause-and-effect diagram in Figure 15-2 presents some possible causes for a flawed survey process in an easy-to-read graphical form. However, the display does not indicate what part of the survey process is most flawed. To determine this, you will recommend another survey, in which respondents will indicate which causes are negative influences on the survey process. You recommend that the results of this survey be presented in the form of a Pareto chart. *Pareto charts* are graphs that include a bar chart, with the bars ordered from the most frequently occurring category to the least frequently occurring category and an *ogive* or *cumulative percentage polygon*.

Because you are graduating, you must train a colleague to conduct the survey and prepare the Pareto chart. Your colleague is not well prepared in statistical methods, so you create a small example to illustrate the use of the Pareto chart. In your example, you assume that 120 students have completed a proposed questionnaire, and that their responses to the seven areas of concern are summarized in the following table.

Area of Concern	Count
1. Too little time for completion	21
2. Ambiguous questions	102
3. Loss of interest because of length	18
4. Little interest in subject	19
5. Inadequate instructions	11
6. Instrument too crowded	25
7. Too many open-ended questions	87

You show your colleague how to enter these data into Minitab and obtain a Pareto chart.

■ Choose **File > New** and click **OK**

■ Name C1 **Concerns** and C2 **Counts**

■ Type the following text values into the first seven rows of C1 to represent the seven areas of concern: **Time**, **Ambiguous**, **Length**, **Subject**, **Instructions**, **Crowded**, and **Open-Ended**

■ Type the corresponding counts into C2: **21**, **102**, **18**, **19**, **11**, **25**, and **87**

To obtain the Pareto chart:

■ Choose **Stat > Quality Tools > Pareto Chart**

The default option for a Pareto chart is raw data. Because your data are in summarized form:

■ Click the **Chart defects table** option button

■ Double-click **C1 Concerns** as the "Labels in" column

- Double-click **C2 Counts** as the "Frequencies in" column

    The completed Pareto Chart dialog box should look as shown in Figure 15-3.

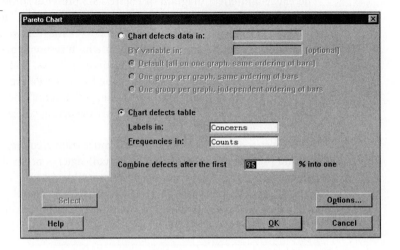

> ◆ **Note** The "Combine defects after the first [ ]% into one" text box instructs Minitab to generate bars until the cumulative percentage exceeds the specified percentage. Then it generates a bar labeled "Others" that contains the rest of the observations. The default cumulative percentage is 95%. ■

- Click **Options**

    In the Pareto Chart - Options dialog box, you can specify labels for each axis and a title for the chart. Alternatively, you may choose not to chart the cumulative percent symbols, connecting lines, and percent scale.

- Type **Concerns** as the "X axis label"
- Type **Counts** as the "Y axis label"
- Type **Pareto Chart for Survey Concerns** as the "Title"
- Click **OK** twice

    The resulting Pareto chart is shown in Figure 15-4.

FIGURE 15-4

Pareto chart for survey concerns

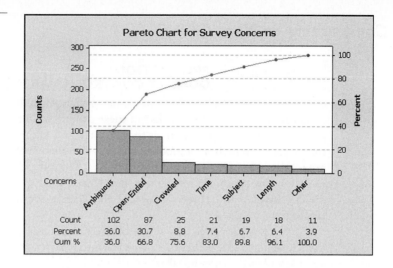

Concerns	Ambiguous	Open-Ended	Crowded	Time	Subject	Length	Other
Count	102	87	25	21	19	18	11
Percent	36.0	30.7	8.8	7.4	6.7	6.4	3.9
Cum %	36.0	66.8	75.6	83.0	89.8	96.1	100.0

The chart shows the bars for each area of concern arranged from the one with the largest count to the one with the smallest count. The ogive (cumulative percent connected line) shows the progress of the cumulative counts. Beneath the plot are the counts for each area of concern and the corresponding percentages and cumulative percentages.

The chart shows that the two largest concerns are ambiguous questions and too many open-ended questions. It shows also that these two occur far more frequently than the others. You can see this by inspecting the bars and by noticing the flattening of the ogive after the second area of concern.

You can point out to your colleague that if these data were real, it would make sense to focus improvement efforts on these two concerns. These results are a good illustration of the general use of Pareto charts to distinguish the "vital few" from the "trivial many".

▶ **Note** Minitab allows you to create Pareto charts from raw data as well as from counts. To see an example entitled "Using Pareto Chart with Raw Data", click Help in the Stat > Quality Tools > Pareto Chart dialog box and then click on examples. ■

Before going on to the next case study, save the work you have done thus far.

■ Click the **Save Project** toolbar button 🖫

PRODUCTION—
QUALITY CONTROL CHARTS

The instructor of your production course brought three bags of candy to class. Each bag contains 19 "fun packs" of these candies for a total of 57 packs. She assigns an identification number to each pack and asks the class to imagine that the 57 packs come from a production line. The class is to evaluate aspects of the production line using various control charts. She informs the class that the weight of each candy should be 2.32 grams, based upon product information. She then divides the class into three groups and asks each to measure different characteristics of these candies.

Group A measures the weight of five candies randomly selected from each of the 57 packs. It enters these 285 weights into *Candya.mtw*, along with the bag and pack identification numbers.

Group B measures the total weight of each pack. It then measures the weight of the pack paper and computes the weight of the candies in each pack. *Candyb.mtw* contains these three variables, along with the bag and pack identification numbers and the number of candies in each pack.

Group C also records the identification number of each bag and pack identification numbers, as well as the number of candies in each pack. In addition, the group counts the number of red, brown, green, orange, and yellow candies in each pack. It enters its data in *Candyc.mtw*.

The instructor then asks the class to determine whether the production line process is "in control" by constructing control charts relevant to their data sets. A process is in control when the mean and variability, or similar characteristics, of the process are stable and any variation in the process is due to random causes.

A *control chart* shows how some aspect of a process varies, usually over time. It normally has statistically determined upper and lower limits. Such charts are among the most widely used of quality tools. Minitab offers 22 control chart types in the Professional version and 14 control chart types in the Student version. Quality control technicians use them to monitor processes because they may show when a process is getting out of control. Control charts help manufacturers and members of service industries achieve a high level of quality by meeting certain specifications. In addition, you can use these charts to discover interesting patterns in a process. In this tutorial, you will construct and interpret five types of Minitab control charts.

Group A sampled five random candies (subgroups of size 5) for 57 packs, weighed each candy, in grams, and placed the results in C3 Weights of *Candya.mtw*.

■ Open the worksheet *Candya.mtw*

To determine whether the mean of the candy-making process is in control, produce an $\overline{X}$ chart to look at the 57 sample means:

- Choose **Stat > Control Charts > Variables Charts for Subgroups > Xbar** (not Xbar-R or Xbar-S)
- Double-click **C3 Weights** as the "All observations for a chart are in one column"
- Type **5** as the "Subgroup sizes"

You decide to use Minitab's default title for all of your control charts in this tutorial.

▶ **Note** Instead of typing 5 as the "Subgroup sizes", you can choose C2 Pack as the ID column. ■

The completed Xbar Chart dialog box should resemble that shown in Figure 15-5.

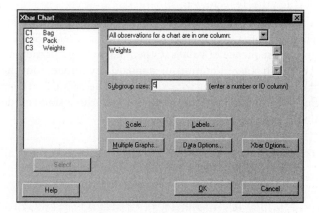

- Click **Xbar Options**

The Xbar Chart - Options dialog box appears. In this box, there are a number of folders—Parameters, Estimate, S Limits, Tests, Stages, Box-Cox (in the Professional version only), Display, and Storage. To enter a mean, instead of having Minitab estimate a mean from the data, for your control chart:

- Verify that the Parameters folder is open
- Type **2.32** as the "Mean"

You will use the default value of the standard deviation that Minitab estimates from the data. The completed Xbar Chart - Options dialog box for the Parameters folder (for the Professional version) should resemble that shown in Figure 15-6.

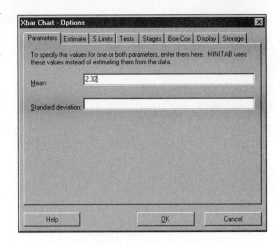

■ Click the **Tests** folder

Minitab can perform eight different tests for special causes on the data. There is a brief explanation for each of these tests. Some are designed to flag an unusual sequence of values. Others flag when a particular number of values fall more than a given number of standard deviations from the *center line* (in this case, the line corresponding to 2.32 grams). The default is to perform a test that identifies individual points that are more than three standard deviations from the center line.

Ask for all of the tests:

■ Click the **drop-down list arrow** and click on **Perform all tests for special causes**

The completed Xbar Chart - Options dialog box for the Tests folder is shown in Figure 15-7.

FIGURE 15-7

The completed Xbar Chart - Options dialog box for the Tests folder

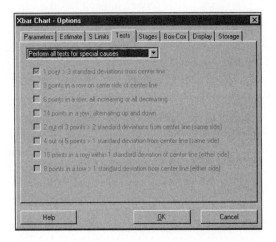

You want Minitab to display information about any test failures in the Session window.

- Click the **Display** folder
- Verify that the "Display test results" option button is clicked

    In the future you will not need to verify this default.
- Click **OK** twice

    Minitab computes the 57 sample means (using subgroups of size 5) and plots them on the $\overline{X}$ chart, as shown in Figure 15-8.

FIGURE 15-8

X-bar chart of Weights

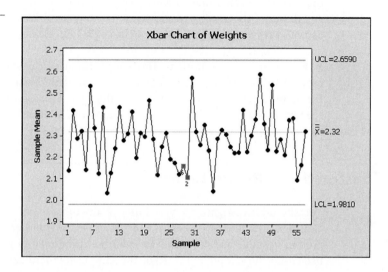

Minitab marks the center line at 2.32. It also places an *upper-control-limit (UCL) line* at 2.6590 and a *lower-control-limit (LCL) line* at 1.9810. Each of these limits is three estimated standard deviations from the center line. Minitab uses a pooled standard deviation estimate for the standard deviation by default; the pooling is over all subgroups. It may not be obvious from the chart, but the two red square symbols in the lower half of the chart indicate that the process is out of control. To interpret the symbols that indicate that the process failed Test 6 and Test 2, you need to look at the output in the Session window shown in Figure 15-9.

- Click the **Session Window** toolbar button 🖳

FIGURE 15-9

Test failures for the Xbar chart

```
MTB > XbarChart 'Weights' 5;
SUBC>    Mu 2.32;
SUBC>    Test 1 2 3 4 5 6 7 8.
```

Xbar Chart of Weights

Test Results for Xbar Chart of Weights

```
TEST 2. 9 points in a row on same side of center line.
Test Failed at points:  29

TEST 6. 4 out of 5 points more than 1 standard deviation from center line (on
        one side of CL).
Test Failed at points:  28, 29

* WARNING * If graph is updated with new data, the results above may no
          * longer be correct.
```

Minitab reports that Test 2 failed at point 29; the nine sample averages before this average were all on one side of the center line. Test 6 failed at points 28 and 29, where Minitab found that four of the five previous sample averages were at least one standard deviation away from, and on the same side of, the center line. Unusual patterns might indicate that the process is drifting away from the target value of 2.32 grams.

You are more than satisfied with this plot. But the instructor reminds you that before you examine an Xbar chart, you first should determine whether the variability of the process of the sample is in control. You remember that the most common way to do this is by constructing a Range chart.

▶ **Note** You can examine methods for estimating the standard deviation with other than the default method of a pooled standard deviation by clicking on the Estimate folder in the Xbar Options dialog box. You can examine methods for specifying the control limits by clicking on the S Limits folder in that dialog box. ■

|15.4| *Constructing a Range Chart*

You should investigate the variability of any process because it determines the control limits for the process mean. To ascertain whether the variance of the process is stable and random, you examine the spread in your samples. You can use either the sample range (R) or the sample standard deviation (S) in Minitab to create R and S charts, respectively. For historical reasons with small samples such as these, the R chart is most commonly used.

To obtain the R chart for your data:

- Choose **Stat > Control Charts > Variables Charts for Subgroups > R**
- Double-click **C3 Weights** as the "All observations for a chart are in one column"
- Type **5** as the "Subgroup sizes"
- Click **R Options**
- Click the **Tests** folder

Minitab can perform four different tests for special causes on the data for this chart. The default is to perform a test that identifies individual points that are more than three standard deviations from the center line. Note the brief explanation for each of the four tests.

Ask for all the tests:

- Click the **drop-down list arrow** and click on **Perform all tests for special causes**

The completed R Chart - Options dialog box for the Tests folder should appear as shown in Figure 15-10.

FIGURE 15-10

*The completed
R Chart - Options
dialog box for
the Tests folder*

- Click **OK** twice

Minitab computes the range of each of the 57 samples of five candies and then plots them on the R chart, as shown in Figure 15-11. Minitab draws the center line at 0.588 (at the average of the sample ranges) and the lower and upper control limits at 0 and 1.243, respectively. It draws a lower control limit (LCL) of 0, which is often called a *zero-valued control limit* because an LCL of 0 is as low as is possible; there are no test failure red square symbols. The R chart indicates that process variability is not out of control. The observed variation in the range of the candy weights in each pack appears to be random. As a consequence, you can use the Xbar chart that you constructed in the previous section.

FIGURE 15-11

R chart of Weights

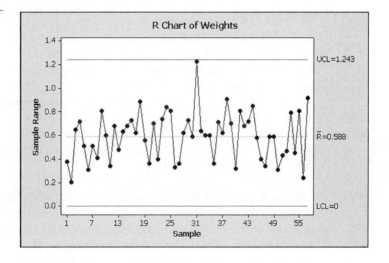

To better present the data, you can place both the Xbar and the R charts together using the Xbar-R chart:

- Choose **Stat > Control Charts > Variables Charts for Subgroups > Xbar-R**

- Double-click **C3 Weights** as the "All observations for a chart are in one column"
- Type **5** as the "Subgroup sizes"
- Click **Xbar-R Options**

    The Xbar Chart - Options dialog box appears.

- Verify that the Parameters folder is open
- Type **2.32** as the "Mean" and click **OK** twice

    The Xbar-R chart appears as is shown in Figure 15-12.

FIGURE 15-12

Xbar-R chart of Weights

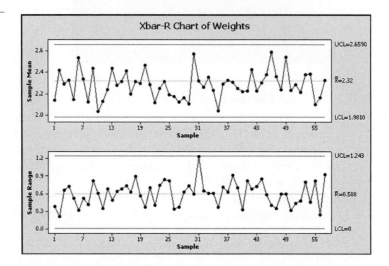

You are pleased with this chart even though no test results are displayed in the Session window. This is because, by default, the Stat > Control Charts > Variables Charts for Subgroups > Xbar-R command supplies only one test—identifying any mean that is more than three standard deviations from the center line—and no mean meets this criteria.

You conclude that the candy-making process is essentially in control; it produces acceptable weights. You tell your instructor that you would recommend that management meet with the process personnel to determine the causes of fluctuation in the process mean. Perhaps together they can identify the causes and improve the process in the future.

◆ **Note** If you decided to use a sample standard deviation instead of the range to investigate the variability of a process, you would use an S chart (Stat > Control Charts > Variables Charts for Subgroups > S) and/or an Xbar-S chart (Stat > Control Charts > Variables Charts for Subgroups > Xbar-S). ■

15.5 | Constructing an Individuals Chart

Group B obtained the weights of the 57 fun packs. These values are stored in C6 NetWgt of *Candyb.mtw*:

- Open the worksheet **Candyb.mtw**

You can investigate each pack individually using an *individuals* chart. Your instructor tells you to specify the historical mean of 20.89 grams for the pack weight to center the chart and to specify the historical standard deviation of 1.41 grams in order to determine what fraction of the production process is within the specification limits of 20.89 ± 3 * 1.41 (or 16.66 and 25.12) grams. Note that for individual values, the control limits are determined by the center line ±3σ.

To obtain the individuals chart:

- Choose **Stat > Control Charts > Variables Charts for Individuals > Individuals**
- Double-click **C6 NetWgt** as the "Variables"

The completed Individuals Chart dialog box should resemble that shown in Figure 15-13.

FIGURE 15-13

The completed Individuals Chart dialog box

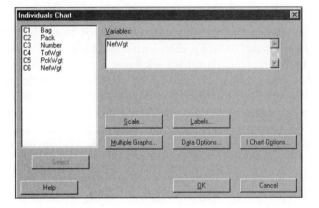

- Click **I Chart Options**

The Individuals Chart - Options dialog box appears. You will enter the historical mean and historical standard deviation and perform all tests for this control chart, instead of having Minitab estimate these statistics from the data.

- Verify that the Parameters folder is open
- Type **20.89** as the "Mean" and **1.41** as the "Standard deviation"

Now, request Minitab to do all eight possible tests.

- Click on the **Tests** folder
- Click the **drop-down list arrow** and click on **Perform all tests for special causes**
- Click **OK** twice

The I (individuals) chart is shown in Figure 15-14.

FIGURE 15-14

I chart of NetWgt

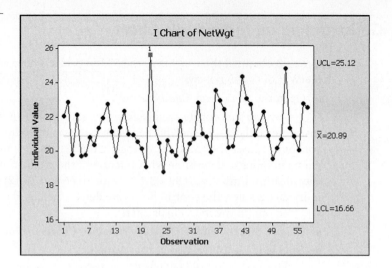

The resulting graph is revealing. The red square symbol indicates that a point failed Test 1.

■ Hold your cursor over the red square symbol below the 1 in the chart

A data tip identifies the point as pack 21 and lists its weight as 25.62 grams. There appears to be a problem with this pack. You contact a member of Group B who informs you that this pack had a number of crushed candies. You suspect that the damage caused this pack to be outside the specification limits—an interesting discovery prompted by examining the control chart. But you wonder if there is a need to explain the variability of the process before you can use this chart. You ask your instructor if this is the case. He informs you that a moving range chart should always be constructed and examined before you study an individuals chart.

15.6   *Constructing a Moving Range Chart*

In the investigation of Group A's data, you selected five randomly sampled candies from each pack. Now because you only have samples of size one—the single net weight from each pack—you cannot estimate process variability as you did before. Instead, you use the *moving range*—the range of two or more consecutive observations—to investigate the variability of the process and construct a moving range chart:

■ Choose **Stat > Control Charts > Variables Charts for Individuals > Moving Range**

■ Double-click **C6 NetWgt** as the "Variables"

■ Click **MR Options**

The Moving Range Chart - Options dialog box appears. You decide to set a standard deviation of 1.41, instead of having Minitab estimate a standard deviation from the data:

- Verify that the Parameters folder is open and type **1.41** as the "Standard deviation"
- Click **OK** twice

The moving range chart is shown in Figure 15-15.

FIGURE 15-15

Moving range chart of NetWgt

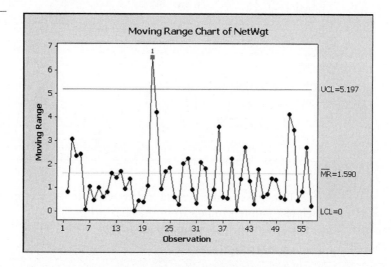

Process variability is not in control, due to the 21st pack. Again, your control chart identifies a problem. Hence, while you should not report the results of the individuals chart (because the Moving Range chart of the process is out of control, you have identified the problem pack by this use of the moving range chart.

While this finding agrees with the problem identified by the individuals chart, this type of occurrence is not always the case. Sometimes it's a good idea to construct both plots on the same graph. Minitab lets you do that with a single command, Stat > Control Charts > Variables Charts for Individuals > I-MR.

- Choose **Stat > Control Charts > Variables Charts for Individuals > I-MR**
- Double-click **C6 NetWgt** as the "Variables"
- Click **I-MR Options**
- Verify that the Parameters folder is open, type **20.89** as the "Mean" and **1.41** as the "Standard deviation"
- Click **OK** twice

The I-MR chart should appear as shown in Figure 15-16.

FIGURE 15-16

Initial I-MR Chart of NetWgt

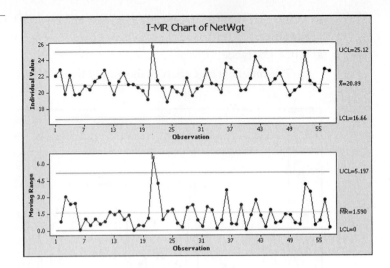

If the test failure symbol is distracting to the viewer, you can remove all such test failure symbols from the chart.

- Click the **Edit Last Dialog** toolbar
- Click **I-MR Options**
- Click on the **Tests** folder
- Click the **drop-down list arrow** and click on **Perform no tests**

The completed Individuals-Moving Range Chart - Options dialog box for the Tests folder should appear as shown in Figure 15-17.

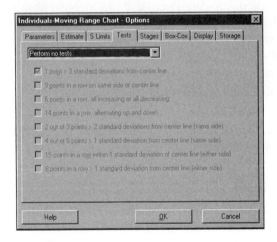

- Click **OK** twice

The resulting chart, with no test failure symbols, is shown in Figure 15-18.

FIGURE 15-18

*Final I-MR chart of
NetWgt*

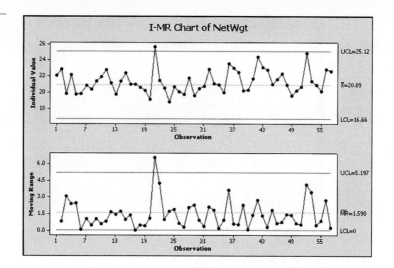

You plan to include this combination chart in your final report. It clearly reveals a problem with the 21st pack. You also plan to explain how you should not use the Xbar chart because the variability of the process is out of control. Because the name of this chart is identical to that of the previous chart, you rename both charts:

- Click the **Show Graphs Folder** 🖻 and rename the previous graph by typing **Initial** before the first graph "I-MR for NetWgt" and rename this graph by typing **Final** before the second "I-MR for NetWgt"

15.7 | *Constructing a Proportion Chart*

Next, you turn to the data in *Candyc.mtw*. You will investigate product quality by studying the proportion of defects (such as crushed candies) in each of the 57 packs. This is a common procedure for monitoring the proportion of defective items coming from a production line.

- Open the worksheet **Candyc.mtw**

Group C entered the number of candies in each pack in C3 Number and the number of defects in each pack in C9 Defects. You can conceptualize this as an *attributes data* problem because each candy is classified as either defective or not. Contrast this type of data to the weight data, which is called *variables data*. Attributes and variables data are terms often used in the context of quality. You may consider attributes data to be *count data* (most often qualitative data) and variables data to be *measurement data* (quantitative data). To analyze attributes data, you use a P (proportion) chart.

- Choose **Stat > Control Charts > Attributes Charts > P**
- Double-click **C9 Defects** as the "Variables"
- Double-click **C3 Number** as the "Subgroup sizes"

The completed P Chart dialog box should look as shown in Figure 15-19.

FIGURE 15-19

The completed P Chart dialog box

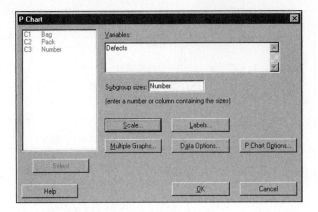

To perform all tests:

- Click **P Chart Options**
- Click on the **Tests** folder
- Click the **drop-down list arrow** and click on **Perform all tests for special causes**
- Click **OK** twice

The P chart is shown in Figure 15-20.

FIGURE 15-20

P chart of Defects

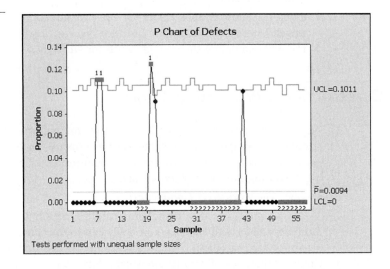

Minitab draws the center line, $\overline{P}$, at 0.0094. This is the proportion of defects over all 57 samples combined and is the default value for the center line. It draws an LCL of 0, as it did with the R Chart. Minitab displays a staircased upper control limit (UCL) at approximately 0.1011. This is a non-constant control limit (that is, not a horizontal straight line) because there are a different number of candies in each pack. The first test, performed with unequal sample sizes, indicates a problem with three packs (seventh, eighth, and 20th) based upon points that exceed the

UCL. The second test failure symbols indicate that there are three occurrences of more than nine points in a row on the same side of the center line. To see the Session window output:

■ Click the **Session Window** toolbar button 🖳

You can conclude that this process is not in control. There are too many defective candies among the 57 packs based upon the test results shown in Figure 15-21.

FIGURE 15-21

Test failures for the P chart

```
MTB > PChart 'Defects' 'Number';
SUBC>    Test 1 2 3 4.
```

P Chart of Defects

Test Results for P Chart of Defects

```
TEST 1. One point more than 3.00 standard deviations from center line.
Test Failed at points:  7, 8, 20

TEST 2. 9 points in a row on same side of center line.
Test Failed at points:  17, 18, 19, 30, 31, 32, 33, 34, 35, 36, 37, 38, 39, 40,
                        41, 51, 52, 53, 54, 55, 56, 57

* WARNING * If graph is updated with new data, the results above may no
        * longer be correct.
```

Based upon this chart you decide to recommend that the company determine the cause of the crushed candies immediately so that fewer defects will be produced in the future.

◈ **Note** If you are using the Student version of Minitab and have been following this tutorial, you will have five open worksheets at this point. This is the maximum number allowed in this version. If you wish to open another data set you will have to close one of the other worksheets. ■

◈ **Note** As mentioned earlier in this book, Minitab comes with a series of tutorials—not to be confused with the tutorials in this book—that provide excellent overviews of different aspects of the software. To reinforce your understanding of some of Minitab's control charts, you may wish to work through the Minitab tutorial called Session Four: Assessing Quality. (The Professional version of this Minitab tutorial contains a tool that we have not mentioned in this book—*process capability analysis*.) You may access the Minitab tutorials with the Help > Tutorials command. ■

You have seen only a small part of what you can do with Minitab's quality tools. Minitab can also chart the number of defectives (instead of the proportion of defectives), counts (instead of proportions), and counts per unit with other attributes charts. In addition, the software allows you to chart moving averages and exponentially weighted moving averages with the Stat > Control Charts > Time-Weighted Charts commands. These five charts are also used routinely in the monitoring and analysis of a process. Furthermore Minitab, Inc., has a sister software package, the Quality Companion, that contains a more powerful cause-and-effect diagram and related productivity tools.

Congratulations! You have completed the last of this book's sixteen tutorials.

■ Click the **Save Project** toolbar button 💾 and exit Minitab

---

MINITAB *at work*

QUALITY MANAGEMENT

In the late 1990s, General Electric (GE) chose to use Minitab statistical software as part of its company-wide "Six Sigma" quality improvement program. GE set for itself the goal of becoming a Six Sigma quality company—one with virtually defect-free manufacturing, service, and business transaction processes—by the year 2000.

Through its Six Sigma program, which received prominent coverage in *The Wall Street Journal* and *Business Week*, GE aims to improve technological, service, and manufacturing quality across all 12 of its key businesses around the world, namely, aircraft engines, appliances, capital services, electrical distribution and control, information services, lighting, medical systems, motors, NBC, plastics, power Systems, and transportation systems.

A continuing component of GE's quality program is its Supplier Training Program, according to which selected suppliers are provided training in Six Sigma quality improvement methods, including the use of Minitab as a statistical problem-solving tool. Minitab training is provided either by GE's internal Minitab experts or by training specialists from Minitab.

As part of GE's worldwide quality improvement program, Minitab training has been provided by both GE and Minitab trainers to GE businesses in many countries, including Canada, China, France, Great Britain, Hungary, India, Japan, Malaysia, Mexico, and the United States.

Minitab Command Summary

This section describes the commands introduced in, or related to, this tutorial. Use Minitab's online Help for a complete explanation of all commands.

Minitab Menu Commands

Menu	Command	Description
Stat ➤	Control Charts	
	➤ Variables Charts for Subgroups	
	➤ Xbar-R	Produces a control chart for subgroup means and a control chart for subgroup ranges.
	➤ Xbar-S	Produces a control chart for subgroup means and a control chart for subgroup standard deviations.
	➤ Xbar	Constructs an $\overline{X}$ chart for variables data stored in a column.
	➤ R	Constructs a range chart using the variables data stored in a column.
	➤ S	Constructs a standard deviation chart using the variables data stored in a column.
	➤ Variables Charts for Individuals	
	➤ I-MR	Produces a chart of individual observations and a moving range chart.
	➤ Individuals	Constructs an individuals chart using the variables data stored in a column.
	➤ Moving Range	Constructs a moving range chart using the variables data stored in a column.
	➤ Attributes Charts	
	➤ P	Constructs a proportion of defectives chart using attributes data stored in a column.
	➤ NP	Constructs a number of defectives chart using attributes data stored in a column.
	➤ C	Constructs a number of defectives per unit chart using attributes data stored in a column for constant subgroup size.
	➤ U	Constructs a chart of the number of defectives per unit sampled using attributes data stored in a column.
	➤ Time-Weighted Charts	
	➤ Moving Average	Constructs a moving average chart using variables data stored in a column.
	➤ EWMA	Constructs an exponentially weighted moving average chart using variables data stored in a column.
	Quality Tools	
	➤ Pareto Chart	Generates a Pareto chart—a bar chart with the bars ordered from largest to smallest usually combined with an ogive.
	➤ Cause-and-Effect	Generates a cause-and-effect (fishbone or Ishikawa) diagram that depicts the potential causes of a problem.

Review and Practice

Matching Problems

Match the following terms to their definitions by placing the correct letter next to the term it describes.

_____ Pooled standard deviation

_____ Individuals

_____ Standard deviation

_____ Mean

_____ Rbar

_____ Tests for special causes

_____ Pareto chart

_____ Cause-and-effect diagram

_____ Non-constant control limit

_____ Zero-valued control limit

a. A dialog box option that fixes the center line at a specified value in a control chart

b. A Minitab dialog box option that fixes the variability value in a control chart at a specified value

c. A non-default estimate of variability in the Xbar chart

d. Default estimate of variability in the Xbar chart

e. Procedures available to determine failures associated with certain control charts

f. Notation that refers to variables data consisting of samples of size 1

g. A staircased limit possible in a P chart

h. A lower control limit that is possible in an MR chart

i. A display that depicts possible causes of a problem

j. A display that includes a bar chart with the bars ordered from largest to smallest

True/False Problems

Mark the following statements with a *T* or an *F*.

a. ____ You can optionally specify the control limits and center line in all of Minitab's control charts.

b. ____ A control chart's center line is always the average of its LCL and UCL.

c. ____ An observation outside the control limits is an indication that a process is out of control.

d. ____ The Xbar command lets you perform nine special-cause tests.

e. ____ There are four tests for special causes available with the P (chart) command.

f. ____ You can specify multiple columns using the R (chart) command.

g. ____ For the P chart, the default center line is the median of the sample proportions.

h. ____ The Session window always displays details of the special-cause test failures.

i. ____ All of Minitab's control chart commands require you to specify the subgroup size.

j. ____ Variables data are measured; attributes data are counted.

Practice Problems

The practice problems instruct you to save your worksheets with a filename prefixed by P, followed by the tutorial number. In this tutorial, for example, use *P15* as the prefix. If necessary, use Help or Appendix A Data Sets, to review the contents of the data sets referred to in the following problems. Interpretations should use the language of the subject matter of the question. If you are using the Student version, be sure to close worksheets when you have completed a problem.

1. There are many reasons why you might be late for class.

    a. List reasons why you might be late in six columns in a Minitab worksheet, one column for each of Minitab's Cause-and-Effect default causes (Personnel, Machines, Material, Methods, Measurements, and Environment).

    b. Create a cause-and-effect diagram using the six columns you just created.

2. This problem asks you to employ two quality tools.

    a. Create a cause-and-effect diagram that depicts the potential reasons an individual may not complete an assignment on time. You may use any or all of Minitab's default causes or make up your own causes.

    b. Based upon your cause-and effect diagram, collect data from at least ten colleagues to determine which of the potential reasons has caused them to not complete an assignment on time. Create a Pareto chart from these data. Write a short paragraph to describe this chart.

3. In 2001, the National Safety Council published a report on United States Vehicle Accidents. Here are the accident causes and their corresponding numbers of occurrences in parentheses: collision between motor vehicles (42,900), collision with a fixed object (18,400), collision with a pedalcycle (800), collision with a railroad train (400), non-collision accidents (5,100), other collision such as animal or animal-driven vehicles (100), and pedestrian accidents (5,800). Create a Pareto chart for these fatalities. What do you learn from this chart?

4. Open *Pubs.mtw*.

    a. Investigate the variability of the process weight per volume by constructing an R chart. The process is not in control. Identify the failed test(s).

    b. The reason that this process is out of control is due to the publication of a 224-page publication. Delete that publication from the worksheet and construct another R-chart. Is the process in control? Explain briefly.

    c. Investigate the process mean weight per volume by constructing an Xbar chart. Is the process in control? Explain briefly.

    d. Construct an Xbar-R chart for the weight per volume. What are the advantages of using this chart instead of separate control charts? What are the disadvantages?

5. Redo the Xbar control chart in Section 15.3 but use the Rbar estimate for the standard deviation instead of the default pooled standard deviation estimate. Are your results similar? Explain briefly.

6. Open *Sleep.mtw*.

   a. Construct R charts for the hours of sleep per week for each of the two brothers. Use seven as the subgroup size. Why is seven an appropriate subgroup size? What do you learn about these two processes from these charts?

   b. Construct Xbar charts for the hours of sleep per week for each of the two brothers. Again use seven as the subgroup size. What do you learn about these two processes from these charts?

   c. Construct Xbar-R charts for the hours of sleep per week for each of the two brothers. Did you learn anything else from these charts that you did not learn in parts a and b? Explain briefly.

   d. Would you present the output from parts a and b or from part c to explain your conclusions? Explain briefly.

   e. Calculate a new variable that is the difference between the hours of sleep per week for the two brothers. Construct an Xbar-R chart (with the sample size again equal to seven) for this variable. What do you learn from this chart?

7. Open *Exh_qc.mtw* that is stored in the Data folder, not the Studnt14 folder. Consider Shift as the ID column for the subgroup sizes for the following charts.

   a. Construct an Xbar-R chart of Faults. Is the process in control? Are there any unusual patterns present in the chart? Explain briefly.

   b. Construct an Xbar-S chart of Faults. Is the process in control? Are there any unusual patterns present in the chart? Explain briefly.

   c. What is the difference between the Xbar-R and Xbar-S charts? Which is most appropriate for this problem?

   d. Modify your most appropriate chart by specifying Shift as the X scale for each chart (*Hint:* Click Edit X Scale and then click on the Time tab) and changing the title to indicate the most unusual pattern that you observed (*Hint:* Click Edit Title.)

8. Open *Election2.mtw*.

   a. Construct a Moving Range chart for Rep%. This process is not in control. Explain why the process is not in control by using the output in the Session window.

   b. Construct a Moving Range chart for Dem%. This process is in control. Explain why the process is in control by using the output in the Session window.

   c. If it is appropriate, construct an I(ndividuals) chart for each of these variables. Explain briefly what historical event produced the similarity between the two out-of-control processes.

   d. Are these processes capable of being brought under "control"? Explain briefly.

9. Open *Pubs.mtw*. Create a new variable that is the sum of the weights for each volume. (*Hint:* Use Stat > Basic Statistics > Store Descriptive Statistics.) Investigate this new variable by constructing an I-MR chart. Is the process in control? Explain briefly.

10. Open *Exh_qc.mtw* that is stored in the Data folder, not the Studnt14 folder. Construct a Moving Range chart for Weight, the weight in pounds of each batch of raw material. Is this process in control? Explain briefly.

11. In the illustration of P charts in this tutorial, the default option (the proportion of defects over all samples combined) was used to determine the center line and control limits. Redo the analysis by letting Minitab use the specified values of .01 for the proportion and 2 for the multiple of the standard deviation to display the control limits to compute the center line and control limits. Are your results similar to those shown in the tutorial? If not, can you explain why?

12. Open *Candyc.mtw*.

    a. Construct a chart for the proportion of red candies in each pack with a specified proportion of .2, not the proportion that Minitab estimates from the data. Is the process in control? Explain briefly.

    b. Modify your P chart so that the upper control limits are two standard deviations above the mean. Is the process now in control? Explain briefly.

On Your Own

As mentioned at the end of this tutorial, Minitab comes with a series of tutorials—not to be confused with the tutorials in this book—that provide excellent overviews of different aspects of the software. To reinforce your understanding of some of Minitab's control charts, it was suggested that you work through the Minitab tutorial called Session Four: Assessing Quality. (The Professional version of this Minitab tutorial contains a tool that has not been mentioned in this book—process capability analysis.)

If you are working with the Professional version, work through this Minitab tutorial. Then choose Stat > Quality Tools > Capability Analysis > Normal and click Help in the dialog box. Examine its normal probability example. Duplicate this example in Minitab and examine its StatGuide information. Write a brief paper comparing and contrasting what you learned from the Minitab tutorial, the example, and the StatGuide information.

If you are working with the Student version, work through this Minitab tutorial. Then choose Stat > Control Charts > Variables Charts for Subgroups > Xbar-R Chart and click Help in the dialog box. Examine its Xbar-R Chart example. Duplicate this example in Minitab and examine its StatGuide information. Write a brief paper comparing and contrasting what you learned from the Minitab tutorial, the example, and the StatGuide information.

Answers to Matching Problems

(d) Pooled standard deviation
(f) Individuals
(b) Standard deviation
(a) Mean
(c) Subgroup range
(e) Tests for special causes
(j) Pareto chart
(i) Cause-and-effect diagram
(g) Non-constant control limit
(h) Zero-valued control limit

Answers to True/False Problems

(a) T, (b) F, (c) T, (d) F, (e) T, (f) T, (g) F, (h) F, (i) F, (j) T

Appendix A

Data Sets

Academe.mtw

This data set contains information about 60 four-year colleges, 15 from each of four regions of the country. Most four-year colleges can be classified as one of the following three types, Doctoral institutions, Comprehensive institutions, or Baccalaureate institutions. From the list of all colleges in each of the 12 region-type combinations, five colleges were randomly selected. For each college, the numbers of instructors at each rank (Professor, Associate, and Assistant) are given by gender. Also given are the average salaries, in thousands of dollars, by rank and gender. This file is used in: On Your Own for Tutorial 9.

Column	Name	Count	Description
C1-T	Region	60	Region of the country; NorthCen (North Central), Northeast, South, or West
C2-T	Type	60	Type of college; Bacc (baccalaureate), Comp (comprehensive), or Doctoral
C3	NumMProf	60	Number of male professors
C4	NumMAssoc	60	Number of male associate professors
C5	NumMAssist	60	Number of male assistant professors
C6	NumFProf	60	Number of female professors
C7	NumFAssoc	60	Number of female associate professors
C8	NumFAssist	60	Number of female assistant professors
C9	SalMProf	60	Average salary (000s) for male professors
C10	SalMAssoc	60	Average salary (000s) for male associate professors
C11	SalMAssist	60	Average salary (000s) for male assistant professors
C12	SalFProf	60	Average salary (000s) for female professors
C13	SalFAssoc	60	Average salary (000s) for female associate professors
C14	SalFAssist	60	Average salary (000s) for female assistant professors

AgeDeath.mtw

These data are for the 151 people who died in an affluent suburb of Boston, Massachusetts during 2001. Information on day and month of death, gender, birthplace, and age at death are included. This file is used in: Tutorials 7, 8, and 13, and Practice Problems for Tutorials 2, 7, 9, and 13.

Column	Name	Count	Missing	Description
C1	Day	151	0	Day of the month at death
C2	Month	151	0	Month of death
C3-T	Gender	151	1	Gender; Female or Male
C4-T	BirthPlace	151	0	State or country of birth
C5	Age	151	0	Age at death

Assess.mtw

Assessors base their home assessments on many different variables. This data set includes a number of those variables, plus the final value of the home and land. This file is used in: Practice Problems for Tutorials 1 and 8.

Column	Name	Count	Missing	Description
C1	Land$	81	2	Assessed value of the land
C2	Total$	81	2	Assessed value of the home and the land
C3	Acreage	81	0	Number of acres
C4-T	Height	81	0	Story height (number and type of floors); 1Story, 1Stryatk (one story plus attic), 1.5Story, 2Stories, 2Storatk (two stories plus attic), SplitLev (split level), or BiLevel
C5	1stFArea	81	0	Area of first floor, in square feet
C6-T	Exterior	81	0	Exterior condition; Excellnt (excellent), Good, or Average
C7-T	Fuel	81	0	Type of fuel; Electric, NatGas (natural gas), Oil, or Solar
C8	Rooms	81	0	Number of rooms
C9	Bedrooms	81	0	Number of bedrooms
C10	FullBath	81	0	Number of full baths
C11	HalfBath	81	0	Number of half baths
C12	Fireplce	81	0	Number of fireplaces
C13-T	Garage?	81	0	Garage or NoGarage

Baby.mtw

Thirty three-month-old infants are randomly divided into five groups of six infants each. The infants in a group are repeatedly shown one of five multicolored designs: A, B, C, D, or E. The median time spent watching the design is recorded for each infant. This file is used in: Tutorials 9 and 13, and Practice Problems for Tutorials 4, 9, and 13.

Column	Name	Count	Description
C1	Time	30	Median time spent watching a design
C2-T	Design	30	One of five designs; A, B, C, D, or E

Backpain.mtw

A nurse completing her master's degree thesis collected the following data for a sample of 279 patients who had received treatment for low-back pain. This file is used in: Tutorials 7 and 8, and Practice Problems for Tutorials 3 and 8.

Column	Name	Count	Description
C1-T	Gender	279	Patient's gender; Female or Male
C2	Age	279	Patient's age
C3	LostDays	279	Number of workdays lost as a result of low-back pain
C4	Cost	279	Cost of treatment for low-back pain

BallparkData.mtw

The following data relate to the 30 Major League Baseball teams that played in the 2001 season. This file is used in: Tutorials 4 and 5, and Practice Problems for Tutorials 3 and 4.

Column	Name	Count	Description
C1-T	Team	30	Name of the team
C2-T	League	30	League; American or National
C3	ParkBlt	30	Year that the ballpark was built
C4	Capacity	30	The ballpark's official capacity
C5	Attend	30	Average home attendance for the 2001 season
C6	WinPct	30	Winning percentage for the 2001 season

BodyTemp.mtw

The body temperatures, in degrees Fahrenheit, were recorded at four points in time for 107 healthy adults. Also recorded were the age, gender, and the smoking status of the participants.

Column	Name	Count	Missing	Description
C1	Age	107	0	Age
C2-T	Gender	107	0	Gender; Female or Male
C3-T	Smoke	107	0	Smoking status; No or Yes
C4	8amDay1	107	69	Body temperature at 8 A.M., Day 1
C5	12amDay1	107	14	Body temperature at 12 A.M., Day 1
C6	8amDay2	107	37	Body temperature at 8 A.M., Day 2
C7	12amDay2	107	1	Body temperature at 12 A.M., Day 2

Candya.mtw – Candyc.mtw

Students in a statistics class opened three bags of candy containing individually wrapped snack packs of candies. They determined the average size of each individual candy (*Candya.mtw*), the weight of the packs (*Candyb.mtw*), and the proportion of candies of one color to the others in the pack and the number of defects (*Candyc.mtw*). These files are used in: Tutorial 15, and Practice Problems for Tutorials 7 and 15.

Candya.mtw

This file is used in: Tutorial 15, and Practice Problems for Tutorial 15.

Column	Name	Count	Description
C1	Bag	285	Bag number from which the individually wrapped packs came
C2	Pack	285	Pack identification number
C3	Weights	285	Weight of the individual candies

Candyb.mtw

This file is used in: Practice Problems for Tutorial 7.

Column	Name	Count	Description
C1	Bag	57	Bag number from which the individually wrapped packs came
C2	Pack	57	Pack identification number
C3	Number	57	Number of candies in each pack
C4	TotWgt	57	Total weight of the pack
C5	PckWgt	57	Weight of the packaging for each pack
C6	NetWgt	57	Total weight minus the package weight

Candyc.mtw

This file is used in: Practice Problems for Tutorial 15.

Column	Name	Count	Description
C1	Bag	57	Bag number from which the individually wrapped packs came
C2	Pack	57	Pack identification number
C3	Number	57	Number of candies in each pack
C4	Red	57	Number of red candies
C5	Brown	57	Number of brown candies
C6	Green	57	Number of green candies
C7	Orange	57	Number of orange candies
C8	Yellow	57	Number of yellow candies
C9	Defects	57	Number of defects

Carphone.mtw

The data in this file are based on a study of the relationship between automobile accidents and the use of cellular phones. This file is used in Practice Problems for Tutorial 12.

Column	Name	Count	Description
C1-T	Phone?	758	Whether or not the subject owned a cellular phone; NoPhone or Phone
C2-T	Accident	758	Whether or not the subject had an automobile accident during the study period; Accident or NoAccid

Chol.mtw

The values in this data set are the low-density lipoprotein (LDL) cholesterol levels for 14 adult males with high levels of cholesterol. The first value for each person was recorded after two weeks on a diet that included corn flakes, and the second, after two weeks on a diet that included oat bran. This file is used in: Tutorials 8 and 13.

Column	Name	Count	Description
C1	CornFlke	14	LDL level after corn flakes
C2	OatBran	14	LDL level after oat bran

CollMass.mtw and CollMass2.mtw

The data in *CollMass.mtw* are various statistics related to 60 four-year colleges in Massachusetts for 2002. *CollMass2.mtw* includes all these variables plus the variables C13 RESI1, C14 SRES1, C15 FITS1, and C16 SATV**2. These files are used in: Tutorials 10 and 11, Practice Problems for Tutorials 3 and 5, and On Your Own for Tutorial 11.

Column	Name	Count	Missing	Description
C1-T	College	60	0	Name of the college
C2-T	PubPriv	60	0	Type of ownership; Private or Public
C3	SFRatio	60	0	Ratio of faculty to students
C4	%FacPhD	60	3	Percentage of faculty with Ph.D.
C5	%Accept	60	1	Percentage of applicants that are accepted
C6	%Enroll	60	1	Percentage of accepted students that enroll
C7	FrClass	60	1	Size of the Freshman class
C8	SATV	60	2	Average SAT Verbal score
C9	SATM	60	2	Average SAT Math score
C10	%Graduate	60	28	Percentage of Freshman class who graduate in six or fewer years

Column	Name	Count	Missing	Description
C11	%Top20%	60	19	Percentage of Freshman class who were in the top 20% of their High School class
C12	Tuition	60	0	Annual tuition
C13	RESI1	60	2	Residuals from Tuition against SATV model
C14	SRES1	60	2	Standardized residuals from Tuition against SATV model
C15	FITS1	60	2	Fitted values from Tuition against SATV model
C16	SATV**2	60	2	Square of SATV

Compliance.mtw

Thirty-four patients who are being treated for heart disease filled out a questionnaire aimed at exploring attitudes toward compliance with a special diet. After answering questions about age, weight, employment status, and education, the respondents answered questions asking about different dimensions of attitude toward the diet based on a seven-point Likert scale. This file is used in: Practice Problems for Tutorials 3, 4, and 9.

Column	Name	Count	Missing	Description
C1	Age	34	0	Age of the patient
C2	Weight	34	0	Weight in pounds
C3-T	Employ	34	1	Employment status; Fulltime, Parttime, Retired, or Unemp (unemployed)
C4-T	Educ	34	1	Highest level of education achieved; EighthGr (eighth grade), HighSch (high school graduate), SomeColl (some college), CollGrad (college graduate), or Masters
C5-T	HistHD	34	0	History of heart disease; No or Yes
C6	Success	34	0	Score on a seven-point scale. The higher the score the more successful the diet is perceived as being
C7	Valuable	34	0	Score on a seven-point scale. The higher the score the more valuable the diet is perceived as being
C8	Helpful	34	0	Score on a seven-point scale. The higher the score the more helpful the diet is perceived as being
C9	Easy	34	0	Score on a seven-point scale. The higher the score the easier the diet is perceived as being
C10	Flexible	34	0	Score on a seven-point scale. The higher the score the more flexible the diet is perceived as being

Congress-Salary.mtw

Congressional salaries are recorded for each year that an increase in salary occurred.

Column	Name	Count	Description
C1	Year	33	Year increase in salary occurred
C2	Senate	33	New salary for Senators, in dollars
C3	House	33	New salaries for Representatives, in dollars

Cotinine.mtw

These data are the levels of serum cotinine, in ng/ml, a metabolite of nicotine, for three groups; subjects with no exposure to tobacco smoke, subjects with some exposure to tobacco smoke, and subjects who report tobacco use. Cotinine is produced as nicotine when it is absorbed by the body. This file is used in: Practice Problems for Tutorial 9.

Column	Name	Count	Description
C1	NoETS	50	Level of serum cotinine for those not exposed to tobacco smoke
C2	ETS	50	Level of serum cotinine for those exposed to tobacco smoke
C3	Smokers	50	Level of serum cotinine for smokers

CPI2.mtw

This data set records percent changes in the Consumer Price Index (CPI) between 1960 and 2001.

Column	Name	Count	Description
C1	Year	42	Year
C2	CPIChnge	42	Percent change in the Consumer Price Index (CPI)

Depth.mtw

Thirty-six people participated in a study of depth perception under different lighting conditions. The people were divided into three groups on the basis of age. Each of the 12 members of an age group was randomly assigned to one of three "treatment" groups. All 36 people were asked to judge how far they were from a number of different objects. An average "error" in judgment, in feet, was recorded for each person. One treatment group was shown the objects in bright sunshine, another under cloudy conditions, and the third, at twilight. This file is used in: Tutorial 9.

Column	Name	Count	Description
C1-T	AgeGroup	36	Age group; Young, MiddleAge, or Older
C2-T	Light	36	Lighting condition; Sun, Clouds, or Twilight
C3	Error	36	Average error in feet

DJC20012002.mtw, DJC20012002a.mtw, and DJC20012002b.mtw

DJC20012002.mtw contains the value of the Dow Jones Composite (DJC) price index for each of the 518 working days in the years 2001 and 2002. Also included is the value for each of the 30 stocks that are included in this index. *DJC20012002a.mtw* contains the values, in dollars, for only the first 15 stocks and *DJC20012002b.mtw*, the values, in dollars, for only the final 15 stocks. These files are used in: Tutorial 5, Practice Problems for Tutorials 5, 11, and 14, and On Your Own for Tutorial 14.

Column	Name	Count	Description
C1-D	Day	518	Date
C2	DJC	518	The Dow Jones Composite price index
C3	AA	518	Value of Alcoa
C4	AXP	518	Value of American Express
C5	T	518	Value of AT&T
C6	BA	518	Value of Boeing
C7	CAT	518	Value of Caterpillar
C8	C	518	Value of Citigroup
C9	KO	518	Value of Coca-Cola
C10	DD	518	Value of DuPont
C11	EK	518	Value of Eastman Kodak
C12	XOM	518	Value of Exxon Mobil
C13	GE	518	Value of General Electric
C14	GM	518	Value of General Motors
C15	HPQ	518	Value of Hewlett-Packard
C16	HD	518	Value of Home Depot
C17	HON	518	Value of Honeywell
C18	INTC	518	Value of Intel
C19	IBM	518	Value of IBM
C20	IP	518	Value of International Paper
C21	JNJ	518	Value of Johnson & Johnson
C22	JPM	518	Value of JP Morgan Chase
C23	MCD	518	Value of McDonald's
C24	MRK	518	Value of Merck
C25	MSFT	518	Value of Microsoft
C26	MO	518	Value of Philip Morris
C27	PG	518	Value of Procter & Gamble
C28	SBC	518	Value of SBC Communications
C29	UTX	518	Value of United Technologies
C30	WMT	518	Value of Wal-Mart Stores
C31	MMM	518	Value of 3M
C32	DIS	518	Value of Walt Disney

Donner.mtw

The Donner party consisted of a group of families traveling toward California in 1846. The party was trapped by heavy snows in the Sierra Nevada Mountains in California and many people died. This data set contains the age, the gender, and the survival status for the 45 members of the party over the age of 15. This file is used in: Practice Problems for Tutorial 11.

Column	Name	Count	Description
C1	Age	45	Age
C2-T	Gender	45	Gender; Female or Male
C3-T	Survive	45	Whether the person survived; No or Yes

Drive.mtw

This data set contains information about a driver education class.

Column	Name	Count	Missing	Description
C1	Max	85	2	Maximum number of students in a class
C2	Init	85	2	Initial number of students who signed up
C3	Dropped	85	2	Number of students who dropped
C4	Replaced	85	2	Number of students who were replaced
C5	Finished	85	2	Number of students who finished the course
C6-T	Type	85	0	Type of class; Auto (automobile) or MCycle (motorcycle)
C7	Order	85	0	Order in which classes were held

DrivingCosts.mtw

These data are the average cost of owning and operating a new car per mile in the United States for 10 years. This file is used in: Practice Problems for Tutorial 10.

Column	Name	Count	Description
C1	Year	10	Year
C2	Cost	10	Average cost per mile, in cents

DrugMarkup.mtw

The price, in dollars, of two patented drugs and their generic equivalents was recorded at 10 outlets. This file is used in: Practice Problems for Tutorial 8.

Column	Name	Count	Description
C1-T	Location	10	Type of location
C2	Prozak	10	Cost of the drug Prozak
C3	PGeneric	10	Cost of the generic version of Prozak
C4	Xanax	10	Cost of the drug Xanax
C5	XGeneric	10	Cost of the generic version of Xanax

Election2.mtw

For every presidential election from 1868 to 2000 this data set contains the percentage of the vote won by both the Republican and the Democratic candidates. This file is used in: Practice Problems for Tutorials 5, 13, and 15.

Column	Name	Count	Description
C1	Year	34	Election year
C2	Rep%	34	Percentage of the vote obtained by the Republican candidate
C3	Dem%	34	Percentage of the vote obtained by the Democratic candidate

EMail.mtw

A faculty member recorded the number of emails he received and the number he sent, for each of 379 consecutive days. This file is used in: Practice Problems for Tutorials 12 and 14.

Column	Name	Count	Description
C1-D	Month	379	Month
C2	Day	379	Day of the month
C3	In	379	Number of emails received
C4	Out	379	Number of emails sent

EmployeeInfo.mtw

The 788 employees at a large college are classified by type of job description and ethnicity. This file is used in: Tutorial 1, and Practice Problems for Tutorial 4.

Column	Name	Count	Description
C1-T	JobStatus	788	Job Status; Admin (administration), Faculty, Security, Service, or Staff
C2-T	Ethnicity	788	Ethnicity; Asian, Black, Hispanic, or White

Endowment.mtw

These data contain endowment information for the 20 private and the 20 public institutions of higher education in the United States that have the largest endowment per student.

Column	Name	Count	Description
C1-T	Type	40	Type of institution; Private or Public
C2-T	Institution	40	Name of institution
C3	Endow(000s)	40	Value of endowment, in thousands of dollars
C4	Enrollment	40	Enrollment at the institution
C5	Ratio	40	Endowment per enrolled student

ExamScores.mtw

The values in the ExamScores data set are for 19 students enrolled in a college physics class. The data consists of the scores (out of 100) for the first three exams and the score on the final exam (out of 200). This file is used in: Practice Problems for Tutorial 2.

Column	Name	Count	Description
C1-T	Student	19	Student's name
C2	Exam1	19	Student's score on the first exam
C3	Exam2	19	Student's score on the second exam
C4	Exam3	19	Student's score on the third exam
C5	Final	19	Student's score on the final exam

Fja.mtw

A statistician named Frank Anscombe developed these data to illustrate the need for regression diagnostics. When you regress combinations of some of these columns, you get the same regression line. However, if you plot the data, each plot is strikingly different. These results show why you can't depend upon traditional measures for regression. This file is used in: Tutorial 11.

Column	Name	Count	Description
C1	X	11	Predictor variable
C2	Y1	11	Response variable
C3	Y2	11	Response variable
C4	Y3	11	Response variable
C5	X4	11	Predictor variable
C6	Y4	11	Response variable

Force.mtw

In an experiment involving 12 undergraduate women, a dynamometer was used to record the strength of the hamstring when the knee is flexed at 5 degrees and then at 45 degrees to the horizontal. The data in C1 and C2 are the force, in pounds, produced in each case. This file is used in: Practice Problems for Tutorials 8 and 13.

Column	Name	Count	Description
C1	Force5	12	Force, in pounds, produced by knee flexed at 5 degrees
C2	Force45	12	Force, in pounds, produced by knee flexed at 45 degrees

GasData.mtw

This data set contains a driver's record of gas usage for maintenance records. This file is used in: Practice Problems for Tutorial 11.

Column	Name	Count	Description
C1	GasDay	201	Day on which driver purchased gas (the first is coded as 1)
C2	Mileage	201	Mileage at the time of gas purchase
C3	Gallons	201	Number of gallons purchased
C4	Cost	201	Cost, in dollars, of gas purchased
C5	MPG	201	Miles per gallon
C6-T	Locale	201	Locale of the gas station; Local or NonLocal

Height.mtw

A professor in the physiology department of a medical school collected data on the self-reported heights of 60 females in an introductory physiology class. Students were asked to provide their heights (in inches) on the first day of class. This file is used in: Tutorial 6, and Practice Problems for Tutorial 3.

Column	Name	Count	Description
C1	ID	60	The student's ID number
C2	Heights	60	The self-reported heights, in inches

Homes.mtw

This data set contains real estate data on 150 randomly selected homes. This file is used in: Practice Problems for Tutorials 7, 10, 11, and 13.

Column	Name	Count	Description
C1	Price	150	Price
C2	Area	150	Area in square feet
C3	Acres	150	Acres
C4	Rooms	150	Number of rooms
C5	Baths	150	Number of baths

Infants.mtw

Sixty-eight low-income, pregnant women participated in a study of breast-feeding and nutritional counseling. This data set contains information about both the mother and the infant. This file is used in: Tutorials 3 and 4, Practice Problems for Tutorials 6, 7, 9, 12, and 13, and On Your Own for Tutorial 4.

Column	Name	Count	Description
C1-T	Ethnic	68	Mother's ethnic origin; Black, Hispanic, or White
C2	Age	68	Age of mother at delivery
C3-T	Smoke	68	Mother's smoking status; NonSmoker, LightSmoker, or HeavySmoker
C4	PreWeight	68	Mother's weight before pregnancy
C5	DelWeight	68	Mother's weight at delivery
C6-T	BreastFed	68	Whether the mother breast-fed; No or Yes
C7	BthWeight	68	Weight of the infant, in grams
C8	BthLength	68	Length of the infant, in centimeters
C9	TimeNut	68	Time spent with a nutritionist, in minutes

Jeans.mtw

A marketing research firm collected data on individuals who last bought jeans from four major stores. In this data set the value 999 indicates a missing value.

Column	Name	Count	Description
C1-T	Size	394	Jean size; Petite, Junior, Misses, or HalfLarg (half sizes or larger sizes)
C2-T	Age	394	Age group; 10to24, 25to34, 35to44, 45to54, 55to64, or 65plus
C3-T	Marital	394	Marital status; DivorSep (divorced or separated), LivAsMar (living as married), Married, Single, or Widowed
C4	Earners	394	Number of full-time earners
C5-T	Employ	394	Employment status; FullTime (full-time: 30 or more hours a week), PartTime (part-time: fewer than 30 hours a week), or NotEmpl (not employed)
C6-T	Educ	394	Education completed; GradeSch (grade school or less), SomeHigh (some high school), HighSchl (high school graduate), BusNurTe (business, nursing, or technical school), SomeColl (some college), CollGrad (college graduate), or GradDeg (graduate degree)
C7-T	Income	394	Total yearly pretax income; Under15 (under $15,000), 15to24.9 ($15,000 to $24,999), 25to34.9 ($25,000 to $34,999), 35to49.9 ($35,000 to $49,999), 50to74.9 ($50,000 to $74,999), 75to99.9 ($75,000 to $99,999), or 100plus ($100,000 or more)
C8	JeanShop	394	Identification number of store where last pair of jeans was purchased
C9	PayJean	394	Price, in dollars, of last pair of jeans purchased
C10-T	Fashion	394	*I consider myself to be fashion/style conscious*; DDagree (definitely disagree), Dagree (disagree), SDagree (somewhat disagree), SAgree (somewhat agree), Agree, or DAgree (definitely agree)
C11-T	Cost	394	*I'm very cost conscious when it comes to clothes*; DDagree (definitely disagree), Dagree (disagree), SDagree (somewhat disagree), SAgree (somewhat agree), Agree, or DAgree (definitely agree)
C12-T	Interest	394	*I take more interest in my wardrobe than most women I know*; DDagree (definitely disagree), Dagree (disagree), SDagree (somewhat disagree), SAgree (somewhat agree), Agree, or DAgree (definitely agree)
C13-T	Spend	394	*I spend a lot of money on clothes and accessories*; DDagree (definitely disagree), Dagree (disagree), SDagree (somewhat disagree), SAgree (somewhat agree), Agree, DAgree (definitely agree)

Lakes.mtw

An extensive study was conducted to determine if the characteristics of lakes in Wisconsin had changed over the last 60 years. Historical data were obtained on 149 lakes between 1925 and 1931. Similar data were obtained on these same lakes between 1979 and 1983.

The collected information included the physical characteristics of the lake, such as maximum depth and surface area, as well as the type of lake. Other variables reported the total area of the lake's watershed and the percentage of the lake that was bog. The data also indicated the amount of development as represented by the number of permanent dwellings within 100 meters of the shore.

In addition, several water quality variables were measured. These included the pH, alkalinity, conductivity, and calcium. Since measurement techniques changed from 1925 to 1979, the data were adjusted to be comparable (the stored data represent the adjusted values).

Column	Name	Count	Missing	Description
C1	LakeID	149	0	An identification number assigned to the lake
C2-T	Type	149	0	The type of lake; Drainage, Drained, Seepage, or Spring
C3	Depth	149	0	Maximum lake depth, in meters
C4	Area	149	0	Lake surface area, in hectares
C5	WSArea	149	0	Watershed area, in hectares
C6	%Bog	149	1	Percentage of lake that is bog
C7	DWHist	149	1	Number of lake shore dwellings, circa 1930
C8	DWCurnt	149	0	Number of lake shore dwellings, circa 1980
C9	PHHist	149	0	Historical pH reading
C10	PHCurnt	149	0	Current pH reading
C11	CondHst	149	0	Historical conductivity reading
C12	CondCur	149	0	Current conductivity reading
C13	AlkHist	149	0	Historical alkalinity reading
C14	AlkCurn	149	0	Current alkalinity reading
C15	CAHist	149	33	Historical calcium reading
C16	CACurnt	149	0	Current calcium reading

Lotto.mtw

A state lotto randomly selects six numbers from the numbers 1 through 47. There are two drawings per week. To check whether the drawings are random, a student monitored the game for several years. Based upon information provided by the local newspaper, the student determined the number of times each of the numbers 1 through 47 had been drawn over the course of the past three years, for a total of 312 drawings. This file is used in: Practice Problems for Tutorial 12.

Column	Name	Count	Description
C1	Observed	47	The observed number of times each number has been drawn

Marathon2.mtw

This data set contains the winning times for the men and for the women running in the Boston Marathon. This file is used in: Practice Problems for Tutorial 5.

Column	Name	Count	Description
C1	YearM	105	Years in which men raced
C2	TimesM	105	The winning times for men, in minutes
C3	YearW	31	Years in which women raced
C4	TimesW	31	The winning times for women, in minutes

Marks.mtw

This data set is the grade summary for a social studies class in a secondary school. Three test scores have been recorded for the 24 students in the class. This file is used in: Tutorial 2.

Column	Name	Count	Description
C1-T	LastName	24	Student's last name
C2-T	FirstName	24	Student's first name
C3	Test1	24	The score (out of 100) on the first test
C4	Test2	24	The score (out of 100) on the second test
C5	Test3	24	The score (out of 100) on the third test

MBASurvey.mtw

These data are the results of a survey given to an MBA class at the beginning of the semester. This file is used in: Practice Problems for Tutorials 2, 3, and 4.

Column	Name	Count	Missing	Description
C1-T	Gender	28	0	Gender; Female or Male
C2-T	HDegree	28	2	Highest degree earned; Bachelors or Masters
C3	GMAT	28	4	Score on the GMAT test
C4	Cash	28	1	Cash on their person, in dollars
C5	AIncome	28	1	Annual income, in dollars
C6	AmEx	28	0	American Express credit card; 1 = yes or 0 = no
C7	Discover	28	0	Discover credit card; 1 = yes or 0 = no
C8	MC	28	0	MasterCard credit card; 1 = yes or 0 = no
C9	Visa	28	0	Visa credit card; 1 = yes or 0 = no
C10	Other	28	0	A credit card other than those above; 1 = yes or 0 = no

Mercedes.mtw

Various characteristics of 15 used Mercedes automobiles are provided. The data on these autos were listed in a display in *The Boston Globe*.

Column	Name	Count	Description
C1-T	Class	15	Class of Mercedes; C, E, or S
C2	Age	15	Age of the car
C3	Miles	15	Mileage on the car
C4	Price	15	Advertised price of the car, in dollars

MLBGame-Cost.mtw

MLBGameCost.mtw contains data about the cost of attending a Major League Baseball game in 2003. It contains the cost of eight items and the size of two small beverages for each of the 30 teams, along with a summary measure of the cost of attending a game. This file is used in: Practice Problems for Tutorials 11 and 12.

Column	Name	Count	Description
C1	Team	30	Team
C2	AvgAdultTicket	30	Cost of average adult ticket price, in dollars
C3	AvgChildTicket	30	Cost of average child ticket, in dollars
C4	Beer	30	Cost of small draft beer, in dollars

Column	Name	Count	Description
C5	BeerOz	30	Number of ounces in small draft beer
C6	Soda	30	Cost of small soft drink, in dollars
C7	SodaOz	30	Number of ounces in small soft drink
C8	HotDog	30	Cost of regular-size hot dog, in dollars
C9	Parking	30	Cost of parking one car, in dollars
C10	Program	30	Cost of game program, in dollars
C11	Cap	30	Cost of least expensive, adult-size adjustable cap, in dollars
C12	GameCost	30	Cost of a game, in dollars

MnWage2.mtw

This data set contains information about the minimum wage from 1950 to 2001. This file is used in: Practice Problems for Tutorials 5, 10, and 11.

Column	Name	Count	Description
C1	Year	52	Year
C2	MinWage	52	Minimum wage, in dollars

MonthlySnow.mtw

Monthly snowfall amounts for Boston from January, 1890, to September, 2001. This file is used in: Practice Problems for Tutorials 2 and 14.

Column	Name	Count	Description
C1	Month	1341	Month, listed numerically beginning with 1 for January
C2	Year	1341	Year
C3	Snowfall	1341	Snowfall, in inches

Movies.mtw

A student in an economics class collected the following data on 32 movies released in the period 1997–1998.

Column	Name	Count	Description
C1-T	Movie	32	Title of the movie
C2	Opening	32	Gross receipts for the weekend after the movie was released, in millions of dollars
C3	Budget	32	The total budget for the movie, in millions of dollars
C4-T	Star?	32	Whether or not the movie has a super star; NoStar or Star
C5-T	Summer?	32	Whether or not the movie was released in the summer; NoSummer or Summer

Murders.mtw and Murderu.mtw

These data include, for each of the 50 states, the murder rate, the state's death penalty status, and the region of the country in which the state lies. The death penalty status is based on whether the state has the death penalty and has used it since 1976 (when the Supreme Court of the United States allowed the re-introduction of the penalty), has the death penalty but not used it since 1976, or does not have the death penalty. In Murders, the data are stacked. In Murderu, they are unstacked by region. These files are used in: Practice Problems for Tutorials 2, 6, 8, 9, and 11.

Murders.mtw

This file is used in: Practice Problems for Tutorials 2, 6, and 11.

Column	Name	Count	Description
C1-T	State	50	State
C2	MurderRate	50	Number of murders per 100,000 people
C3-T	DPStatus	50	Death penalty status; NoDP (the state did not have the death penalty), DPNoEx (the state had the death penalty but had not used it since 1976), or DPEx (the state had the death penalty and had at least one execution since 1976)
C4-T	Region	50	Region of the country; NorthCen (North Central), Northeast, South, or West

Murderu.mtw

This file is used in: Practice Problems for Tutorials 2, 8, and 9.

Column	Name	Count	Description
C1-T	NCState	12	North Central state
C2	NCMurder	12	North Central murder rates
C3-T	NCStatus	12	Death penalty status of North Central states
C4-T	NEState	9	Northeast state
C5	NEMurder	9	Northeast murder rates
C6-T	NEStatus	9	Death penalty status of Northeast states
C7-T	SState	16	Southern state
C8	SMurder	16	Southern murder rates
C9-T	SStatus	16	Death penalty status of Southern states
C10-T	WState	13	Western state
C11	WMurder	13	Western murder rates
C12-T	WStatus	13	Death penalty status of Western states

MusicData.mtw

MusicData.mtw contains a randomly simulated sample of 1,000 musical sales in which the genre numbers and names of music types (Country, Jazz, and so on) are recorded based on the probabilities of each genre's being selected. The probabilities are based on music sales in 2001. This file is used in: Tutorial 12, and Practice Problems for Tutorial 6. The data from this file are used in: Tutorial 6.

Column	Name	Count	Description
C1	GenreN	9	Numerical designation (1, 2, ..., 9) of the musical genre
C2-T	Genre	9	Name of the Genre; Classical, Country, Gospel, Jazz, Pop, R&B/Urban, Rap/Hip-Hop, Rock, or Other
C3	Probability	9	The probability of each genre of music being selected
C4	Sample	1000	The sample of 1,000 genre numbers
C5-T	TxSample	1000	The sample of 1,000 genre names

NHL2003.mtw

These data are for the 30 teams in the National Hockey League for the 2002–2003 season. This file is used in: Practice Problems for Tutorial 5. A subset of the data from this file is used in: Practice Problems for Tutorial 8.

Column	Name	Count	Description
C1-T	Team	30	Team
C2	Won	30	Number of games won
C3	Lost	30	Number of games lost
C4	Tie	30	Number of games tied
C5	OverLoss	30	Number of games ending in an overtime loss
C6	Points	30	Number of points
C7	GoalsFor	30	Number of goals scored
C8	GoalsAgainst	30	Number of goals conceded
C9-T	Division	30	Division; East or West
C10	Payroll	30	Player payroll, in millions of dollars

Note02.mtw

This data set contains information about a selection of notebook (portable) computers that were available in 2002. This file is used in: Practice Problems for Tutorials 1 and 4.

Column	Name	Count	Description
C1-T	Processor	7	Type of processor
C2	Speed	7	Speed, in megahertz
C3	BatteryHrs	7	Hours before battery needs recharging
C4-T	PDevice	7	Pointing device; SP (stick pointer), TP (touchpad), or TP and SP (stick pointer and touchpad)
C5	Display	7	Size of screen, in inches
C6	Weight	7	Weight, in pounds
C7	USBPorts	7	Number of USB ports
C8	CardSlots	7	Number of card slots
C9-T	Floppy	7	Floppy drive; E (external), M (modular), M and E (external and modular), or None
C10	Price	7	Price, in dollars

OldFaithful.mtw

The data in this worksheet are the durations, the time intervals to the next eruption, and the heights of eruptions of the Old Faithful geyser in Yellowstone National Park in California. This file is used in: Tutorial 0, and Practice Problems for Tutorials 3 and 7.

Column	Name	Count	Description
C1	Duration	50	Duration of the eruption, in seconds
C2	Interval	50	Interval before the next eruption, in minutes
C3	Height	50	Height of the eruption, in feet

OpenHouse.mtw

OpenHouse.mtw is based on a listing of 105 open houses found in the Real Estate section of *The Boston Sunday Globe*. This file is used in: Practice Problems for Tutorials 10 and 11.

Column	Name	Count	Missing	Description
C1	Price	105	0	Price, in dollars
C2	Bedrooms	105	1	Number of bedrooms
C3	Bath	105	3	Number of bathrooms
C4-T	Style	105	6	Style of home; Condo (condo minimum) or Single (single family)
C5-T	Location	105	0	Location of home; City, InSuburb (inner suburb), MidSuburb (middle distant suburb), or OutSuburb (outer suburb)

PayData.mtw

This data set contains the salaries and related information for each of ten employees in the sales department of the Tayco company. This file is used in: Tutorial 1.

Column	Name	Count	Description
C1	Salary	10	The salary of an employee in the sales department
C2	YrsEm	10	The number of years employed at Tayco
C3	PriorYr	10	The number of prior years' experience
C4	Educ	10	Years of education after high school
C5	Age	10	Current age
C6	ID	10	The company identification number for the employee
C7-T	Gender	10	Gender; Female or Male

PhoneRates.mtw

These data are the weekend and weekday per-minute telephone rates from the United States to 13 countries from a telecommunication company.

Column	Name	Count	Description
C1-T	Country	13	Name of country
C2	Weekend	13	Weekend per-minute rate, in cents
C3	Weekday	13	Weekday per-minute rate, in cents

Pizza2.mtw

This data set contains campus newspaper ratings for 13 pizza shops for two semesters. This file is used in: Practice Problems for Tutorial 2. The data from this file are used in: Practice Problems for Tutorial 1.

Column	Name	Count	Missing	Description
C1-T	Pizzeria	13	0	Pizza shop
C2-T	Type	13	0	Chain or Local
C3	FallRank	13	0	Ranking (first is best)
C4	FallScore	13	0	Score (highest is best)
C5	SpringRank	13	3	Ranking from previous spring
C6	SpringScore	13	3	Scores from previous spring

Process.mtw

The process of creating, administering, and tallying the student government surveys, whose results are in the file *Process.mtw*, involves a number of causes that could contribute to process problems. This file contains six columns corresponding to the six variables used to create a Minitab cause-and-effect diagram. This file is used in: Tutorial 15.

Column	Name	Count	Description
C1-T	Personnel	3	Potential process problems caused by people
C2-T	Machines	5	Potential process problems caused by machines
C3-T	Materials	2	Potential process problems caused by materials
C4-T	Methods	2	Potential process problems caused by methods
C5-T	Measurements	2	Potential process problems caused by the measuring process
C6-T	Environment	3	Potential process problems caused by the environment

Prof.mtw

The student government randomly distributed surveys to 15 students in each of 146 sections in a variety of disciplines. The survey asked students to evaluate the course and the instructor. Each participating section received 15 surveys. Some sections returned all 15 surveys, whereas others returned fewer. The results for each section were averaged for use in this worksheet. There are nine columns and 146 rows. Each row presents a summary of the information taken from the 1 to 15 surveys returned by a particular class. This file is used in: Practice Problems for Tutorials 7 and 12.

Column	Name	Count	Missing	Description
C1-T	Dept	146	0	The academic department of the course to which the 15 survey were distributed
C2	Number	146	0	The course number
C3	Interest	146	0	The section average of the surveyed students' responses to *The course stimulated your interest in this area*; 0 = strongly disagree, 1 = disagree, 2 = neutral, 3 = agree, or 4 = strongly agree
C4	Manner	146	0	The section average of the surveyed students' responses to *The instructor presented course material in an effective manner*; 0 = strongly disagree, 1 = disagree, 2 = neutral, 3 = agree, or 4 = strongly agree

Column	Name	Count	Missing	Description
C5	Course	146	0	The section average of the surveyed students' responses to *Overall, I would rate this course as ...*; 0 = poor, 1 = below average, 2 = average, 3 = above average, or 4 = excellent
C6	Instrucr	146	0	The section average of the surveyed students' responses to *Overall, I would rate this instructor as...*; 0 = poor, 1 = below average, 2 = average, 3 = above average, or 4 = excellent
C7	Responds	146	0	The number of completed surveys returned out of the 15 surveys distributed
C8	Size	146	16	The number of students in the section
C9-T	Year	146	0	The level of the course; Freshman, Soph (sophomore), Junior, or Senior

Pubs.mtw

This data set includes information about a series of publications over time. This file is used in: Practice Problems for Tutorial 15.

Column	Name	Count	Description
C1	Volume	84	Volume of publication
C2	Number	84	Number of publication within the volume
C3	Pages	84	Number of pages
C4	Weight	84	Weight, in grams
C5-T	Cover	84	Cover type; PlnBrown (plain brown), PlnGreen (plain green), GlsyBrwn (glossy brown), or GlsyGren (glossy green)
C6-T	Binder	84	Binding status; Binder or NoBinder

PulseA.mtw

Ninety-two students in a statistics class performed the following experiment. Each student recorded his or her pulse rate and then flipped a coin. If the coin came up heads, the student ran in place for a minute. Otherwise, the student did not. The student then took his or her pulse rate again. The pulse rates and related information about the students are recorded in the following chart. This file is used in: Practice Problems for Tutorials 1, 2, 3, 4, 5, 7, 8, 9, 10, and 12.

Column	Name	Count	Missing	Description
C1	Pulse1	92	0	Initial pulse rate, in beats per minute
C2	Pulse2	92	0	Second pulse rate, in beats per minute
C3-T	Ran	92	0	Whether or not the student ran in place; Ran or Still
C4-T	Smokes	92	0	Smoking status; Smoke or NonSmoke

PulseA.mtw
(continued)

Column	Name	Count	Missing	Description
C5-T	Gender	92	0	Student's gender; Female or Male
C6	Height	92	0	Height, in inches
C7	Weight	92	0	Weight, in pounds
C8-T	Activity	92	1	Usual level of activity; Slight, Moderate, or ALot

Radlev.mtw

A television station in southwestern Ohio did a survey to determine the radon levels in homes in its viewing area. Questionnaires and radon detection kits were sent to all viewers who requested them. This file includes the results of only certain questions from the full questionnaire. This file is used in: Practice Problems for Tutorial 2.

Column	Name	Count	Missing	Description
C1	Radon	543	30	Radon measurement, in pico curies
C2	Age	543	60	Age of the house, in years
C3	Days	543	65	Number of days the kit was exposed
C4-T	Insulate	543	9	Amount of insulation in the house; Excellnt (excellent), Average, Poor, or DontKnow
C5-T	Sump?	543	17	Was the sample taken near a sump pump?; Yes or No

Random-Integers.mtw

Twenty-four undergraduate students and 28 graduate students were asked to provide a "random" integer between 0 and 999. The gender and the integers chosen are provided in this data set. This file is used in: Practice Problems for Tutorial 8.

Column	Name	Count	Missing	Description
C1	UGender	24	0	Gender of undergraduate student; 1 = male or 2 = female
C2	URandomI	24	1	"Random" integer chosen by an undergraduate student
C3	GGender	28	0	Gender of graduate student; 1 = male or 2 = female
C4	GRandomI	28	0	"Random" integer chosen by a graduate student

During the summer of 1988, one of the hottest on record in the Midwest, a graduate student in environmental science conducted a study for a state Environmental Protection Agency. She studied the impact of an electric generating plant along a river by observing several water characteristics at five sites along the river. Site 1 was approximately four miles up river from the electrical plant, about six miles down river from a moderately large Midwestern city, and directly down river from a large suburb of this city. Site 2 was directly up river from the cooling inlets of the plant. Site 3 was directly down river from the cooling discharge outlets of the plant. Site 4 was approximately three-quarters of a mile down river from site 3, and site 5 was approximately six miles down river from site 3.

The Agency was most concerned with the plant's use of river water for cooling. Scientists feared that the plant was raising the temperature of the water and hence endangering the aquatic species that lived in the river.

The student anchored data sounds (battery-operated and self-contained canisters that float in the river) at the five sites and took hourly measurements of the temperature, pH, conductivity, and dissolved oxygen content. She left the devices there for five consecutive days.

The data sets Rivera through Rivere contain the recordings for each separate site; they therefore do not include the columns C6 Site and C7 Hour1 described below. Rivers.mtw contains the recordings from all five sites and includes all of the columns described in the following chart (it also contains the three missing values from Riverb). These files are used in: Tutorial 14, and Practice Problems for Tutorials 4, 13 and 14.

Column	Name	Count	Missing	Description
C1	Hour	+	0	The hour of the measurement, using a 24-hour clock
C2	Temp	+	0	The temperature, in °C, of the river at the given hour
C3	PH	+	0	The pH of the river at the given hour
C4	Cond	+	0	The conductivity, in electrical potential, of the river at the given hour
C5	DO	+	0	The dissolved oxygen content of the river at the given hour
C6	Site	478	0	Site
C7	Hour1	478	0	The hour recorded as a number from 1 through 100

+ The counts for the different data sets are Rivera = 95; Riverb = 95; Riverc = 95; Riverd = 96; Rivere = 97; Rivers = 478. Note that Riverb and Rivers have three missing values.

Riverc2.mtw

Riverc2 contains six additional variables.

Column	Name	Count	Missing	Description
C1	Hour	95	0	The hour of the measurement, using a 24-hour clock
C2	Temp	95	0	The temperature, in °C, of the river at the given hour
C3	PH	95	0	The pH of the river at the given hour
C4	Cond	95	0	The conductivity, in electrical potential, of the river at the given hour
C5	DO	95	0	The dissolved oxygen content of the river at the given hour
C6	FORE1	48	0	Forecasts for the next 48 hours
C7	Lags	95	1	Lags
C8	Diff	95	1	Differences
C9	AForecst	48	0	ARIMA forecasts
C10	LAForcst	48	0	Lower limits for ARIMA forecast
C11	UAForcst	48	0	Upper limits for ARIMA forecast

Salary02.mtw

This data set contains 2001 and 2002 salary information for 60 faculty members teaching at a small New England College. Other variables include gender, rank, and years in rank. This file is used in: Practice Problems for Tutorials 2, 5, and 13.

Column	Name	Count	Description
C1-T	GenderProf	20	Gender of Professors; F (female)or M (male)
C2	YearsInRankProf	20	Number of years as Professors
C3	Salary01Prof	20	Salary for 2001–2002 academic year for Professors
C4	Salary02Prof	20	Salary for 2002–2003 academic year for Professors
C5-T	GenderAssoc	15	Gender of Associate Professors; F (female) or M (male)
C6	YearsInRankAssoc	15	Number of years as Associate Professors
C7	Salary01Assoc	15	Salary for 2001–2002 academic year for Associate Professors
C8	Salary02Assoc	15	Salary for 2002–2003 academic year for Associate Professors
C9-T	GenderAssist	23	Gender of Assistant Professors; F (female) or M (male)

Column	Name	Count	Description
C10	YearsInRankAssist	23	Number of years as Assistant Professors
C11	Salary01Assist	23	Salary for 2001–2002 academic year for Assistant Professors
C12	Salary02Assist	23	Salary for 2002–2003 academic year for Assistant Professors
C13-T	GenderInstr	2	Gender of Instructors; F (female) or M (male)
C14	YearsInRankInstr	2	Number of years as Instructors
C15	Salary01Instr	2	Salary for 2001–2002 academic year for Instructors
C16	Salary02Instr	2	Salary for 2002–2003 academic year for Instructors

SBP.mtw

A drug designed to lower systolic blood pressure (SBP) was administered to 12 subjects. The data set contains their SBP before and after the drug was taken.

Column	Name	Count	Description
C1	Before	12	Systolic blood pressure before taking the drug
C2	After	12	Systolic blood pressure after taking the drug

SchoolsData.mtw

This data set contains information for 135 public high schools in the Greater Boston area.

Column	Name	Count	Description
C1-T	School	135	School district
C2	Enrollment	135	Enrollment
C3	Cost/Pupil	135	Cost per pupil
C4	AveTeach$	135	Average teacher salary
C5	SATV	135	Average SAT Verbal score
C6	SATM	135	Average SAT Math score
C7	SATPartRate	135	SAT participation rate
C8	10GMCASEng	135	Average 10th grade MCAS English score
C9	10GMCASMth	135	Average 10th grade MCAS Math score
C10	S/TRatio	135	Student/teacher ratio
C11	S/CounselRatio	135	Student/Counselor ratio
C12	DropoutRate	135	Dropout rate
C13	%College	135	Percentage going to college
C14	%2YrPub	135	Percentage going to two-year public colleges

Column	Name	Count	Description
C15	%4YrPub	135	Percentage going to four-year public colleges
C16	%2YrPri	135	Percentage going to two-year private colleges
C17	%4YrPri	135	Percentage going to four-year private colleges
C18	%Military	135	Percentage going into the military
C19	%Work	135	Percentage going to work
C20	%Other	135	Percentage choosing other activities

Sleep.mtw

For 55 consecutive nights two brothers recorded their number of hours of sleep. This file is used in: Practice Problems for Tutorials 14 and 15.

Column	Name	Count	Description
C1	Sleep1	55	Hours of sleep for one brother
C2	Sleep2	55	Hours of sleep for the other brother

SP5002.mtw

A professor in a business school monitored the stock market. This data set contains 523 consecutive closing values for the Standard and Poor's 500 Cash (Spot) Index. This file is used in: Practice Problems for Tutorial 14.

Column	Name	Count	Description
C1-D	Date	523	Date
C2	Value	523	Closing value

SPCarData.mtw

This data set consists of fuel economy ratings for 2002, two-seater sports cars.

Column	Name	Count	Missing	Description
C1-T	Model	13	0	Manufacturer and model
C2-T	Transmission	13	0	Type of transmission; A (automatic) or M (manual)
C3	Speeds	13	1	Number of speeds
C4	EngineCap	13	0	Engine capacity
C5	Cylinders	13	0	Number of cylinders
C6	CityMPG	13	0	Miles per gallon for city driving
C7	HwyMPG	13	0	Miles per gallon for highway driving
C8	PremiumGas	13	0	Requires premium gas?; 1 = yes or 0 = no
C9	GasGuzzle	13	0	The automobile is rated as a "Gas Guzzler"?; 1 = yes or 0 = no
C10	Turbo	13	0	The engine is turbo-charged?; 1 = yes or 0 = no

SpeedCom.mtw

Computers are becoming faster each year! For each of the years 1991 to 2002 here are the speeds of the fastest supercomputers in gigaflops (billions of mathematical operations per second), as well the number of processors in each computer. This file is used in: Practice Problems for Tutorial 10.

Column	Name	Count	Description
C1	Year	12	Year
C2-T	Computer	12	Name and type of computer
C3	NProcessors	12	Number of processors
C4	SPDGigaflops	12	Speed of the computer, in gigaflops

Stores2.mtw

A marketing research firm collected data on the number of individuals who last bought jeans from four major stores, classified according to their approximate annual household income before taxes, measured to the nearest hundred dollars. This file is used in: Practice Problems for Tutorial 12.

Column	Name	Count	Description
C1	Store	4	Store identification
C2	Under25K	4	The number of individuals earning less than $25,000 who purchased jeans at each store
C3	25KTo34.9K	4	The number of individuals earning from $25,000 to $34,900 who purchased jeans at each store
C4	35Kto49.9K	4	The number of individuals earning from $35,000 to $49,900 who purchased jeans at each store
C5	50KTo74.9K	4	The number of individuals earning from $50,000 to $74,900 who purchased jeans at each store
C6	75KPlus	4	The number of individuals earning $75,000 or more who purchased jeans at each store

Survey.mtw

This data set contains the results of a survey of 100 students in a statistics class. This file is used in: Practice Problems for Tutorials 2, 4, and 9.

Column	Name	Count	Missing	Description
C1-T	Gender	100	0	Gender; Female or Male
C2	Age	100	0	Age, in years
C3	Height	100	0	Height, in inches
C4	Coins	100	0	Value of coins, in cents, in possession of student
C5	Keys	100	0	Number of keys in possession of student
C6	Credit	100	0	Number of credit cards in possession of student
C7	Pulse	100	6	Pulse rate, in beats per minute
C8	Exercise	100	0	Student engages in vigorous exercise?; 1 = yes or 2 = no
C9	Smoke	100	0	Student smokes?; 1 = yes or 2 = mo
C10	ColorBlind	100	0	Student is color-blind?; 1 = yes or 2 = no
C11	Hand	100	0	Student handedness?; 1 = left-handed, 2 = right-handed, or 3 = ambidextrous

TBill2.mtw

This data set lists the weekly 6-month Treasury Bill values for 2001 and 2002. This file is used in: Practice Problems for Tutorial 1.

Column	Name	Count	Description
C1	High	104	High value for week
C2	Low	104	Low value for week
C3	Close	104	Closing value for week

Temco.mtw

This data set contains the salary information for all salaried employees in four departments at the Temco company. This file is used in: Practice Problems for Tutorials 9, 10, and 11.

Column	Name	Count	Description
C1	Salary	46	The salary of an employee
C2	YrsEm	46	The number of years employed at Temco
C3	PriorYr	46	The number of prior years' experience
C4	Educ	46	Years of education after high school

Column	Name	Count	Description
C5	ID	46	The company identification number for the employee
C6-T	Gender	46	The gender of the employee; Female or Male
C7-T	Dept	46	The employee's department; Advertse (advertising), Engineer (engineering), Purchase (purchasing), or Sales
C8	Super	46	The number of employees supervised by this employee

Textbooks.mtw

This data set contains information on 36 introductory, applied statistics textbooks published between 1998 and 2002. This file is used in: Practice Problems for Tutorials 2, 5, 8, 10, and 13.

Column	Name	Count	Missing	Description
C1	Type	36	0	Type of intended audience; BusEcon (business/economics) or GenEd (general education)
C2	Edition	36	0	Edition
C3	List$	36	1	Publisher's list price, in dollars
C4	Amazon$	36	4	Price on Amazon.com, in dollars
C5	NumChap	36	0	Number of chapters in the book
C6	NumPages	36	0	Number of pages in the book
C7	ReadEase	36	0	A measure of readability. The higher the score the more readable the text
C8	GradeLev	36	0	Estimated grade level for readability
C9	MTBUse	36	0	Intensity of Minitab use; 0 = text does not use Minitab, 1 = text provides Minitab output only, or 2 = text provides both output and instruction in Minitab
C10	GraphUse	36	0	Text makes use of graphs through the book?; 1 = yes or 0 = no
C11	NumDSFiles	36	0	Number of data set files that are included with the text
C12	NumExer	36	0	Number of exercises in the text

Top25Stars.mtw

The top 25 box office stars in 2002 are listed with the amount of money their movies have made and the number of movies in which they have appeared.

Column	Name	Count	Description
C1-T	Star	25	Name of star
C2	TBoxOffice	25	Total box office proceeds from movies, in dollars
C3	Movies	25	Number of movies in which the star has appeared

Tvhrs.mtw

This file includes the results of a study of TV-viewing patterns.

Column	Name	Count	Description
C1-T	ID	120	Subject identification number
C2-T	AgeGrp	120	Age group; GradeSch (grade school), CollStu (college student), or 50plus (50 years of age or older)
C3	Age	120	Ages
C4-T	Gender	120	Gender; Female or Male
C5-T	Sesame	120	Whether they watched Sesame Street; No or Yes
C6	HrsTV	120	Hours spent watching TV in a week
C7	HrsMTV	120	Hours spent watching MTV in a week
C8	HrsNews	120	Hours spent watching news in a week
C9-T	Educ	120	Highest level of education achieved; SomGrade (some grade school), SomHigh (some high school), HighSchl (high school diploma), SomeColl (some college), College (college degree), <=2Grad (two or fewer years of graduate school), or >2Grad (more than 2 years of graduate school)

These data are for those batters on the New York Yankees and the Minnesota Twins baseball teams who played for that team in the 2001 baseball season and who were on the same team at the start of the 2002 season. Performance statistics are for the 2001 season. This file is used in: Practice Problems for Tutorial 6.

Column	Name	Count	Description
C1-T	Team	18	Team; Twins or Yankees
C2	Num	18	Player's number
C3-T	Name	18	Player's name
C4	2002Salary(000s)	18	Salary for the 2002 season, in $1,000s
C5-T	Pos	18	Position; C (catcher), 1B (first base), 2B (second base), 3B (third base), SS (short stop), LF (left field), CF (center field), RF (right field), or DH (designated hitter)
C6-T	Bat	18	Bats; Both (both left-handed and right-handed), L (left-handed), or R (right-handed)
C7-T	Throw	18	Throws; L (left-handed) or R (right-handed)
C8	Age	18	Age, in years
C9	HtFeet	18	Height, in feet (rounded down)
C10	HtInches	18	Additional inches of height
C11	Weight	18	Weight, in pounds
C12-T	Born	18	State or country of birth
C13	GP	18	Number of games played
C14	AB	18	Number of at-bats
C15	R	18	Number of runs scored
C16	H	18	Number of singles
C17	2B	18	Number of doubles
C18	3B	18	Number of triples
C19	HR	18	Number of home runs
C20	RBI	18	Number of runs-batted-in
C21	BB	18	Number of base-on-balls
C22	SO	18	Number of strikeouts
C23	BAvg	18	Batting average
C24	Inn	18	Number of innings played
C25	PO	18	Number of putouts
C26	A	18	Number of assists
C27	E	18	Number of errors
C28	FPct	18	Fielding percentage

TwoTowns.mtw

These data are for 138 homes listed with a real estate company in a suburb of Boston. This file is used in: Practice Problems for Tutorials 4, 5, and 11.

Column	Name	Count	Description
C1-T	Town	138	Town; Framingham or Natick
C2	Price	138	Asking price, in dollars
C3	Rooms	138	Number of rooms
C4	Bedrooms	138	Number of bedrooms
C5	Baths	138	Number of bathrooms
C6	HalfBaths	138	Number of half-bathrooms
C7	Garages	138	Number of garage spaces

UGradSurvey.mtw

These data are from a survey on Internet use of the 24 undergraduate students in a fall 2001 statistics class.

Column	Name	Count	Missing	Description
C1-T	Gender	24	0	Gender; Female or Male
C2-T	HSGrade	24	2	Average Grade in High School; B, B+, B+/A-, A-, or A
C3	IYear	24	0	Year of first Internet use
C4	ITimes	24	4	Number of times logged onto the Internet each day
C5	IMinutes	24	4	Total minutes logged onto the Internet each day
C6	EMail	24	0	Email use of Internet?; 1 = yes or 0 = no
C7	IM	24	0	Instant messenger use of Internet?; 1 = yes or 0 = no
C8	Songs	24	0	Song downloading use of Internet?; 1 = yes or 0 = no
C9	Chats	24	0	Chat room participation use of Internet?; 1 = yes or 0 = no
C10	Photos	24	0	Photograph exchange use of Internet?; 1 = yes or 0 = no
C11	Other	24	0	Other use of Internet?; 1 = yes or 0 = no

USAArrivals.mtw

These data are on-time statistics for flights of US Airways into Boston's Logan Airport on the Sunday after Thanksgiving in 2000 and 2001.

Column	Name	Count	Description
C1-T	Carrier	155	US Airways; US
C2-D	Date	155	Date of flight
C3	FlightNum	155	Flight number
C4-T	Origin	155	Origin of flight
C5-D	SchArrival	155	Scheduled arrival time
C6-D	ActArrival	155	Actual arrival time
C7	SchElapsed	155	Scheduled flying time, in minutes
C8	ActElapsed	155	Actual flying time, in minutes
C9	ArrvlDelay	155	Arrival delay, in minutes

USDemData.mtw

This data set displays census values for the United States, including the estimated population and population density for the years 1790, 1800, ..., 2000.

Column	Name	Count	Description
C1	Year	22	Year
C2	Population	22	Population of the United States, in millions
C3	Density	22	Density of the United States population, in people per square mile

WastesData.mtw

New and abandoned hazardous waste sites are being discovered across the United States. This data set lists the number of hazardous waste sites found in each state (and the District of Columbia). Also given is the region of the country that the state is located.

Column	Name	Count	Description
C1-T	State	51	State
C2	NumSites	51	Number of hazardous waste sites
C3-T	Region	51	Region of the country; ENCentrl (East North Central), ESCentrl (East South Central), MAtlantic (Mid-Atlantic), Mountain, NEngland (New England), Pacific, SAtlantic (South Atlantic), WNCentrl (West North Central), or WSCentrl (West South Central)

YearlySnow.mtw

This data set contains the total snowfall amounts and equivalent rainfall (in inches) for Boston recorded for the 109 years between 1892 and 2000. This file is used in: Tutorial 13 and Practice Problems for Tutorials 13 and 14. A subset of the data from this file is used in: Practice Problems for Tutorial 1.

Column	Name	Count	Description
C1	Year	109	Year
C2	Snowfall	109	Snowfall, in inches
C3	Rain	109	The equivalent rainfall, in inches

YogurtData.mtw

These data, collected by a research company, show the results of testing 17 brands of plain yogurt. The tests evaluated overall nutritional value, cost per ounce, and the number of calories per eight-ounce serving. This file is used in: Practice Problems for Tutorial 9. The data from this file are used in: Tutorial 1.

Column	Name	Count	Description
C1-T	Rating	17	Nutritional rating; Excellent, Very Good, Good, or Fair
C2	Cost	17	Cost per ounce, in cents
C3	Calories	17	Calories per 8-oz. serving

▶ **Note** The following Data folder files were used: *Exh_qc.mtw* in Practice Problems for Tutorial 15, *Grades.mtw* in Practice Problems for Tutorial 6, and *Pancake.mtw* in Practice Problems for Tutorial 9. ■

Minitab Menus and Toolbars

Menus

The 17 menus below are for the Student version of Minitab. They differ little from, and form the core of, those in the Professional version. Professional version users may activate the Student version menus by using the Tools > Manage Profiles > Manage command.

FIGURE B-1

File Menu for the Data window

FIGURE B-2

File Menu for the Session window

FIGURE B-3

File Menu for the Graph window

FIGURE B-4

File Menu for the History window

FIGURE B-5

*Edit Menu for the
Data window*

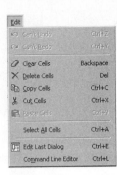

FIGURE B-8

Calc Menu

FIGURE B-6

*Edit Menu for the
Session window*

FIGURE B-9

Stat Menu

FIGURE B-10

Graph Menu

FIGURE B-7

Data Menu

FIGURE B-11

Editor Menu for the Data window

FIGURE B-12

Editor Menu for the Session window

FIGURE B-13

Editor Menu for the Graph window

FIGURE B-14

Editor Menu for the Brushing window

FIGURE B-15

Tools Menu

FIGURE B-16

Window Menu

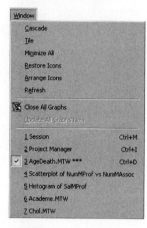

FIGURE B-17

Help Menu

Toolbars

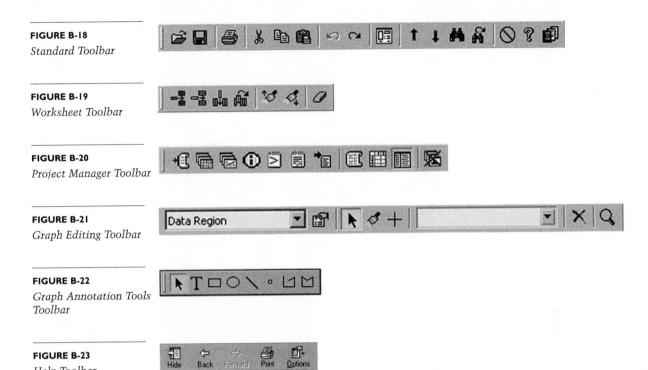

FIGURE B-18
Standard Toolbar

FIGURE B-19
Worksheet Toolbar

FIGURE B-20
Project Manager Toolbar

FIGURE B-21
Graph Editing Toolbar

Data Region

FIGURE B-22
Graph Annotation Tools Toolbar

FIGURE B-23
Help Toolbar

Hide   Back   Forward   Print   Options

#, allows for internal documentation on Exec macro, 385

Academe.mtw data set, 311
summary, 484

active cell, 28

actual power, 271

Add > Footnote button, 150

Addison-Wesley Publishing Web site, 4, 14

AgeDeath.mtw data set, 88–89, 229–241, 251, 256–265, 309, 413–416, 423
summary, 484

? (level of significance), 232

alternative hypothesis, 231

Analysis of Variance, *See* ANOVA

Anderson-Darling test, 268, 351, 368

ANOVA (Analysis of Variance), 283
checking assumptions for one-way, 289–291
comparing means of separate populations, 284–289
comparing means of several populations with responses in separate columns, 294–295
Tukey's multiple comparisons test, 291–294
two-way, 296–306

ANOVA > Balanced ANOVA command (Stat Menu), 303

ANOVA > General Linear Model command (Stat Menu), 303

ANOVA > Interactions Plot command (Stat Menu), 303

ANOVA > Interval Plot command (Stat Menu), 288

ANOVA > One-Way command (Stat Menu), 286, 289–290, 292

ANOVA > One-Way (Unstacked) command (Stat Menu), 294–295

ANOVA > Test for Equal Variances command (Stat Menu), 276, 306

ANOVA > Two-Way command (Stat Menu), 298

Append Graph to Report button, 122, 150

Append Selected Lines to Report button, 122

ARIMA (autoregressive integrated moving average)
time series forecasting, 447–450
time series modeling, 435–438, 441–446

AR 2 model, 441

ASCII text files (American Standard Code for Information Interchange), 83–84

Assess.mtw data set, 57, 279
summary, 485

associated variables, 145

Assume equal variances check box, 258–259

attributes data, 473

autocorrelation plots, 435–437

Autofill, 200, 306

autoregressive (AR) component, 436

autoregressive integrated moving average, *See* ARIMA

Baby.mtw data set, 155, 284–295, 308, 417–420, 425
summary, 485

Backpain.mtw data set, 128, 130–131, 241–248, 272–276, 281
summary, 486

balanced two-way ANOVA, 303

BallparkData.mtw data set, 129, 140–151, 155–156, 159–177, 182–191
summary, 486

bar chart
cluster, 136–138
creating, 97–102
Pareto charts, 459–461
pie charts compared, 104
stack, 138–139
summarized data, 100–101

Bar Chart > A function of a variable, One Y, Simple command (Graph Menu), 146–147

Bar Chart > Counts of unique values, Cluster command (Graph menu), 137–138

Bar Chart > Counts of unique values, Stack command (Graph Menu), 139

Bar Chart > Simple command (Graph Menu), 97–98, 104

Bartlett's test, 306

base, for random number generation, 208

Basic Statistics > Correlation command (Stat Menu), 172–173
checking for collinearity using, 345
simple linear regression, 314

Basic Statistics > Covariance command (Stat Menu), 171–172

Basic Statistics > Display Descriptive Statistics command (Stat Menu)
checking data for normality, 212
comparing means of several populations, 284
group comparison, 143–145, 146
single variable data analysis, 115–118
two-way ANOVA, 301

Basic Statistics > Normality Test command (Stat Menu), 213–215
one-way ANOVA, 290

Basic Statistics > Normality Test
    command (cont.)
    two-sample inference, 267–268
    verifying normality assumption,
        351
Basic Statistics > Paired t command
    (Stat Menu), 265–266, 268
Basic Statistics > 1 Proportion
    command (Stat Menu),
    242–243
Basic Statistics > 2 Proportions
    command (Stat Menu),
    272–273
Basic Statistics > 1-Sample t
    command (Stat Menu),
    237–241
Basic Statistics > 2-Sample t
    command (Stat Menu), 258,
    262
Basic Statistics > 1-Sample Z
    command (Stat Menu),
    231–234
Basic Statistics > Store Descriptive
    Statistics command (Stat
    Menu), 118, 145
bell-shaped curve, 210, 212. *See also*
    normal distribution
best subsets regression, 362–368
bimodal distribution, 20
binary logistic regression, 368–371
binary variables, 368
binomial probabilities, 200–206
binomial probability function, 200
bins, histograms, 105–107
bivariate outliers, 162
BodyTemp.mtw data set, summary,
    486
box-and-whisker plot, 113
Box-Jenkins analysis, 441–446. *See*
    also ARIMA
Box-Pierce chi-square statistics, 443
boxplot
    for comparing means of several
        populations, 284–285
    creating simple, 113–115
    group comparison, 142–143
    marginal plots with, 170
    two-sample inference, 259–260
Boxplot > One Y, With Groups
    command (Graph Menu),
    142–143, 419

Boxplot > Simple command (Graph
    Menu), 113–115
brief k command, 338
Brush button, 161
Brush command (Editor Menu), 161
brushing palette, 161
Brushing window, 524
built-in graphs, 20, 118
button commands, 25, 51

Calc Menu, 25, 523
Calculator command (Calc Menu),
    34–36, 102
    for calculating Mean Absolute
        Deviation, 120–121
    for summarizing cases, 60–61, 63
Cancel button, 11
Candya.mtw data set, 462–468
    summary, 487
Candyb.mtw data set, 252, 462,
    469–473
    summary, 487
Candyc.mtw data set, 462, 473–475,
    481
    summary, 487
Carphone.mtw data set, 400
    summary, 488
cases, 26
    summarizing, 60–64
categorical variables, 94, 134
cause-and-effect diagrams, 457–458
ceiling function, 236
cells, 9
    data entry, 26–30
    formulas not allowed in, 35
    maximum number of, 9, 26
centered variable, 347
centering, 74
center line, 464
Central Limit Theorem, 233
Change Data Type command (Data
    Menu), 28
character graphs, 108
chi-square distribution, 380
chi-square goodness-of-fit test
    comparing observed distribution
        of counts to hypothesized
        distribution, 379–384
    Exec macro for, 384–388

chi-square test for independence
    two qualitative variables in
        contingency table, 388–393
    two qualitative variables using
        raw data, 394–398
Chol.mtw data set, 265–268,
    408–413
    summary, 488
classical decomposition, 430–435
    ARIMA model forecast compared,
        449–450
Close All Graphs button, 168
Close button, 11, 107
Close Worksheet command (File
    Menu), 39
cluster bar charts, 136–138
clusters, 141
cluster variables, 136–139
Code > Numeric to Text command
    (Data Menu), 69–71, 209
Code > Text to Text command (Data
    Menu), 394
Code > Use Conversion Table
    command (Data Menu), 209
coding data, 69–71
coefficient of determination, 175
collinearity, 345–348
CollMass.mtw data set, 130, 195,
    313–330
    summary, 488
CollMass2.mtw data set, 345–371,
    377
    summary, 488–489
column factor, 298
column name cell, 26
column name row, 30
column operations, 65
columns, 9
    copying, 36–40
    deleting, 45
    maximum number of, 26
    naming, 30–31
    sampling from, 218–219
    stacking, 76–78
    summarizing, 65–66
    unstacking, 78–80
Column Statistics command (Calc
    Menu), 65–66, 69, 115
Column > Value Order command
    (Editor Menu), 95, 133, 209

Command Line Editor command
(Edit Menu), 169, 384

commands, *See* Minitab commands

commas, ignored by Minitab during
data entry, 28, 44

Compliance.mtw data set, 128, 154,
310–311
summary, 489

confidence bands, 323

confidence intervals, 228
estimate of population mean with
1-sample sign, 410–411
estimating population median
with Wilcoxon, 412–413
one-sided, 234
for population mean when
population standard deviation
is known, 234–235
for population slope of simple
linear regression line, 317–318
summarized data, 239–241
two-sided, 235

CongressSalary.mtw data set,
summary, 490

context-sensitive help, 10–11

contingency tables, 133–136
for independence for two
qualitative variables in,
388–393

control chart, 462

Control Charts > Attributes Charts >
C command (Stat Menu), 476

Control Charts > Attributes Charts >
NP command (Stat Menu), 476

Control Charts > Attributes Charts >
P command (Stat Menu),
473–475

Control Charts > Attributes Charts >
U command (Stat Menu), 476

Control Charts command (Stat
Menu), 458

Control Charts > Time-Weighted
Charts > EWMA command
(Stat Menu), 476

Control Charts > Time-Weighted
Charts > Moving Average
command (Stat Menu), 476

Control Charts > Variables Charts
for Individuals > I-MR
command (Stat Menu),
471–473

Control Charts > Variables Charts
for Individuals > Individuals
command (Stat Menu),
469–470

Control Charts > Variables Charts
for Individuals > Moving
Range command (Stat Menu),
470–471

Control Charts > Variables Charts
for Subgroups > R command
(Stat Menu), 466–468

Control Charts > Variables Charts
for Subgroups > S command
(Stat Menu), 468

Control Charts > Variables Charts
for Subgroups > Xbar
command (Stat Menu), 463

Control Charts > Variables Charts
for Subgroups > Xbar-R
command (Stat Menu),
467–468

Control Charts > Variables Charts
for Subgroups > Xbar-S
command (Stat Menu), 468

Cook's distance, 357

Copy button, 51

Copy Cells button, 52

Copy Cells command (Edit Menu),
50

Copy > Columns to Columns
command (Data Menu), 37–39

Copy Graph button, 151

copying
columns, 36–40
text output into ReportPad and
Microsoft Word, 121–123

correlation, 172–173. *See also* linear
regression

correlation coefficient, 172
simple linear regression, 314

correlation matrix, 173

Cotinine.mtw data set, 309
summary, 490

count data, 473

counts, 95
comparing observed distribution
to hypothesized, 379–384
expected and observed, 380

covariance, 171–172

CPI2.mtw data set, summary, 490

creating
bar charts, 97–102
boxplots, 113–115
cause-and-effect diagrams,
457–458
dotplots, 111–112
Exec macros, 384–388
graphs, 19–20
individuals charts, 469–470
individual value plots, 112–113
marginal plots, 170–171
moving range charts, 470–473
Pareto charts, 459–461
pie charts, 103
projects, 93–94
proportion charts, 473–476
range charts, 466–469
scatterplots, 159–165
stem-and-leaf displays, 107–111
subsets of data, 75–76
text files, 83–84
Xbar charts, 462–466

critical value, 233

crosshairs, 164–165

Crosshairs button, 164

Crosshairs command (Editor Menu),
164

cross-tabulation, 133

cumulative binomial probabilities,
204–206

cumulative counts, 96

cumulative percentage polygon, 459

cumulative percents, 96

cumulative probabilities, normal
distribution, 215–218

Current Data Window button, 25, 51

Cut button, 42, 51

Cut Cells button, 52

Cut Cells command (Edit Menu), 50

cutpoint, 107

data
checking for normality, 212–215
coding, 69–71
combining using Stack option,
76–78
creating subsets, 75–76
entry, 27–28
exporting, 182–184
exporting formatted, 184–185
exporting to Excel, 46–48,
186–187

data, (cont.)
  importing from Excel, 48–50,
    190–191
  manipulating using Calculator
    command, 34–36
  ranking, 71
  retrieving, 33–34
  rounding, 63–64
  saving, 29–30
  separating using Unstack option,
    78–80
  sorting, 72–74
  stacking, 76–78
  standardizing, 74–75
  summarized, 100–101
  summarizing cases, 60–63
  types of, 17, 28
  unstacking, 78–80
data analysis
  group comparison, 133–151
  nonparametric methods, 404–420
  one variable, 92–124
  printing results, 80–83
Data and Story Library (DASL;
    Carnegie-Mellon University),
    131
Data command (Window Menu), 51
data-entry arrow, 27
data files, 6. See also specific date
    files
Data folder (location of data sets), 5,
    14
data folders, 6, 14
Data Menu, 25, 523
data sets, 6
  getting help with, 12–14
  getting information about, 16–19,
    41
  number that can be open at once,
    25
  opening, 40–41
  summarized, 484–520
  viewing information about, 45–46
data subsets, 75–76
data tips, 98
Data window, 7–8, 24–26
  Edit Menu, 523
  Editor Menu, 524
  entering data, 27–30
  File Menu, 522
  moving within, 32–33
  printing from, 31–32
  switching to Session window, 25
  and worksheets, 9

date/time data, 17, 28
decomposition, 430–435
degrees of freedom (DF), 382–383,
    393
Delete Cells button, 52
Delete Cells command (Edit Menu),
    43, 45
Delete Rows command (Edit Menu),
    43
deleting
  columns, 45
  rows, 42–44
dependent samples, 255, 265
dependent variables, 312
Depth.mtw data set, 296–306
  summary, 490
depths, in stem-and-leaf display, 108
detrended values, 428
DF (degrees of freedom), 318,
    382–383, 393
dialog boxes, 31
directory, 14
discrete distribution, generating
    random data from, 207–210
Display Data command (Data Menu),
    69
Display Descriptive Statistics
    command (Stat Menu), 17–18
distributions. See also normal
    distribution
  binomial probabilities, 200–206
  checking data for normality,
    212–215
  cumulative probabilities and
    inverse cumulative
    probabilities for normal,
    215–218
  generating random data from
    discrete, 207–210
  generating random data from
    normal, 210–212
DJC20012002.mtw data set,
    178–181, 198, 374
  summary, 491
DJC20012002a.mtw data set, 455
  summary, 491
DJC20012002b.mtw data set, 454
  summary, 491
dollar sign ($), ignored by Minitab
    during data entry, 44
Donner.mtw data set, 376
  summary, 492

dotplot, 113
  creating simple, 111–112
  group comparison, 140–141
  marginal plots with, 170
Dotplot command > Multiple Y's,
    Simple command (Graph
    Menu), 416
Dotplot command > One Y, with
    Groups command (Graph
    Menu), 140–141
Dotplot command > Simple
    command (Graph Menu),
    111–112
Drive.mtw data set, summary, 492
DrivingCosts.mtw data set, 334
  summary, 492
DrugMarkup.mtw data set, 280
  summary, 493
dummy variables, 242, 357–362
Dunnett multiple comparisons test,
    291
Durbin-Watson statistic, 353–354

Edit Last Dialog button, 36, 51
Edit Last Dialog command (Edit
    Menu), 36
Edit Menu, 25, 523
Editor Menu, 25, 524
Edit Title button, 98
Edit X Scale button, 106
Edit Y Axis button, 149
Edit Y Scale button, 147–148
education, 124
effect size, 246, 269
Election2.mtw data set, 197, 423,
    480
  summary, 493
EMail.mtw data set, 401, 455
  summary, 493
EmployeeInfo.mtw data set, 153,
    388–398
  summary, 493
Enable Commands command (Editor
    Menu), 66, 67, 83
end command, 385
Endowment.mtw data set, summary,
    494
endpoint, 107
Entering and Exploring Data
    Tutorial, 198

Erase Variables command (Data Menu), 45, 182

estimated median, 412

estimating
  population mean with 1-sample sign confidence interval estimate, 410–411
  population median with Wilcoxon confidence interval estimate, 412–413
  response variable for multiple linear regression, 328–330
  response variable for simple linear regression, 321–323

ETA1-ETA2, 415

ExamScores.mtw data set, 87–88
  summary, 494

Excel, *See* Microsoft Excel

Exec macro, for chi-square goodness-of-fit test, 384–388

Exh_qc.mtw data set, 480, 481, 520

Exit command (File Menu), 14, 21, 39–40

expected counts, 380
  small, 391

explanatory variables, 283, 312

exponential growth model, 428

exporting
  data, 182–184
  data to Excel, 46–48, 186–187
  graphs, 149–151

factor column, Kruskal-Wallis test, 417–418

factors, 286
  interaction between, 297
  levels in two-way ANOVA, 304–306

File Menus, 25, 522

Find button, 51

Find command (Editor Menu), 45

First Time Alert, 24

fishbone diagram, 457

Fisher multiple comparisons test, 291

fitted line plots, 176–177
  simple linear regression, 320–321

fitted value, 290

five-number summary, 114

Fja.mtw data set, 338–344
  summary, 494

floating point, 184

footnotes, to graphs, 149–150

Force.mtw data set, 281, 423–424
  summary, 495

forecasting
  classical decomposition, 430–434
  using ARIMA, 447–450

formatted data
  exporting, 184–195
  importing, 188–190

FORTRAN format, 184

Four in one graph, 365

F-tests
  one-way ANOVA, 286–287
  two-way ANOVA, 298, 301–303

functions, 64

GasData.mtw data set, 376
  summary, 495

General Electric, Six Sigma program, 476

Global macros, 384

Glossary, 120

Go To... command (Editor Menu), 32–33

Go To command (Editor Menu), 32

Grades.mtw data set, 225, 520

Graph Annotation Tools Toolbar, 525

Graph Editing Toolbar, 525

Graphing Data Tutorial, 198

Graph Menu, 25, 523

graphs
  creation, 19–20
  exporting, 149–151
  Four in one, 365
  importance in regression, 338–344
  naming, 98–99
  overlaying, 180–181
  printing, 20
  saving, 101–102

Graphs folder (Project Manager), 99

Graph window
  Editor Menu, 524
  File Menu, 522

group comparison
  boxplots, 142–143
  cluster and stack bar charts, 136–139
  contingency tables, 133–136
  dotplots, 140–141

exporting graphs, 149–151
  individual value plots, 141
  subgroups, 143–146
  using charts to display descriptive statistics, 146–149

grouping variables, 165–168

groups, 141

G-statistic, 371

Height.mtw data set, 131, 211–219
  summary, 496

Help button, 10–11, 51

Help command (Help Menu), 51

Help Menu, 25, 524
  to find a formula, 260–262
  using, 11–12

Help Toolbar, 525

Help window, 10–14

High Influence (HI) value, 356

histogram, 19–20
  for checking data for normality, 212
  marginal plots with, 170
  two-sample inference, 260, 266

Histogram command (Graph Menu), 19

Histogram command > Simple command (Graph Menu), 105–107

History folder (Project Manager), 99
  viewing, 168–170

History window, 522

Homes.mtw data set, 251, 333, 374–375, 424
  summary, 496

homoscedasity, 349, 351–353

Hsu multiple comparisons test, 291

H-test, Kruskal-Wallis, 417

human resources, 331

hypothesis testing, 228
  comparing population means from two independent samples, 256–265
  comparing population proportions from two independent samples, 272–275
  confidence interval when population standard deviation is known, 234–235
  mean of paired data, 265–268
  one-way ANOVA, 286

hypothesis testing, (cont.)
   population mean using sign test, 408–410
   population mean when population standard deviation is known, 229–234
   population mean when population standard deviation is unknown, 237–241
   population median using Wilcoxon test, 411–412
   population proportion, 241–245
   power of a test, 245–248
   sample size and power for comparing means of two independent sample, 269–271
   sample size and power for comparing two independent proportions, 275–276
   two-way ANOVA, 297–298

importing
   data from Excel, 48–50, 190–191
   formatted data, 188–190
increment, 108
independent populations
   comparing median of *K* using Kruskal-Wallis test, 417–420
   comparing medians of two, 413–416
independent samples, 255
independent variables, 312
indicator variables, 242
   incorporating in regression model, 357–362
   interpreting regression coefficient, 361–362
individuals charts, 469–470
individual value plot
   creating simple, 112–113
   group comparison, 141
   two-sample inference, 259, 260
Individual Value Plot > One Y, with Groups command (Graph Menu), 141
Individual Value Plot > Simple command (Graph Menu), 112–113
Infants.mtw data set, 93–118, 133–139, 157, 225, 251, 308, 401, 425
   summary, 496
inference, *See* statistical inference
influential points, 317

info session command, 67
Info window, 17
inserting rows, 42–44
insertion cursor, 28
Insert Rows button, 43, 51, 52
Insert Rows command (Editor Menu), 43
integer, 184
interaction, between factors, 297
interactions plot, 302–303
interval plot, 261–262, 288
Interval Plot > One Y, With Groups command (Graph Menu), 261–262, 288
intervals, histograms, 105–107
inverse cumulative probabilities, 215–218
Ishikawa diagram, 457

Jeans.mtw data set, summary, 497

k time periods (lags), 435
keyboard shortcuts, 33
Kruskal-Wallis test, 417–420

lags, 435, 439–440
Lakes.mtw data set, summary, 498
layering variables, 157
least squares/regression line
   computing, 174–175
   displaying, 175–177
level of significance, 232–234
Levene's test, 306
leverage points, 317
leverage value, 356
linear regression, 312
   multiple linear, 326–328
   response variable estimates for multiple linear, 328–330
   response variable estimates for simple, 321–323
   simple linear, 313–323
   verifying assumptions, 349–355
linear trend analysis, 427–429
links, 11
Linux operating system, Minitab doesn't support, 2
Ljung-Box Q (LBQ) statistics, 437, 443
Local macros, 384

Lotto.mtw data set, 400
   summary, 499
Lotus 1-2-3, 9
   Minitab cells contrasted, 35
lower-control limit (LCL), 465

Macintosh operating system, Minitab doesn't support, 2
macros, 384
MAD (Mean Absolute Deviation), 120–121, 429
Make Indicator Variable command (Calc Menu), 357–362
Make Patterned Data > Simple Set of Numbers command (Calc Menu), 201, 306
Make Patterned Data > Text Values command (Calc Menu), 305
Mallows C-p criterion, 364
Mann-Whitney test, 17, 414–416
MAPE (Mean Absolute Percentage Error), 429
Marathon2.mtw data set, 198
   summary, 499
marginal plot, 170–171
Marginal Plot > With Boxplots command (Graph Menu), 170–171
Marginal Plot > With Dotplots command (Graph Menu), 170
Marginal Plot > With Histograms command (Graph Menu), 170
margin of error, 235
Marks.mtw data set, 60–84
   summary, 499
Matrix Plot > Matrix of Plots, Simple command (Graph Menu), 362–363
Maximize button, 7–8, 26
maximizing windows, 7–9
maximum, 146
MBASurvey.mtw data set, 88, 129, 154
   summary, 500
mean, 146, 463–465
   comparing separate populations with one-way ANOVA, 284–289
   comparing several populations with responses in separate columns with one-way ANOVA, 294–295

hypothesis testing about
population using sign test,
408–410
of paired data, 265–268
with 1-sample sign confidence
interval estimate, 410–411
sample size and power for
comparing means of two
independent samples, 269–271
trimmed, 118, 120
Mean Absolute Deviation (MAD),
120–121
time series, 429
Mean Absolute Percentage Error
(MAPE), 429
Mean Squared Deviation (MSD), 429
measurement data, 473
median
comparing *K* independent
populations using Kruskal-
Wallis test, 417–420
comparing two independent
populations, 413–416
estimating population with
Wilcoxon, 412–413
hypothesis testing about
population using Wilcoxon
test, 411–412
Mood's Median Test, 419
medical diagnostics, 420
Meet MINITAB, 5
Menu Bar, 7
Mercedes.mtw data set, summary,
500
Merge Worksheet command (Data
Menu), 76
.mgf file type, 102
Microsoft Calculator command
(Tools Menu), 102
Microsoft Excel, 9
exporting data to, 46–48, 186–187
importing data from, 48–50,
190–191
Minitab cells contrasted, 35
Microsoft Word, copying text output
into, 122–123
MiniGuide window, 119–120
Minimize button, 7–8
minimizing windows, 7–9
minimum, 146
Minitab. *See also* cases; columns;
data; data analysis; rows
context-sensitive help, 10–11

exiting, 14, 21, 33
installing, 4–6
introduction to, 2–6
opening, 6–9, 14, 23–26
sample session, 14–21
technical support, 4
use in Six Sigma program
(General Electric), 476
Minitab commands
issuing, 4, 10
organization within menus, 25
pictorial aids for, 30
right-hand mouse button, 52
summary of, 50–51
Toolbar button commands, 51
Minitab Help, 5
Minitab menus, 5–6. *See also*
specific menus
summary, 522–524
Minitab Portable format, 183
MINITAB Release 14, 2, 5
Minitab Session Command Help, 5
MINITAB Student Release 14, 2
Minitab toolbars, 7, 25, 525
button commands, 25, 51
Minitab Tutorials, 5
Minitab tutorials (not this book's),
475
Minitab Web site, 2, 4
Minitab window, 7–8, 24–25
MLBGameCost.mtw data set, 376,
401
summary, 500–501
MnWage2.mtw data set, 197, 334,
374
summary, 501
modal category, 136
MonthlySnow.mtw data set, 91, 455
summary, 501
Mood's Median Test, 419
Move Columns command (Editor
Menu), 182
Movies.mtw data set, summary, 501
moving averages, classical
decomposition, 430
moving range charts, 470–473
.mpj file type, 94
MSD (Mean Squared Deviation), 429
.mtb file type, 386
.mtp file type, 183
.mtw file type, 13, 30

multicollinearity, 345
multiple comparisons test, 291–294
multiple linear regression, 326–328
response variable estimates for,
328–330
multiplicative model, 430
Murders.mtw data set, 224, 375
summary, 502
Murderu.mtw data set, 90, 280, 309
summary, 502
MusicData.mtw data set, 207–210,
224–225, 379–388
summary, 503

N, number of observations in
column, 18, 117
naming
columns, 30–31
files, 29
graphs, 98–99
New > Worksheet command (File
Menu), 50
NHL2003.mtw data set, 196–197
summary, 503
95% confidence bands, 323
95% prediction bands, 323
non-constant control limit, 474
nonlinear interpolation (NLI), 410
nonparametric methods, 407
comparing medians of *K*
independent populations using
Kruskal-Wallis test, 417–420
comparing medians of two
independent populations,
413–416
estimating population mean with
1-sample sign confidence
interval estimate, 410–411
estimating population median
with Wilcoxon confidence
interval estimate, 412–413
runs test for randomness, 404–407
testing hypothesis about
population mean using sign
test, 408–410
testing hypothesis about
population median using
Wilcoxon test, 411–412
Nonparametrics > Friedman
command (Stat Menu), 419
Nonparametrics > Kruskal-Wallis
command (Stat Menu), 418

Nonparametrics > Mann-Whitney command (Stat Menu), 414–416

Nonparametrics > Mood's Median Test command (Stat Menu), 419

Nonparametrics > Runs Test command (Stat Menu), 406

Nonparametrics > 1-Sample Sign command (Stat Menu), 408–410

Nonparametrics > 1-Sample Wilcoxon command (Stat Menu), 411–413

non-stationary time series, 439, 441

normal distribution
    checking data for normality, 212–215
    cumulative probabilities, 215–218
    generating random data from, 210–212
    inverse cumulative probabilities, 215–218

normality
    checking assumption in regression models, 349, 351
    checking data for, 212–215

normal probability plot (NPP), 213–215
    of one-way ANOVA residuals, 290
    two-sample inference, 267
    of two-way ANOVA residuals, 298–299, 300

note command, 385

Note02.mtw data set, 56, 156
    summary, 504

Notepad command (Tools Menu), 83–84, 185

Not equal, default alternative hypothesis, 275

null hypothesis, 231

number of lags, 435, 437

numeric data, 17, 28, 94

observations, 26
    unusual, 355–357

observed counts, 380

ogive, 459, 461

OldFaithful.mtw data set, 14–21, 129, 252
    summary, 504

one-sample Z-test, 230–231

one-sided confidence interval, 234

one variable data analysis, See single variable data analysis

one-way ANOVA, 283
    checking assumptions, 289–291
    comparing means of separate populations, 284–289
    comparing means of several populations with responses in separate columns, 294–295
    Tukey's multiple comparisons test, 291–294

one-way tables, 157

Open Graph command (File Menu), 102

OpenHouse.mtw data set, 335–336, 375–376
    summary, 504

Open Project button, 33, 123

Open Project command (File Menu), 123

Open Worksheet command (File Menu), 15, 32–33, 40, 93, 187, 190

option buttons, 31

order 1 autoregressive component, 437

order 2 autoregressive component, 437, 438, 441

order 3 autoregressive component, 437

order 1 seasonal autoregressive component, 441

Other Files > Export Special Text command (File Menu), 184–185

Other Files > Import Special Text command (File Menu), 189

Other Files > Run an Exec command (File Menu), 387

outliers, 162, 342

Output Editable command (Editor Menu), 67

overall error rate, 294

overlaid scatterplot, 181

paired samples, 255, 265–266

pairwise averages, 412, 419

pairwise differences, 419

pairwise slopes, 419

Pancake.mtw data set (in Data Folder), 310, 520

Panel button, 167

Panel command (Editor Menu), 167

paneling, 167–168

Pareto charts, 459–461

partial autocorrelation plots, 435, 437–438

Paste button, 47, 48, 51

Paste Cells button, 52

Paste Cells command (Edit Menu), 50

PayData.mtw data set, 40–50
    summary, 505

Pearson's correlation coefficient, 172

PhoneRates.mtw data set, summary, 505

pictorial gallery, 19

pie chart, 103–104

Pie Chart command (Graph Menu), 103–104

Pizza2.mtw data set, 54, 89
    summary, 505

pizza producers, 249

Poisson distribution, 202, 220

pooled proportion, 274–275

pooled standard deviation, 465

population mean
    comparing two independent samples, 256–265
    confidence interval when population standard deviation is known, 234–235
    estimating with 1-sample sign confidence interval estimate, 410–411
    hypothesis testing using sign test, 408–410
    hypothesis testing when population standard deviation is known, 229–234
    inferences about when population standard deviation is unknown, 237–241
    sample size for estimating when population standard deviation is known, 235–237

population median
    comparing K independent populations using Kruskal-Wallis test, 417–420

comparing two independent, 413–416
estimating with Wilcoxon confidence interval estimate, 412–413
hypothesis testing using Wilcoxon test, 411–412
population proportion
comparing two independent, 272–275
sample size and power for comparing two independent, 275–276
population standard deviation
inferences about population mean when known, 229–234
inferences about population mean when unknown, 237–241
sample size for estimating population mean when known, 235–237
power, of a test, 245–248
sample size and power for comparing means of two independent sample, 269–271
sample size and power for comparing two independent proportions, 275–276
target and actual, 271–272
Power and Sample Size > One-Way ANOVA command (Stat Menu), 248
Power and Sample Size > 1 Proportion command (Stat Menu), 242–243
Power and Sample Size > 2 Proportions command (Stat Menu), 248, 275–276
Power and Sample Size > 1-Sample t command (Stat Menu), 248
Power and Sample Size > 2-Sample t command (Stat Menu), 248, 269–270
Power and Sample Size > 1-Sample Z command (Stat Menu), 248
prediction bands, 323
prediction interval, 321
predictors, 312
collinearity, 345–348
predictor table, 316
predictor variable, 174
Print Graph command (File Menu), 20, 102

printing
from Data window, 31–32
graphs, 20
results of data analysis, 80–83
print session command, 67–68
Print Session Window command (File Menu), 81–83
Print Window button, 51
Print Worksheet command (File Menu), 31
Probability Distributions > Beta command (Calc Menu), 222
Probability Distributions > Binomial command (Calc Menu), 201, 205, 409
Probability Distributions > Cauchy command (Calc Menu), 222
Probability Distributions > Chi-Square command (Calc Menu), 222, 382–384
Probability Distributions > Discrete command (Calc Menu), 222
Probability Distributions > Exponential command (Calc Menu), 222
Probability Distributions > F command (Calc Menu), 222
Probability Distributions > Gamma command (Calc Menu), 222
Probability Distributions > Hypergeometric command (Calc Menu), 222
Probability Distributions > Integer command (Calc Menu), 222
Probability Distributions > Laplace command (Calc Menu), 222
Probability Distributions > Largest Extreme value command (Calc Menu), 222
Probability Distributions > Logistic command (Calc Menu), 222
Probability Distributions > Loglogistic command (Calc Menu), 222
Probability Distributions > Lognormal command (Calc Menu), 222
Probability Distributions > Normal command (Calc Menu), 216
Probability Distributions > Poisson command (Calc Menu), 222

Probability Distributions > Smallest Extreme value command (Calc Menu), 222
Probability Distributions > t command (Calc Menu), 222, 241, 318
Probability Distributions > Triangular command (Calc Menu), 222
Probability Distributions > Uniform command (Calc Menu), 222
Probability Distributions > Weibull command (Calc Menu), 222
process capability analysis, 475
process monitoring, 462
Process.mtw data set, 457–461
summary, 506
professional graphs, 108
Prof.mtw data set, 253, 400
summary, 506–507
Project Description command (File Menu), 94
Project Manager, 98–100
Project Manager button, 98
Project Manager command (Window Menu), 98, 168
Project Manager Toolbar, 525
projects, 9, 29, 84
creating, 93–94
saving and reopening, 123–124
proportion charts, 473–476
public safety, 220
Pubs.mtw data set, 479, 480
summary, 507
PulseA.mtw data set, 55–56, 90, 127, 130, 154–155, 194, 253, 281, 282, 308–309, 333–334, 401
summary, 507–508
p-value, 233

quadratic model, 428
quadratic regression, 324–326
qualitative variables
chi-square test for independence for two qualitative variables in contingency table, 388–393
chi-square test for independence for two qualitative variables using raw data, 394–398
comparing observed distribution of counts to hypothesized distribution, 379–384

qualitative variables (cont.)
  Exec macro for chi-square
    goodness-of-fit test, 384–388
  summarizing, 95–97
Quality Companion, 476
Quality Tools command (Stat Menu),
  458
Quality Tools > Cause-and-Effect
  command (Stat Menu),
  457–458
Quality Tools > Pareto Chart
  command (Stat Menu),
  459–461
quantitative response, 312
quantitative variables, 94. *See also*
  scatterplot
  examining relationships between
    two, 159–191
  summarizing numerically,
    115–118
  summarizing with displays,
    105–115

R standardized residuals, 317
Radlev.mtw data set, 89
  summary, 508
random data
  generating from discrete
    distributions, 207–210
  generating random data from
    normal distributions, 210–212
Random Data > Bernoulli command
  (Calc Menu), 221
Random Data > Beta command (Calc
  Menu), 221
Random Data > Cauchy command
  (Calc Menu), 221
Random Data > Chi-Square
  command (Calc Menu), 221
Random Data > Discrete command
  (Calc Menu), 208
Random Data > Exponential
  command (Calc Menu), 221
Random Data > F command (Calc
  Menu), 221
Random Data > Gamma command
  (Calc Menu), 221
Random Data > Hypergeometric
  command (Calc Menu), 221
Random Data > Integer command
  (Calc Menu), 221

Random Data > Laplace command
  (Calc Menu), 221
Random Data > Largest Extreme
  Value command (Calc Menu),
  221
Random Data > Logistic command
  (Calc Menu), 221
Random Data > Loglogistic
  command (Calc Menu), 221
Random Data > Lognormal
  command (Calc Menu), 221
Random Data > Multivariate Normal
  command (Calc Menu), 221
Random Data > Normal command
  (Calc Menu), 211
Random Data > Poisson command
  (Calc Menu), 221
Random Data > Sample From
  Columns command (Calc
  Menu), 218
Random Data > Smallest Extreme
  Value command (Calc Menu),
  221
Random Data > t command (Calc
  Menu), 221
Random Data > Triangular command
  (Calc Menu), 221
Random Data > Uniform command
  (Calc Menu), 221
Random Data > Weibull command
  (Calc Menu), 221
RandomIntegers.mtw data set, 279
  summary, 508
randomness, runs test for, 404–407
range (R) charts, 466–469
Rank command (Data Menu), 71
ranking data, 71
raw data, 100
Rbar, 478
R charts, 466–469
ReadMe file, 5
read-only output, 67
Redo button, 43, 51, 52
Redo command (Edit Menu), 43
regression coefficient, 175
  for indicator variable, 361–362
regression line
  computing, 174–175
  displaying, 175–177

regression models. *See also* linear
  regression
  best subsets, 362–368
  binary logistic, 368–371
  checking independence
    assumption using time series
    plot, 353–355
  checking normality assumption,
    351–353
  collinearity, 345–348
  graphs in regression, 338–344
  indicator variable, 357–362
  stepwise, 368
  unusual observations, 355–357
Regression > Best Subsets command
  (Stat Menu), 363–368
Regression > Binary Logistic
  Regression command (Stat
  Menu), 369–371
Regression > Fitted Line Plot
  command (Stat Menu),
  176–177
  simple linear regression, 320, 322
Regression > Regression command
  (Stat Menu), 174–175, 338–344
  quadratic regression, 324
  simple linear regression, 315, 318,
    321
  variance inflation factor option,
    346–348
  verifying assumptions, 349–350
Regression > Stepwise command
  (Stat Menu), 368
Related Documents folder (Project
  Manager), 99
Rename Worksheet button, 188
Replace command (Editor Menu), 45
ReportPad folder (Project Manager),
  99
  copying text output into, 121–123
residuals, 317
  obtaining, 318–319
  storing, 349–351
residual value, 290
response column, Kruskal-Wallis
  test, 417–418
response variables, 174, 312
  estimates for multiple linear
    regression, 328–330
  estimates for simple linear
    regression, 321–323
Restore button, 8

retailing, 249

right-hand mouse button commands, 52

Rivera.mtw data set, 453–454
   summary, 509

Riverb.mtw data set, 455
   summary, 509

Riverc.mtw data set, 427–450, 454
   summary, 509

Riverc2.mtw data set, 455
   summary, 510

Riverd.mtw data set, summary, 509

Rivere.mtw data set, summary, 509

Rivers.mtw data set, 155, 424
   summary, 509

robust procedures, 239

Round function, 63

rounding, 63–64

row factor, 298

row header, 42

row operations, 65

rows, 9, 26
   deleting and inserting, 42–44

Row Statistics command (Calc Menu), 61–63

R-Sq(adj), 175, 287, 316, 328

R-Sq (coefficient of determination), 175, 287, 316, 328

.rtf file type, 122

run, 406

runs test for randomness, 404–407

Salary02.mtw data set, 90–91, 196, 423, 424
   summary, 510–511

Samples in different columns option, 263

Samples in one column option, 263

sampling from columns, 218–219

sampling without replacement, 218

sampling with replacement, 218

SAR 1 model, 441

Save Current Worksheet As command (File Menu), 29, 41, 183

Save Current Worksheet command (File Menu), 33, 39, 93

Save Graph As command (File Menu), 101–102

Save Project As command (File Menu), 111

Save Project button, 93–94, 111, 123

Save Project command (File Menu), 94

Save Report As button, 122

Save Report As command (File Menu), 122

saving
   current worksheet, 25
   data, 29–30
   graphs, 101–102
   projects, 123–124

SBP.mtw data set, summary, 511

scatterplot, 159–165
   adding grouping variable, 165–168
   for binomial probabilities, 204, 206
   importance in regression, 341–344
   least squares/regression line, 175–177
   overlaid, 181
   paneling, 167–168
   for paneling, 167
   for simple linear regression, 314
   verifying homoscedasity assumption, 352

Scatterplot command (Graph Menu), 429

Scatterplot command > Simple command (Graph Menu), 159–160, 162, 179
   verifying homoscedasity, 159–160, 162, 179

Scatterplot command > With Connect Line command (Graph Menu), 175–178, 236–237, 271

Scatterplot command > With Groups command (Graph Menu), 165–167

S charts, 466, 469

SchoolsData.mtw data set, summary, 511–512

scientific research, 277

S-curve model, 428

Search tab, 12

seasonal component, 429

seed number, for random number generation, 208

Select button, 165

separating data, 78–80

Seq SS table, 325, 328

Session command language, 66
   disabling, 82–83

session commands, 66–69

Session command (Window Menu), 51, 66

Session folder (Project Manager), 99

Session Four: Assessing Quality, 475, 481

Session window, 7–9, 24
   Edit Menu, 523
   Editor Menu, 524
   File Menu, 522
   switching to Data window, 25

Session Window button, 25, 51, 66

Set Base command (Calc Menu), 208, 211, 218

Set ID Variables button, 161

Set ID Variables command (Editor Menu), 161

shortcut menu, 33

Show Graphics Folder button, 107

Show History button, 168

Show Info button, 16, 45, 51

Show ReportPad button, 122

Show Session Folder button, 169

side-by-side bar charts, 138

sign test, hypothesis testing about population mean, 408–410

simple linear regression, 313–323
   response variable estimates for, 321–323

simulation, 207

single variable data analysis
   bar chart creation, 97–102
   boxplot creation, 113–115
   dotplot creation, 111–112
   individual value plot creation, 112–113
   Mean Absolute Deviation, 120–121
   pie chart creation, 103–104
   project creation, 93–94
   stem-and-leaf display creation, 107–111
   summarizing qualitative variables, 95–104
   summarizing quantitative variables numerically, 115–118
   summarizing quantitative variables with displays, 105–115

single variable data analysis (cont.)
using StatGuide, 119–120

Six Sigma program (General Electric), 476

Sleep.mtw data set, 454, 480
summary, 512

small expected cell counts, 391

Sort command (Data Menu), 72–74, 80

sorting data, 72–74

SPCarData.mtw data set, summary, 512

Spearman's Rank Correlation Coefficient (rho), 425

SpeedCom.mtw data set, 334–335
summary, 513

split stems, 109

Split Worksheet command (Data Menu), 76

SP5002.mtw data set, 453
summary, 512

SRES against FITS plot, 352

stack bar charts, 138–139

Stack > Columns command (Data Menu), 76–78

stacking data, 76–78

standard deviation, 464–465
pooled, 465

standard deviation (S) charts, 466

standard error of the estimate, 175

standard error of the mean (SE Mean), 232

Standardize command (Calc Menu), 74–75, 347

standardized residuals, 317

standardizing data, 74–75

standard normal, 232

Standard Toolbar, 525

StatGuide, 5
using, 119–120

StatGuide button, 119

StatGuide window, 119

stationary time series, 439, 441

statistical inference, 228
about population mean when population standard deviation is known, 229–234
about population mean when population standard deviation is unknown, 237–241
about population proportion, 241–245
comparing population means from two independent samples, 256–265
comparing population proportions from two independent samples, 272–275
confidence interval for population mean when population standard deviation is known, 234–235
mean of paired data, 265–268
one sample, 229–249
power of a test, 245–248
sample size and power for comparing means of two independent sample, 269–271
sample size and power for comparing two independent proportions, 275–276
sample size for estimating population mean when population standard deviation is known, 235–237
two sample, 256–276

statistics, 65

Stat Menu, 25, 523

status bar, 9, 25

stock market analysis, 451

Stem-and-Leaf command (Graph Menu), 107–111

stem-and-leaf display
creating, 107–111
for normality testing of data, 215
two-sample inference, 260

stemplot, 107

stems, 108

stepwise regression, 368

Stores2.mtw data set, 400–401
summary, 513

Studnt 14 folder, 5, 14

subcommands, 112

subdirectory, 14

subgroups, 143–146

subscripts, 257

subsets, of data, 75–76

Subset Worksheet command (Data Menu), 75–76, 366

summarized data
bar graphs using, 100–101
comparing two proportions, 274–275
percentages using, 102
t-test, 239–241
two sample t-tests, 264

summarizing
cases, 60–64
columns, 65–66

sums, 146

supernova, 277

Supplier Training Program (General Electric), 476

Survey.mtw data set, 90, 153–154, 310
summary, 514

Tables > Chi-Square Test (Table in Worksheet), 390–393

Tables > Cross Tabulation and Chi-Square command (Stat Menu), 95–96
contingency tables, 134
one-way table, 157
qualitative data analysis, 394–396

Tables > Descriptive Statistics command (Stat Menu), 145–146

Tables > Tally Individual Values command (Stat Menu), 95–96, 210, 380

target power, 270–271

TBill2.mtw data set, 57
summary, 514

technical support, 4

Temco.mtw data set, 308, 334, 335, 375
summary, 514–515

tests for special causes, 464, 466

Textbooks.mtw data set, 87, 194–195, 279, 336, 424
summary, 515

text boxes, 31

text data, 17, 28

text files
  copying into ReportPad and
    Microsoft Word, 121–123
  creating, 83–84
time series, 428
  creating, 178–181
  stationary and non-stationary,
    439, 441
  transforming, 439–441
time series analysis
  autocorrelation plots, 435–437
  Box-Jenkins analysis, 441–446
  classical decomposition, 430–435
  forecasting with ARIMA, 447–450
  partial autocorrelation plots, 435,
    437–438
  trend analysis, 427–429
Time Series > ARIMA command
  (Stat Menu), 441–446
Time Series > Autocorrelation
  command (Stat Menu),
  435–437
Time Series > Cross Correlation
  command (Stat Menu), 438
Time Series > Decomposition
  command (Stat Menu),
  430–435
Time Series > Differences command
  (Stat Menu), 440–441
Time Series > Double Exp
  Smoothing command (Stat
  Menu), 450
Time Series > Lag command (Stat
  Menu), 439–440
Time Series > Moving Average
  command (Stat Menu), 450
Time Series > Partial
  Autocorrelation command
  (Stat Menu), 437–438
time series plot, 178, 429
  checking independence
    assumption using, 353–355
Time Series Plot command (Graph
  Menu), 429
Time Series Plot > Multiple
  command (Graph Menu), 448
Time Series Plot > Simple command
  (Graph Menu), 404–405
  verifying independence
    assumption, 353–354

Time Series > Single Exp Smoothing
  command (Stat Menu), 450
Time Series > Time Series Plot
  command (Stat Menu), 429
Time Series > Trend Analysis
  command (Stat Menu),
  428–429
Time Series > Winter's Method
  command (Stat Menu), 450
time series tools, 426
time-shifting, 439
Title Bar, 7, 8, 9
toggles, 67
toolbars, 7, 25, 525
Tools Menu, 25, 524
Top25Stars.mtw data set, summary,
  516
total quality management (TQM),
  456, 476
total quality management tools
  cause-and-effect diagrams,
    457–458
  individuals charts, 469–470
  moving range charts, 470–473
  Pareto charts, 459–461
  proportion charts, 473–476
  range charts, 466–469
  Xbar charts, 462–466
transformation
  quadratic regression, 326
  time series, 439–441
trend analysis, 427–429
trend line, 429
trimmed mean, 118, 120
t-statistics, 437
t-test
  one-sample, 237–239
  summarized data, 239–241
Tukey multiple comparisons test,
  291–294
Tvhrs.mtw data set, summary, 516
TwinsYankees.mtw data set, 226–227
  summary, 517
two-sample t-tests
  using stacked data, 256–260
  using summarized data, 264
  using unstacked data, 262–264
two-sided confidence interval, 235

TwoTowns.mtw data set, 156,
  195–196, 375
  summary, 518
two variable data analysis, See group
  comparison
two-way ANOVA, 283, 296–306
two-way tables, 157
Type I error, 232
Type II error, 245
typographical conventions, used in
  this book, 3–4

UGradSurvey.mtw data set
  summary, 518
unbalanced two-way ANOVA, 303
Undo button, 43, 51, 52
Undo command (Edit Menu), 43
Unstack Columns command (Data
  Menu), 78–80
  with one-way ANOVA, 294
  with two-sample t-tests, 262–263
unstacking data, 78–80
unusual observations, 317, 355–357
Up One Level icon, 15
upper-control limit (UCL), 465
USAArrivals.mtw data set,
  summary, 519
USDemData.mtw data set,
  summary, 519

variables, 17, 26
variables data, 473
variance inflation factor (VIF), 345

Walsh averages, 412
WastesData.mtw data set, summary,
  519
What's New, 5
whiskers, 114
Wilcoxon Signed Ranks test, 411
Wilcoxon test
  estimating population median
    with confidence interval
    estimate, 412–413
  hypothesis testing about
    population median, 411–412

Window Menu, 25, 524

windows, 7–9. *See also* specific windows

Worksheet > Description command (Editor Menu), 94

worksheets, 9
    exploring existing, 15–21
    maximum number open at one time in Student version (5), 39
    opening, 7
    previewing contents when opening, 124

renaming, 188
    saving current, 25

Worksheets folder (Project Manager), 99

Worksheet Toolbar, 525

X (influential points), 317
    marking, 343

Xbar charts, 462–466

YearlySnow.mtw data set, 404–407, 423, 453
    summary, 520

YogurtData.mtw data set, 23–40, 308
    summary, 520

zero-valued control limit, 467